Hydraulic Components Volume B

Hydraulic Filters

Dr. Medhat Kamel Bahr Khalil, Ph.D, CFPHS, CFPAI.
Director of Professional Education and Research Development,
Applied Technology Center, Milwaukee School of Engineering,
Milwaukee, WI, USA.

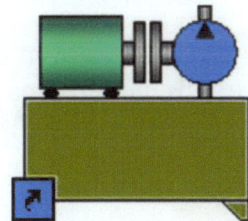

CompuDraulic LLC
www.CompuDraulic.com

CompuDraulic LLC

Hydraulic Components Volume B

Hydraulic Filters

ISBN: 978-0-9977634-0-9

Printed in the United States of America
First Published by Sept. 2022
Revised by Sept. 2023

Disclaimer

It is always advisable to review the relevant standards and the recommendations from the system manufacturer. However, the content of this book provides guidelines based on the author's experience.

Any portion of information presented in this book might not be suitable for some applications due to various reasons. Since errors can occur in circuits, tables, and text, the author/publisher assumes no liability for the safe and/or satisfactory operation of any system designed based on the information in this book.

The author/publisher does not endorse or recommend any brand name products by including such brand name products in this book. Conversely the author/publisher does not disapprove any brand name product not included in this book. The publisher obtained data from catalogs, literatures, and material from hydraulic components and systems manufacturers based on their permissions. The author/publisher welcomes additional data from other sources for future editions. This disclaimer is applicable for the workbook (if available) for this textbook.

Hydraulic Components Volume B
Hydraulic Filters

PREFACE

Keeping the oil clean is an essential requirement for reliable and efficient machine operation. This book introduces knowledge foundation about hydraulic filters. The book introduces about various types of filters constructions, configurations, accessories. The book also introduces the various concepts of filtration mechanisms and filter media. The book overviews various types of materialistic contaminations such a fluidic, chemical, and particulate contamination. The book discusses filter selection criteria, maintenance, troubleshooting, and failure analysis of filters including the standard test methods for filter performance.

Dr. Medhat Kamel Bahr Khalil

ACKNOWLEDGEMENT

All praise is to Allah who granted me the knowledge, resources, and health to finish this work.

To the soul of my parents who taught me the values of ISLAM

To my family: wife, sons, daughters in law, and grandchildren

To my best teachers and supervisors

The author wishes to thank the following gentlemen for their effective support in developing this book:

- Kamara Sheku, Dean of Applied Research at Milwaukee School of Engineering.
- Tom Wanke, CFPE, Director of Fluid Power Industrial Consortium and Industry Relations at Milwaukee School of Engineering.
- Paul Michael, Research Chemist, Fluid Power Institute at MSOE.

The author thanks the following companies (listed alphabetically) for permitting him to use portions of their copyrighted literatures in this book.

- American Technical Publishers
- Assofluid
- Bosch Rexroth
- C.C. Jensen Inc
- Donaldson
- Hydac
- Hydraulic and Pneumatic Magazine
- Lightening Reference Handbook (IFPS)
- MP Filtri
- MSOE
- Noria Corporation
- Pall Corporation
- Parker Hannifin
- Schroeder
- Spectro Scientific

Lastly, the author extends his thanks to the following sources of public information used to enrich the contents of the book.

www.ohfab.com
www.tricocorp.com
www. mecoil.net
www.metrohm.com
www.centerlinedistribution.com
www.oilmax.com
www.gallagherseals.com
www.capsnplugs.com
www.magneticfiltration.com

ABOUT THE BOOK

Book Description:

The book is targeting students and professionals who are looking to advance their fluid power careers. The book is colored and has the size of standard A4. This book is the second in a series that the author plans to publish to offer separate book for every hydraulic component. This book introduces knowledge foundation about hydraulic filters. The book introduces about various types of filters constructions, configurations, accessories. The book also introduces the various concepts of filtration mechanisms and filter media. The book overviews various types of materialistic contaminations such a fluidic, chemical, and particulate contamination. The book discusses filter selection criteria, maintenance, troubleshooting, and failure analysis of filters including the standard test methods for filter performance.

Book Objectives:

Chapter 1: Introduction to Hydraulic Filters
This chapter presents an overview of hydraulic filters including the contribution of filters in hydraulic systems, ISO1219 symbols, construction, and operating principles. The chapter also presents various types of filters based on application in which the filter is used, type of connection to the circuit, body style of the filter, placement in the hydraulic circuit. The chapter also discusses the added accessories to the filter such as bypass valve and clogging indicators. Examples from industry are presented.

Chapter 2: Filter Media and Filtration Mechanisms
This chapter presents an overview of filter elements including the construction and material of the filter media. This chapter discusses surface filters versus depth filters. The chapter discusses also the principles of various filtration mechanisms that are applicable in hydraulic filters such as direct interception, absorption, adsorption, and magnetic separation.

Chapter 3: Hydraulic Fluid Analysis
This chapter discusses standard methods for hydraulic fluid analysis including methods for particle and material analysis. The chapter covers the various standard cleanliness classes used to evaluate the contamination level in hydraulic fluids. The chapter also provides examples for interpretation of hydraulic fluid analysis reports.

Chapter 4: Fluidic Contamination
This chapter covers the sources of hydraulic fluids fluidic contamination. For each source, the chapter explains how the system performance will be affected and possible recommendations to minimize such consequences.

Chapter 5: Chemical Contamination

This chapter presents the sources of chemical contamination. For each source, the chapter explains how the system performance will be affected and possible recommendations to minimize such consequences.

Chapter 6- Particulate Contamination

This chapters presents the sources of particulate contamination. For each source, the chapter explains how the system performance will be affected and possible recommendations to minimize such consequences.

Chapter 7- Maintenance of Filters

This chapter provides guidelines for **Filters** selection, replacement, maintenance scheduling, installation, testing, storage and transportation. This chapter is supported by examples and figures granted by leading fluid power manufacturers.

Chapter 8- Filter Selection Criteria

This chapter presents a selection checklist as a guide for selecting proper filters. The chapter also discusses briefly the concepts for cost-effective filtration and selecting a filter cleanliness level based on system requirements. This chapter presents several examples of filtration solution for hydraulic systems.

Chapter 9- Troubleshooting and Failure Analysis of Filters

This chapter discusses hydraulic filters inspection, troubleshooting, and failure analysis. In this chapter, a troubleshooting chart for filter faults is presented. This chapter also presents examples of defective filters.

Note: you may notice that there are some duplications in the figures and body text. The reason is that the author wants to make each subject is a standalone chapter that can be taught independent from the other chapters.

Book Statistics:

The table shown below contains interesting statistical date about the textbook:

Chapter #	Pages	Figures	Tables	Words	Editing Time (Hours)
Chapter 1	56	69	0		189
Chapter2	20	26	0		181
Chapter 3	59	57	18		181
Chapter 4	30	22	3		172
Chapter 5	18	22	1		168
Chapter 6	41	43	5		194
Chapter 7	31	25	5		120
Chapter 8	12	8	2		165
Chapter 9	5	4	2		89
	272				1,459Hour = 61 Days

ABOUT THE AUTHOR

Medhat Khalil, Ph.D. is Director of Professional Education & Research Development at the Applied Technology Center, Milwaukee School of Engineering, Milwaukee, WI, USA. Medhat has consistently been working on his academic development through the years, starting from bachelor's and master's Degrees in Mechanical Engineering in Cairo Egypt and proceeding with his Ph.D. in Mechanical Engineering and Post-Doctoral Industrial Research Fellowship at Concordia University in Montreal, Quebec, Canada. He has been certified and is a member of many institutions such as: Certified Fluid Power Hydraulic Specialist (CFPHS) by the International Fluid Power Society (IFPS); Certified Fluid Power Accredited Instructor (CFPAI) by the International Fluid Power Society (IFPS); Member of Center for Compact and Efficient Fluid Power Engineering Research Center (CCEFP); Listed Fluid Power Consultant by the National Fluid Power Association (NFPA); and Listed Professional Instructor by the American Society of Mechanical Engineers (ASME). Medhat has balanced academic and industrial experience. Medhat has vast working experience in Fluid Power teaching courses for industry professionals. Being quite aware of the technological developments in the field of fluid power,

Medhat had worked for several world-wide recognized industrial organizations such as Rexroth in Egypt and CAE in Canada. Medhat had designed several hydraulic systems and developed several analytical and educational software. Medhat also has considerable experience in modeling and simulation of dynamic systems using Matlab-Simulink. Medhat has been selected among the inductees for Pioneers in fluid Power by NFPA (2012) and Hall of Fame in fluid Power by IFPS (2021).

Chapter 1

Introduction to Hydraulic Filters

Objectives

This chapter presents an overview of hydraulic filters including the contribution of filters in hydraulic systems, ISO1219 symbols, construction and operating principles. The chapter also presents various types of filters based on application in which the filter is used, type of connection to the circuit, body style of the filter, placement in the hydraulic circuit. The chapter also discusses the added accessories to the filter such as bypass valve and clogging indicators. Examples from industry are presented.

Brief Contents

1.1 - Contribution of Filters in hydraulic Systems

1.2 - Types of Filters Based on Application

1.3 - Types of Filters Based on Types of Contamination

1.4 - Interpretation of ISO 1219 Symbols for Hydraulic Filters

1.5 - Basic Construction and Operation of Hydraulic Filters

1.6 - Types of Filters Based on Hydraulic Connections

1.7 - Types of Filters Based on the Filter Body Style

1.8 - Filter Clogging Indicators

1.9 - Types of Filters Based on their Placement in the Circuit

Chapter 1 – Introduction to Hydraulic Filters

1.1 - Contribution of Filters in hydraulic Systems

Every hydraulic system has suspended particles in its fluid. Maintaining clean hydraulic fluid is 80% of the effort required to maintain a reliable hydraulic system.

Hydraulic filters are classified based on:
- Application.
- Place in a hydraulic circuit.
- Body style.
- Types of contamination.

Filters are also available in:
- Different sizes.
- Dirt holding capacity.
- Contamination removal efficiency.

A properly selected filter must perform the following tasks:
- Remove particulate contaminants from the hydraulic fluid.
- Maintain required cleanliness level as determined by the system manufacturer.
- Remove chemical contaminants and their products (varnish, sludge, etc.) from the fluid.
- Prevent aging of the hydraulic fluid due to chemical contaminants.
- Maintain the lubricity of the fluid.
- Extend the life of the hydraulic fluid.
- Remove water content in the fluid.
- Absorb moisture from breathing reservoirs.
- Permit preventive maintenance.
- Increase component life and system reliability.
- Increase intervals between scheduled maintenance.
- Prevent unexpected failures and consequent costs of unplanned shutdown.

1.2 - Types of Filters Based on Application

As shown in Fig. 1.1, hydraulic filters serving hydraulic systems in a wide range of both industrial and mobile applications. Wherever a hydraulic system is used, at least one filter is required.

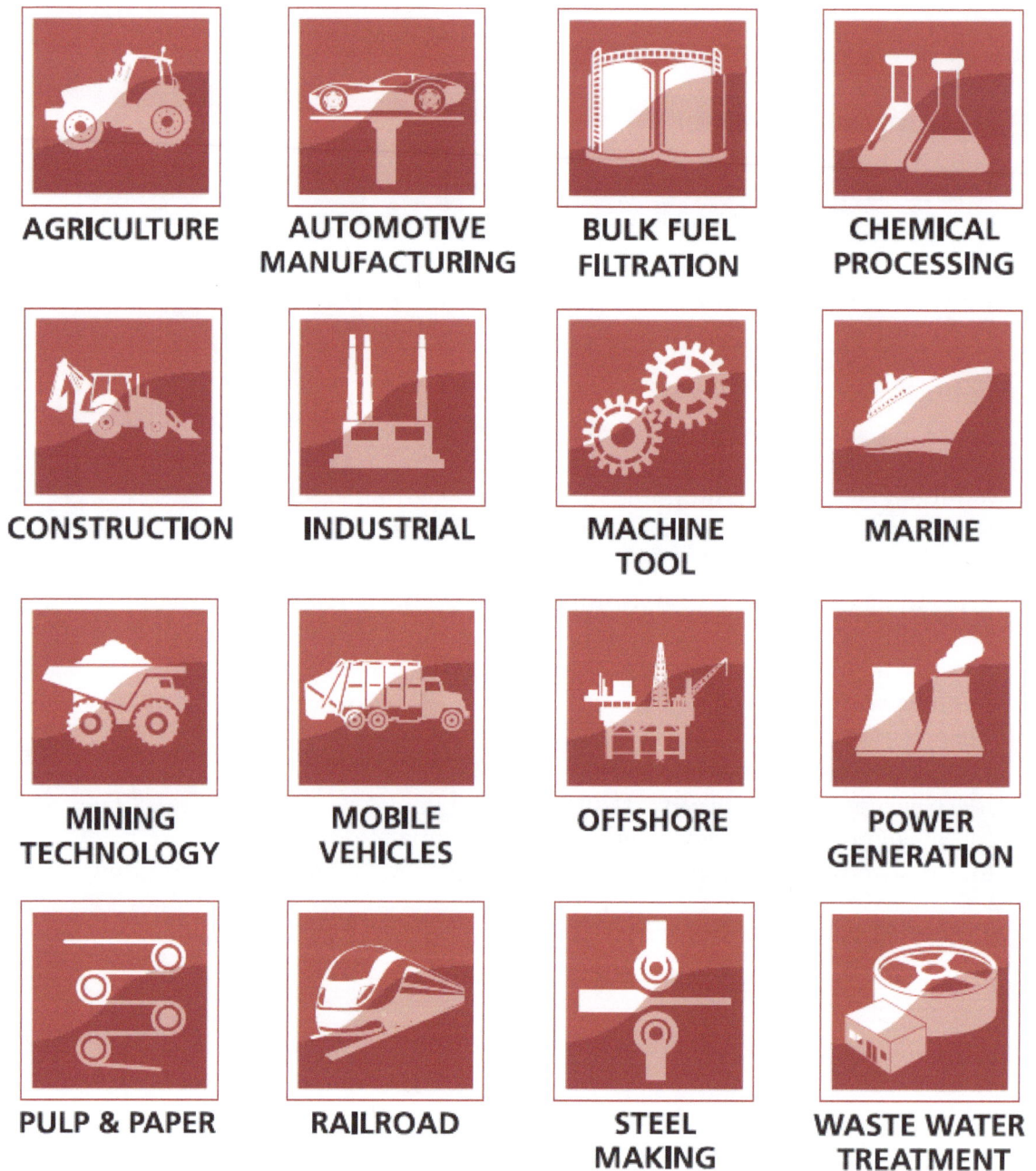

AGRICULTURE   AUTOMOTIVE MANUFACTURING   BULK FUEL FILTRATION   CHEMICAL PROCESSING

CONSTRUCTION   INDUSTRIAL   MACHINE TOOL   MARINE

MINING TECHNOLOGY   MOBILE VEHICLES   OFFSHORE   POWER GENERATION

PULP & PAPER   RAILROAD   STEEL MAKING   WASTE WATER TREATMENT

Fig. 1.1- Applications of Hydraulic Filters (Courtesy of Schroeder)

Figures 1.2 and 1.3 show hydraulic fluid filters among filtration solutions for a tractors and excavators; respectively.

Fig. 1.2- Filtration Solutions for Tractors (Courtesy of Donaldson)

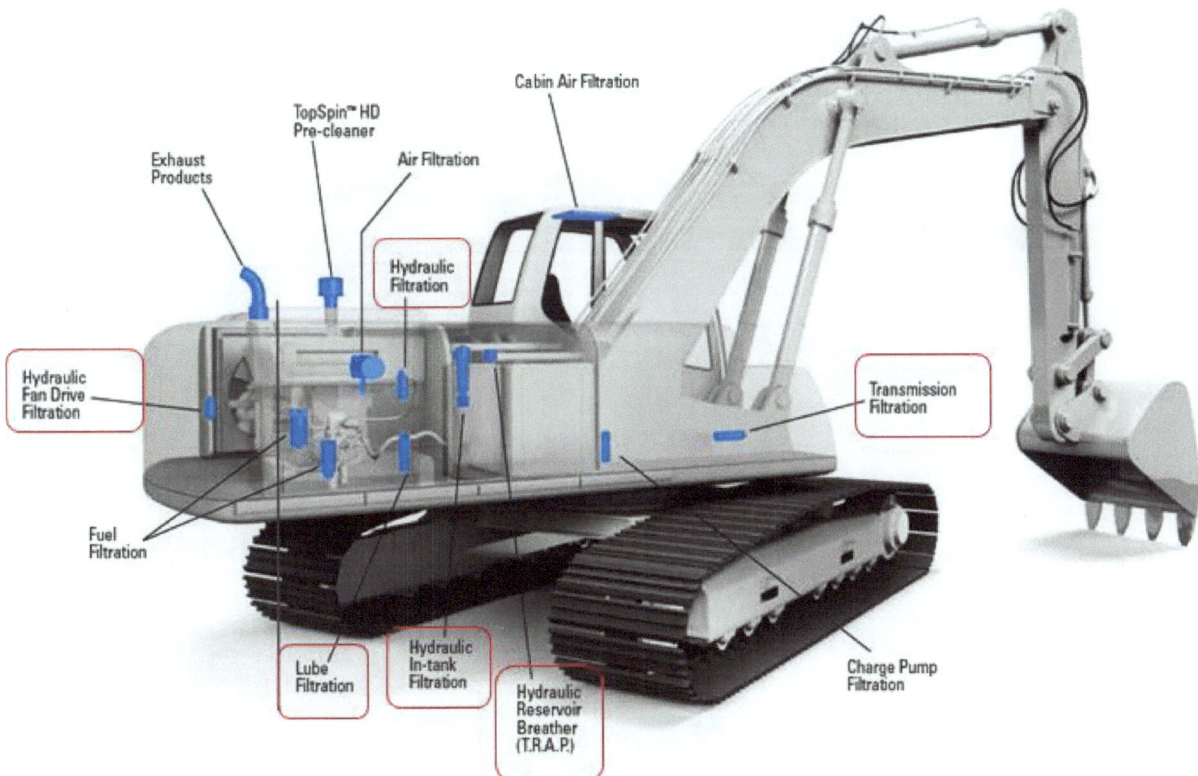

Fig. 1.3- Filtration Solutions for Excavators (Courtesy of Donaldson)

Figures 1.4 and 1.5 show hydraulic fluid filters among filtration solutions for a Dump Trucks and industrial hydraulic power units; respectively.

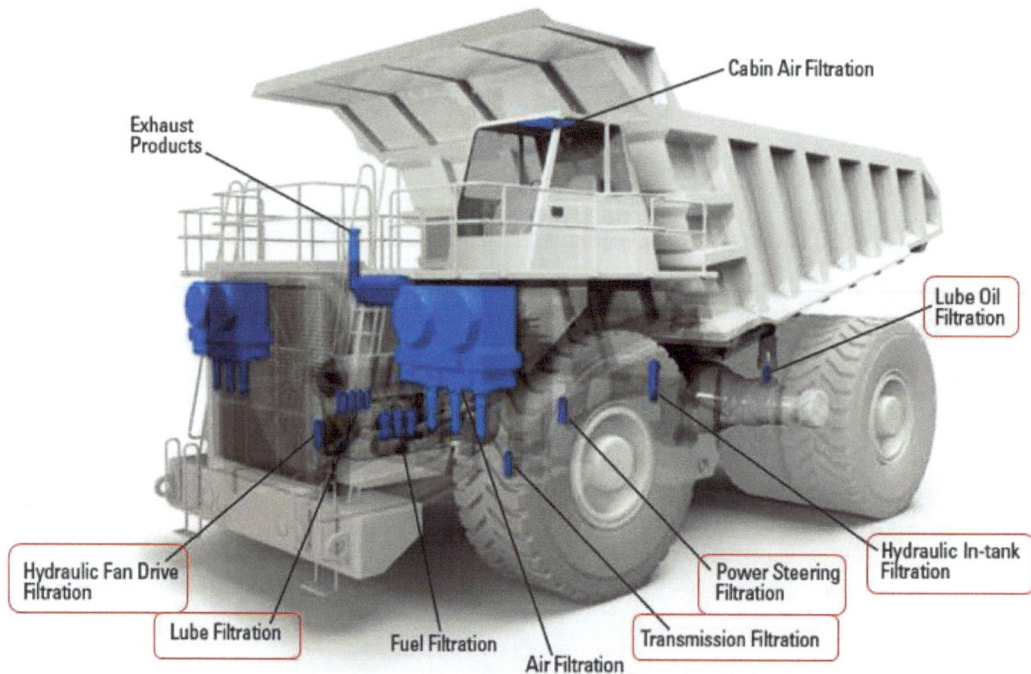

Fig. 1.4- Filtration Solutions for Dump Trucks (Courtesy of Donaldson)

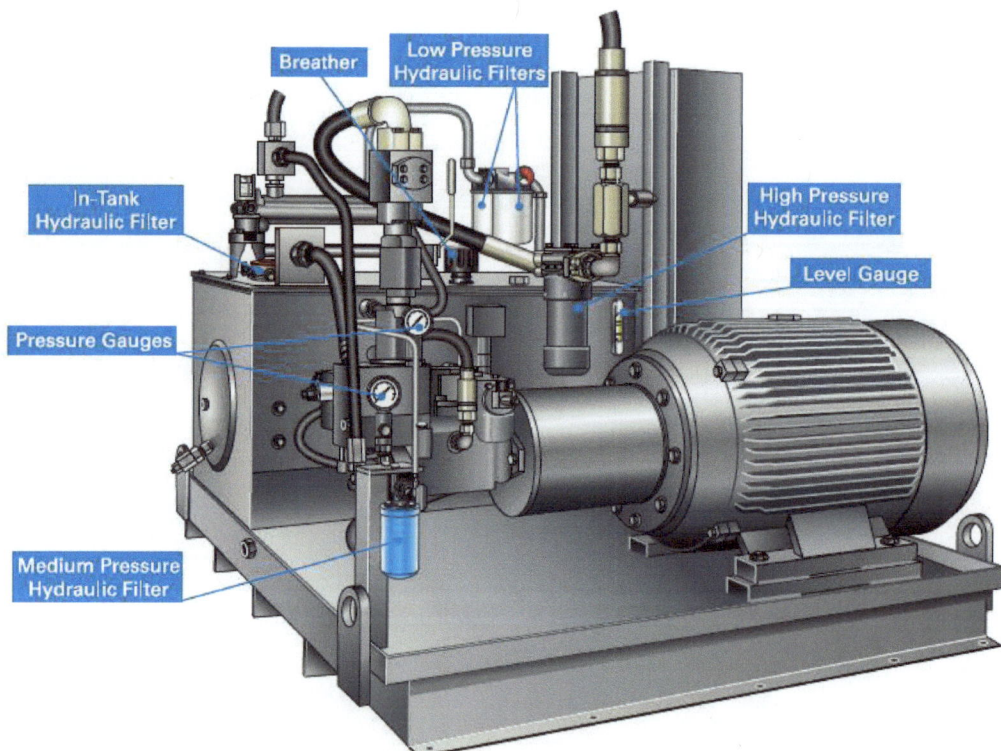

Fig. 1.5- Filtration Solutions for Industrial Hydraulic Power Units (Courtesy of Donaldson)

1.3 - Types of Filters Based on Types of Contamination

Hydraulic filters are classified based on the types of contamination as follows:

- Filters for *Fluidic Contaminants*. These filters are used to remove water content in the hydraulic fluids.

- Filters for *Chemical Contaminants*. These filters are used to remove products of hydraulic fluids breaking down, such as oxidations, varnish and sludges.

- Filters for *Particulate Contaminants*. These filters are used to remove particulate contaminates whether solid or elastic, abrasive or nonabrasive, and metallic or nonmetallic.

1.4 - Interpretation of ISO 1219 Symbols for Hydraulic Filters

Like all hydraulic components, hydraulic filters are presented in symbols that reflect the basic construction and function of the filter. Figure 1.6 shows the symbols for hydraulic filters as follows:

1. Non-Bypass-Filter.
2. Bypass-Filter is a hydraulic filter with built-in bypass valve.
3. Bypass-Filter with visual clogging indicator based on inlet pressure.
4. Bypass-Filter with visual clogging indicator based on differential pressure.
5. Sandwich-mounted non-bypass-Filter with visual clogging indicator based on differential pressure.
6. Bypass-Filter with Self-Cleaning feature. Self-Cleaning filter is a large filter that is equipped with a manual or an automatic self-cleaning mechanism. This mechanism wipes the outer surface of the filter element and the inner surface of the filter housing without stopping the machine. Such a filter contains surface type filter media.
7. Non-bypass-filter with electrical clogging indicator.
8. Bypass-to-Tank Filter. Used when element collapse Pressure < max working pressure. It protects sensitive components by directing dirty fluid to tank.

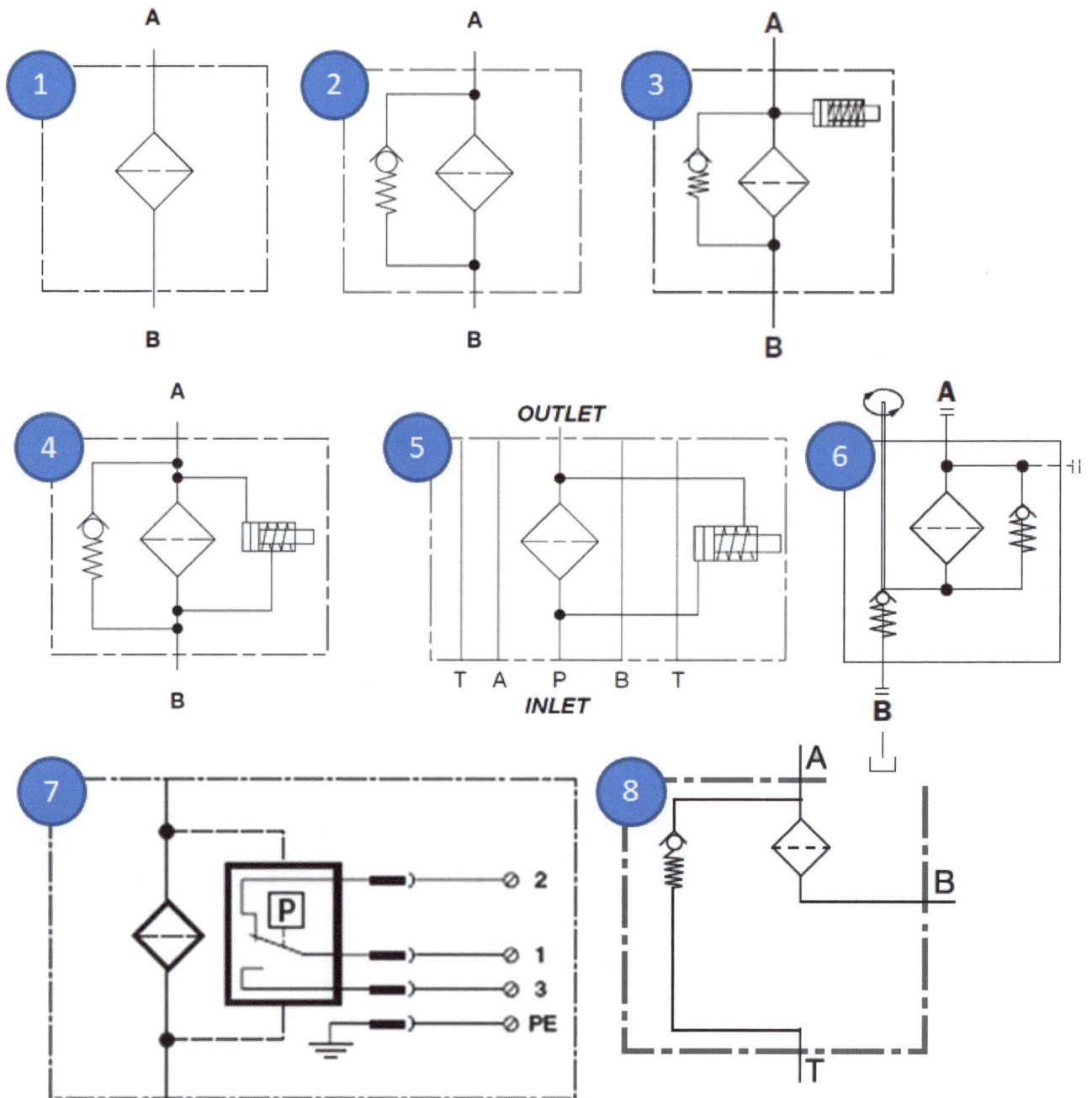

Fig. 1.6- Examples of Hydraulic Filters Symbols

1.5- Basic Construction and Operation of Hydraulic Filters

As shown in Fig. 1.7, basic construction of hydraulic filters contains the following elements:

Filter Housing: Is also referred to as "*Bowl*". It is the pressure vessel of the filter, secures the element and creates a seal between the element inlet and outlet areas. It also has a drain plug for draining before disassembling the fitter.

Filter Head: The head contains the port and possibly other options such as a bypass valve and clogging indicator. It also contains the filter mounting method. The primary concern in selecting the housing is the pressure rating. This should be determined before the housing style is selected.

Filter Element: Is also referred to as "*Filter Cartridge*". It consists of a central tube, media to remove specific type of contaminants, and end caps.

Clogging Indicator: Is to show the level of clogging of the filter media. It provides an alarm to replace the element before the bypass valve setting is reached. Replacing the element before the bypass opens prevents significant reduction in filtering efficiency.

Bypass Valve: A bypass valve is an optional feature in hydraulic filters. So, a filter may be specified as follows:

- <u>Bypass-Filters:</u> The *bypass-filter* allows the fluid to flow directly to the system when the filter media is clogged in order to limit the differential pressure across the media and prevents media collapse. Standard filter assemblies normally have a bypass valve cracking pressure between 1.5 – 7 bar (25 - 100 psi). When specifying a bypass-filter, it can generally be assumed that the manufacturer has designed the element to withstand the bypass valve differential pressure when the bypass valve opens. It is to be noted that, when the bypass valve opens, all contaminants will get into the system. Some systems can withstand short term operation with this contamination.
- <u>Non-Bypass Filters:</u> When sensitive components are used, such as servo and proportional valves, *Non-Bypass-Filter* prevents any unfiltered flow from going downstream. A low collapse pressure element should never be used in a non-bypass type housing. When contaminate buildup causing excessive differential pressure, the element may rupture or collapse. Hence, large quantities of contamination are immediately induced into the system, with severe problems. Therefore, when specifying a non-bypass filter design, make sure that the element has a differential pressure rating greater than maximum operating pressure on the filtration line.
- <u>Bypass-to-Tank Filters:</u> Alternative to non-bypass filters, and to avoid using high collapse pressure filter elements, a *Bypass-to-Tank* is another option. This allows the unfiltered bypass flow to return to tank through a third port, preventing unfiltered bypass flow from entering the system.

Figure 1.8 shows that hydraulic fluid introduced to the inlet port of the filter. The fluid is then forced into the cartridge from the outside surface to inside surface in order to utilize the larger area at the outside surface to retain maximum amount of dirt. Clean fluid is then pass through the central tube to the outlet port. In case of clogged filter media, bypass valve opens.

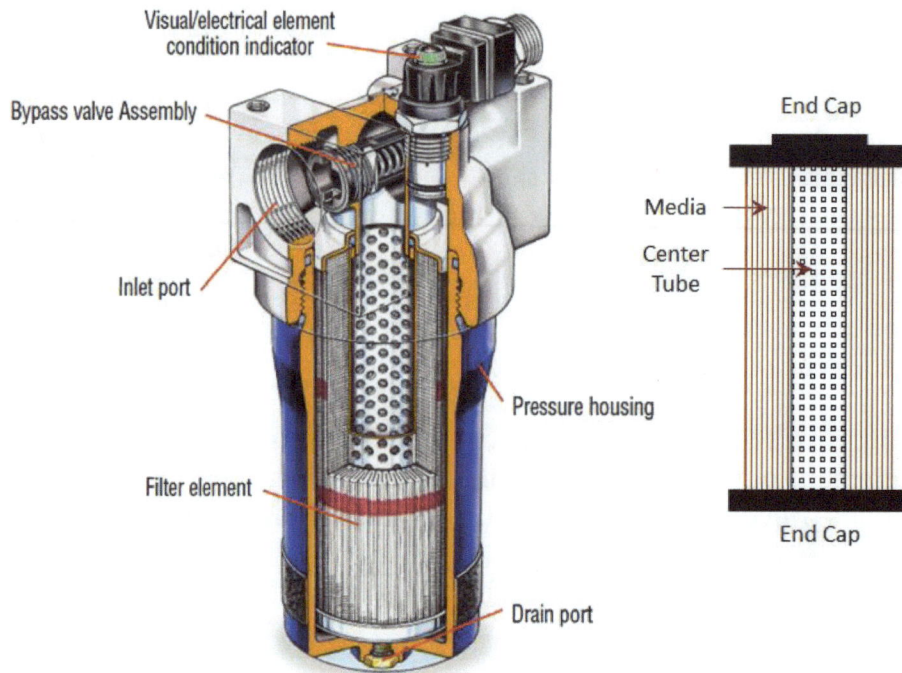

Fig. 1.7- Basic Construction of Hydraulic Filter (Courtesy of Parker)

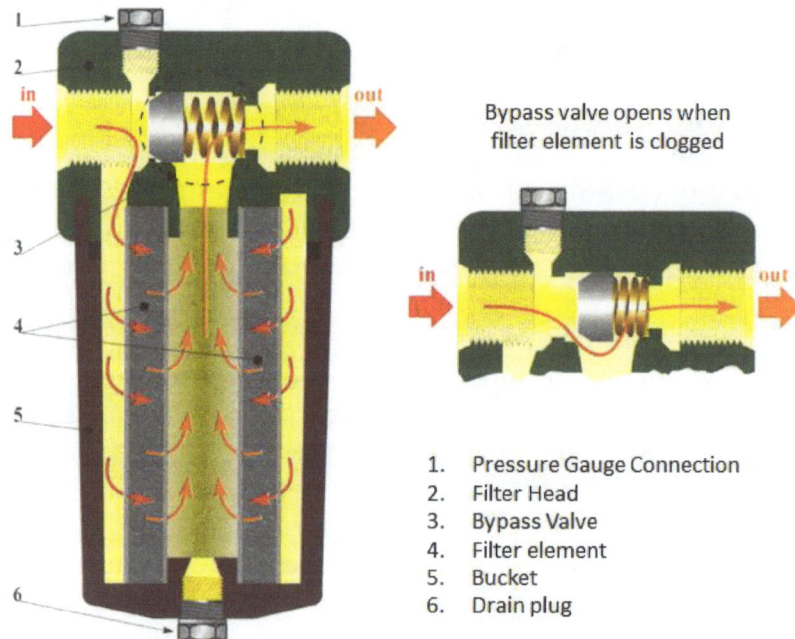

1. Pressure Gauge Connection
2. Filter Head
3. Bypass Valve
4. Filter element
5. Bucket
6. Drain plug

Fig. 1.8- Basic Construction of Hydraulic Filter (Courtesy of Assofluid)

1.6 – Types of Filters Based on Hydraulic Connections

As shown in Fig. 1.9, hydraulic filters can be connected to hydraulic system by one of the following methods:

Line-Mounted Filters: *Line Mounted* filters are assembled directly on a hydraulic line.

Flange-Mounted Filters: *Flange-Mounted* filters are assembled on one side of a manifold.

Sandwich-Mounted Filters: *Sandwich-Mounted* filters are assembled in between two standard manifolds or subplates.

Depending on the filter body style, a filter element in any filter can be replaced without disassembling the filter head from the system.

| Line-Mounted | Flange-Mounted | Sandwich-Mounted |

Fig. 1.9- Types of Filters Based on Hydraulic Connections

1.7 – Types of Filters Based on the Filter Body Style

As shown in Fig. 1.10, hydraulic filters are configured in various body styles in order to ease the assembly and to make it accessible during maintenance. The following sections shows examples of hydraulic filters body styles. Both pressure and return filters are available in a duplex version.

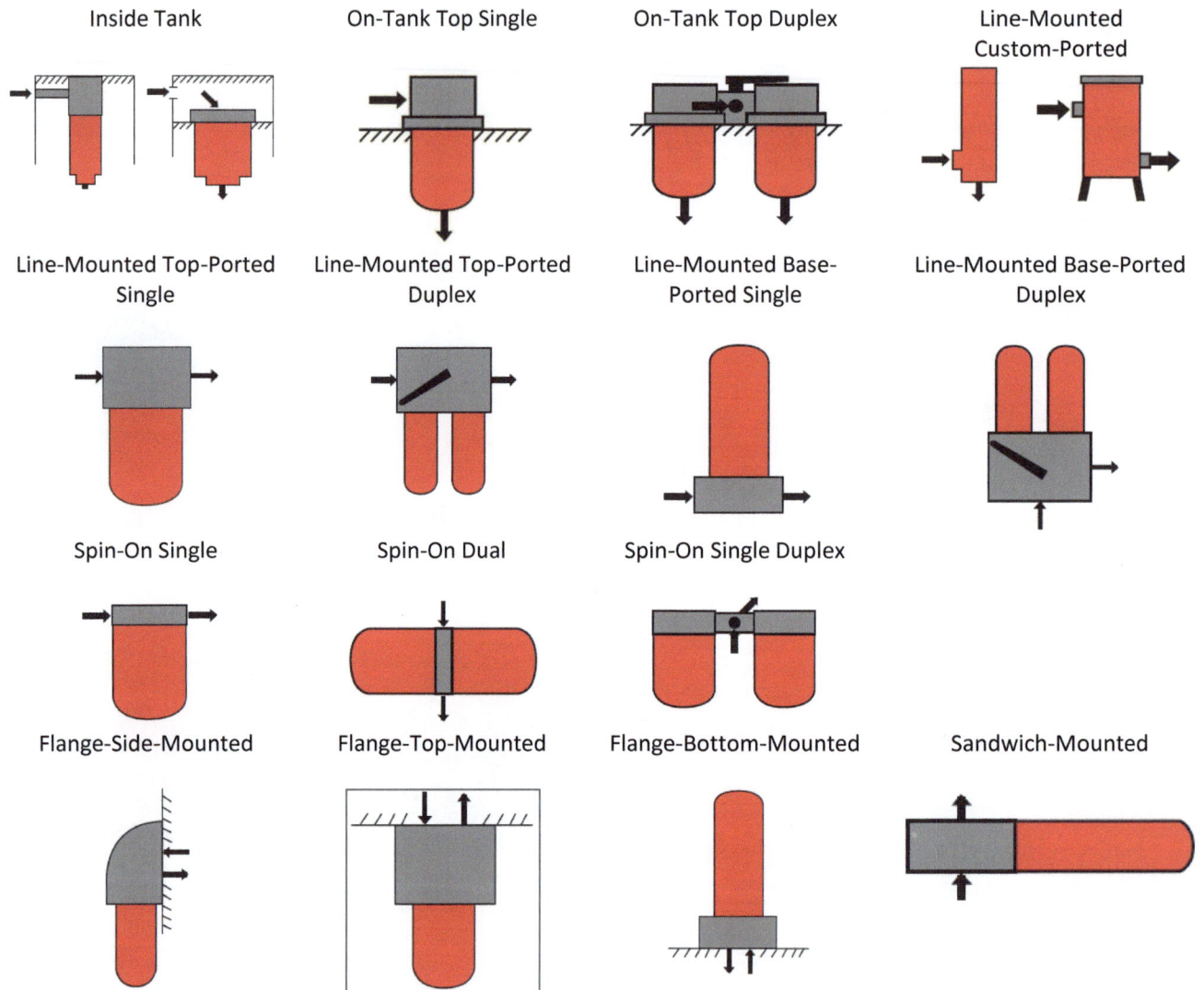

Fig. 1.10- Types of Filters Based on the Filter Body Style (Courtesy of Hydac)

1.7.1 – Inside Tank Filters

Figure 1.11 shows a unique design of *Inside Tank* filters that allow filter to be completely installed inside the reservoir. This saves space, protects the filter, and reduces leak points.

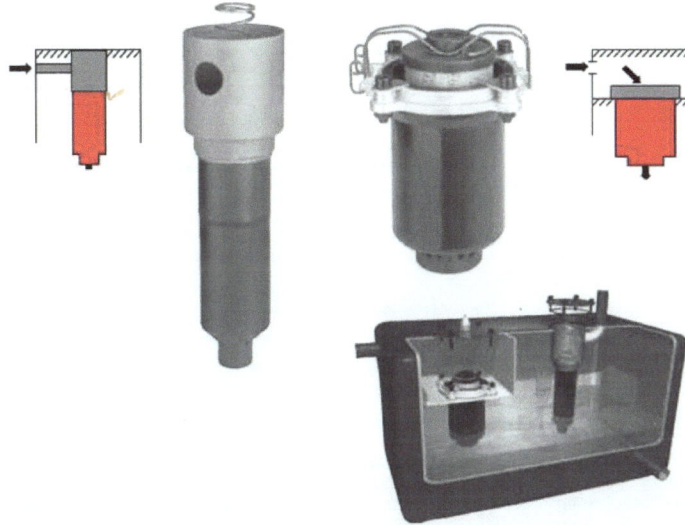

Fig. 1.11 Examples of Inside Tank Filters RFM-Series (Courtesy of Hydac)

1.7.2 – On-Tank Top Single Filters

Figure 1.12 shows various brands of *On-Tank Top Single* filters. Such filters installed where the head and the inlet/outlet ports are accessible from above the tank, while the housing remains inside the tank, offering system design flexibility.

| SF Series | ST Series | UT610 Series |
| (Courtesy of Hydac) | (Courtesy of Schroeder) | (Courtesy of Pall) |

Fig. 1.12- Examples of On-Tank Top Single Filters

Figure 1.13 shows a model of an *On-Tank Top Single* filter. This model offers a variety of options including aluminum or plastic access covers. Optional air breather featuring T.R.A.P.™ technology is available eliminating the cost associated with an additional penetration to the hydraulic tank for breather installation.

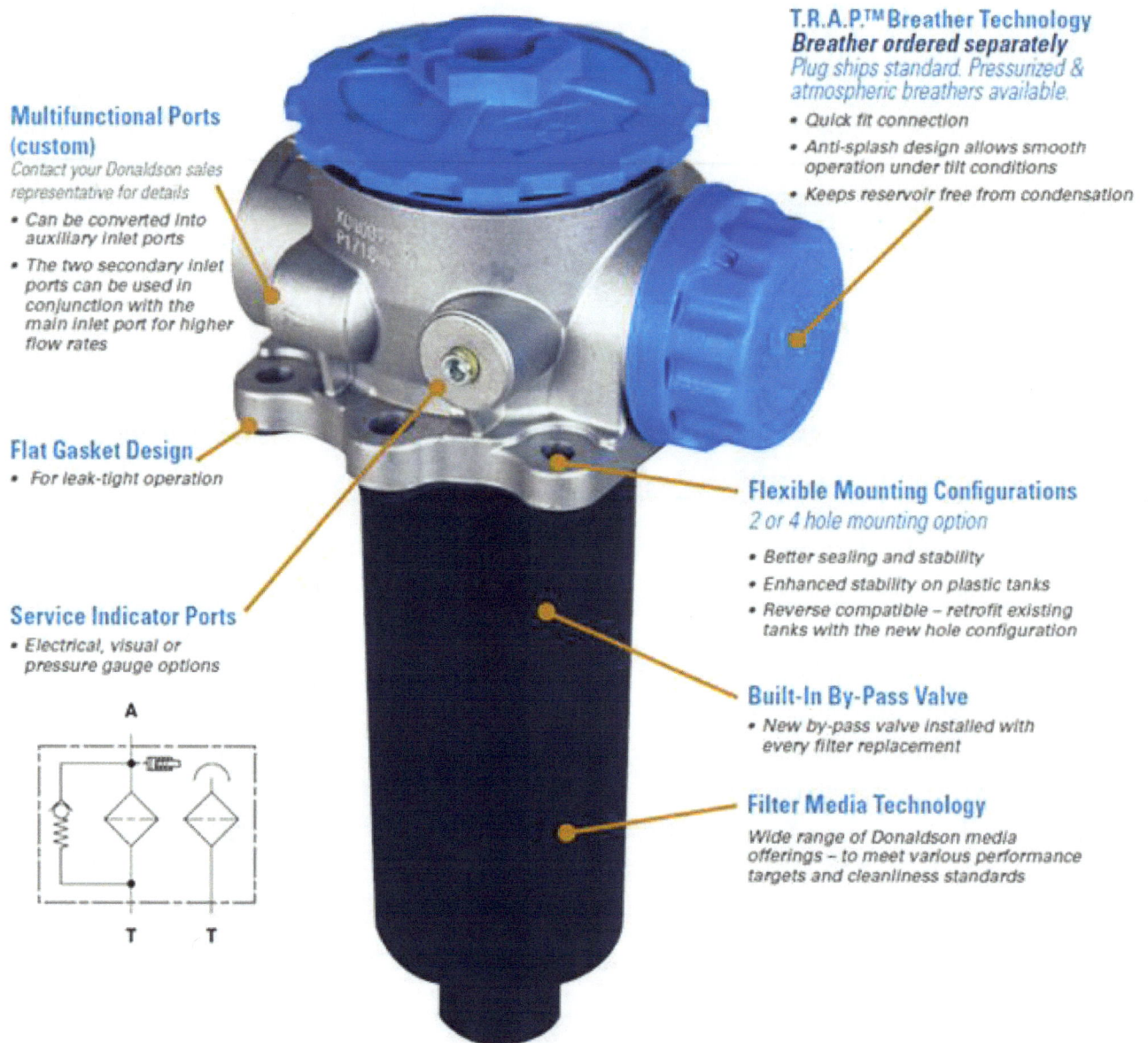

T.R.A.P.™ Breather Technology
Breather ordered separately
Plug ships standard. Pressurized & atmospheric breathers available.
- *Quick fit connection*
- *Anti-splash design allows smooth operation under tilt conditions*
- *Keeps reservoir free from condensation*

Multifunctional Ports (custom)
Contact your Donaldson sales representative for details
- *Can be converted into auxiliary inlet ports*
- *The two secondary inlet ports can be used in conjunction with the main inlet port for higher flow rates*

Flat Gasket Design
- *For leak-tight operation*

Service Indicator Ports
- *Electrical, visual or pressure gauge options*

Flexible Mounting Configurations
2 or 4 hole mounting option
- *Better sealing and stability*
- *Enhanced stability on plastic tanks*
- *Reverse compatible – retrofit existing tanks with the new hole configuration*

Built-In By-Pass Valve
- *New by-pass valve installed with every filter replacement*

Filter Media Technology
Wide range of Donaldson media offerings – to meet various performance targets and cleanliness standards

Fig. 1.13- Example of On-Tank Top Single Filter - FIK Series (Courtesy of Donaldson)

1.7.3 – On-Tank Top Duplex Filters

Both pressure and return filters are available in a duplex version. Such filters are made of two filter chambers and includes the necessary valving. One filter works at a time to allow for continuous, uninterrupted filtration. When a filter element needs servicing, the duplex valve is shifted, diverting flow to the opposite filter chamber. The dirty element can then be changed, while filtered flow continues to pass through the filter assembly. The duplex valve typically is an open cross-over type, which prevents any flow blockage. Figure 1.14 shows various models of *On-Top Tank Duplex* filters.

www.behringersystems.com   www.hydraulicoilfilters.com

Fig. 1.14- Examples of On-Tank Top Duplex Filters

1.7.4 – Line-Mounted Top-Ported Single Filters

Figure 1.15 shows various models for *Line-Mounted Top-Ported Single* filters.

LF Series
(Courtesy of Hydac)

LMP900 - 901 Series
(Courtesy of MP Filtri)

CTF60 Series
(Courtesy of Schroeder)

Fig. 1.15- Examples of Line-Mounted Top-Ported Single Filters

Figure 1.16 and 1.17 show detailed construction of line-mounted top-ported single filters.

Integrated By-pass Valve
Robust, proven design

Unique Head to Cartridge Interface Connection

RadialSeal™ Sealing Technology
- *No metal-to-metal contact – downstream flow*
- *Robust, reliable seal on clean side of filter – prevents cross contamination of oil*

Filter Cartridge
- *Double wire mesh support on outside of cartridge maintains pleat spacing under high pressure differential*
- *Locking grab handles makes for cleaner servicing and simplifies filter position during servicing*

Industrial Hand Grips
No special servicing tools needed

Locking Grab Handles
Cleaner, easier servicing

RadialSeal™ Sealing Technology
- *No metal-to-metal contact – upstream flow*
- *Easy-to-torque, mistake-proof sealing*
- *Robust, reliable seal*

Anti-dust Seal
- *Keeps threads free from contamination*
- *Easier to remove and reassemble during service*

Synteq XP Media Technology
Delivers high performance – lower pressure drop, superior cold-start filtration and extended filter life

Closed End Cap
Eliminates the possibility of contamination to clean side of assembly during servicing

Oil Drain Port
Oil drain port used to drain oil during servicing

IMPORTANT SERVICE INSTRUCTIONS:
To prevent thread damage when installing new filter, fully lubricate the entire thread and o-ring surface with a Molybdenum-containing gear oil or anti-seize paste such as Schaeffer #214S Supreme One 80W-140 gear oil or Dow Corning Molykote P-37 anti-seize past.

Fig. 1.16- Example of Line-Mounted Top-Ported Single Filters - FLK Series (Courtesy of Donaldson)

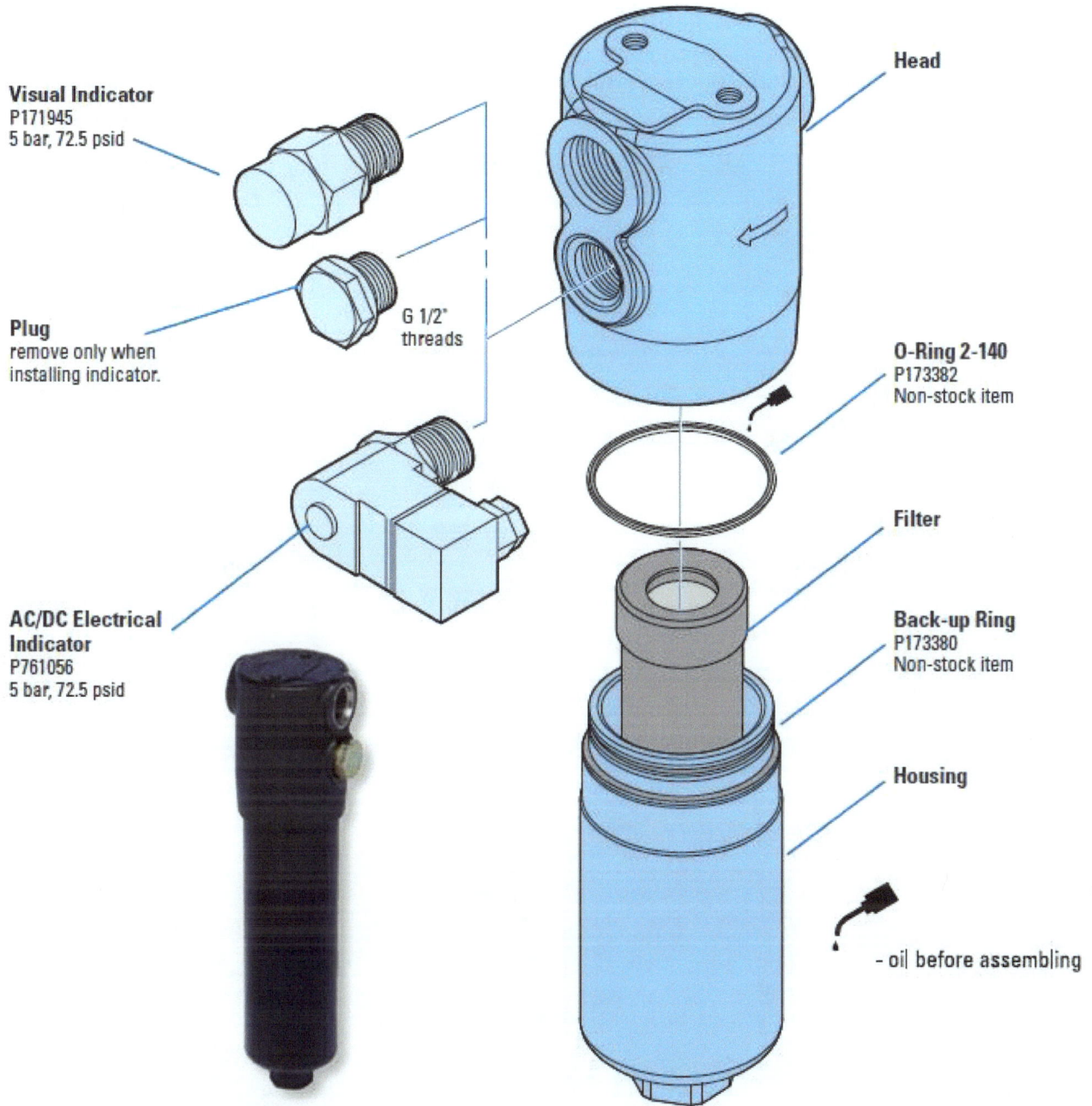

Visual Indicator
P171945
5 bar, 72.5 psid

Plug
remove only when
installing indicator.

G 1/2"
threads

**AC/DC Electrical
Indicator**
P761056
5 bar, 72.5 psid

Head

O-Ring 2-140
P173382
Non-stock item

Filter

Back-up Ring
P173380
Non-stock item

Housing

- oil before assembling

**Fig. 1.17- Example of Line-Mounted Top-Ported Single Filters - FPK02 Series
(Courtesy of Donaldson)**

1.7.5 – Line-Mounted Top-Ported Duplex Filters

Figure 1.18 shows various models of *Line-Mounted Top-Ported Duplex* filters.

LMD Series
(Courtesy of MP Filtri)

PLD Series
(Courtesy of Schroeder)

DPK2400 Series
(Courtesy of Donaldson)

Fig. 1.18- Examples of Inline Top-Ported Duplex Filters

1.7.6 – Line-Mounted Base-Ported Single Filters

Figure 1.19 shows various models for *Line-Mounted Base-Ported Single* filters.

HF4P Series
(Courtesy of Hydac)

Series Athalon™ UH210
(Courtesy of Pall)

KF30 Series
(Courtesy of Schroeder)

Fig. 1.19- Examples of Line-Mounted Base-Ported Single Filters

1.7.7 – Line-Mounted Base-Ported Duplex Filters

Figure 1.20 shows an example of *Line-Mounted Base-Ported Duplex* filters.

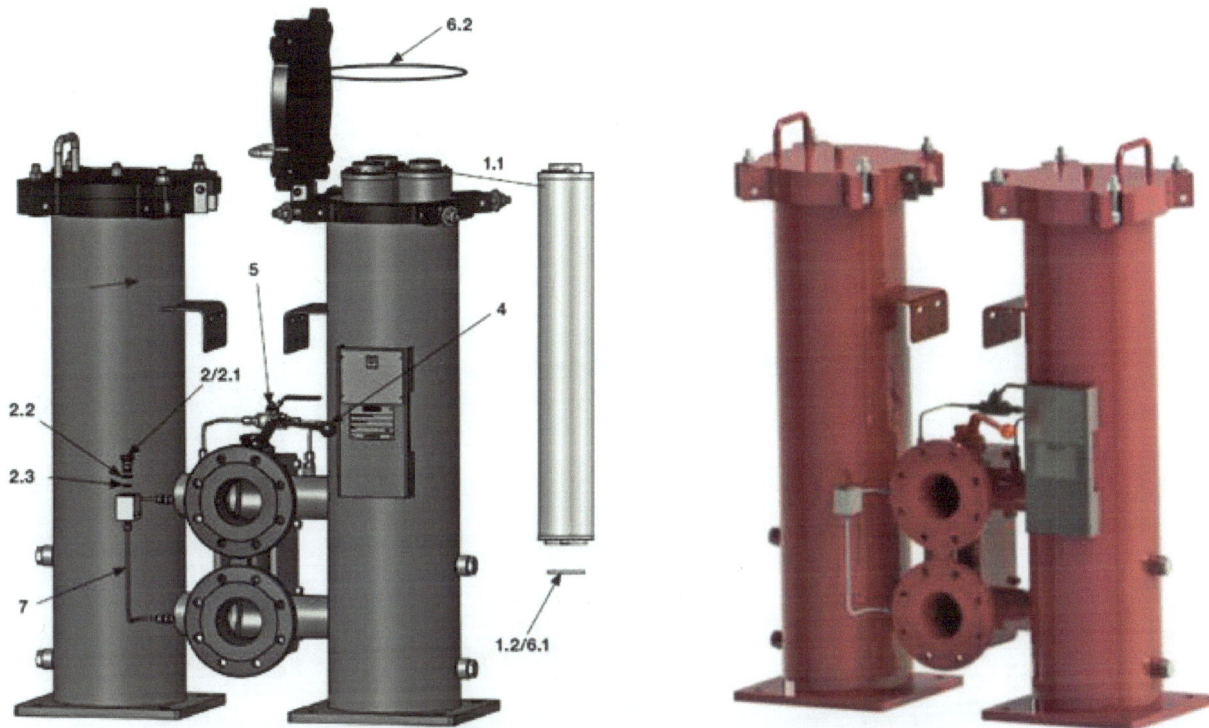

Item	Consists of	Designation
1.		**Filter element**
	1.1	Filter element
	1.2	O-ring
		No. of elements per filter side / size
2.		**Indicator plug VD 0 A 1.0 /-V**
	2.1	Clogging indicator or indicator plug
	2.2	Profile seal ring
	2.3	O-ring
3.		**SEAL KIT VD/VM/VR/VR FKM**
4.		**Lever for change-over valve**
5.		**Equalization line ball valve**
6.		**SEAL KIT RFLD...FKM**
	6.1	O-ring *(element)*
	6.2	Lid seal
7.		**Indicator and equalization line pipe and plumbing**

**Fig. 1.20- Example of Line-Mounted Base-Ported Duplex Filters - RFLDH Series
(Courtesy of Hydac)**

Figure 1.21 shows another example of *Line-Mounted Base-Ported Duplex* filters.

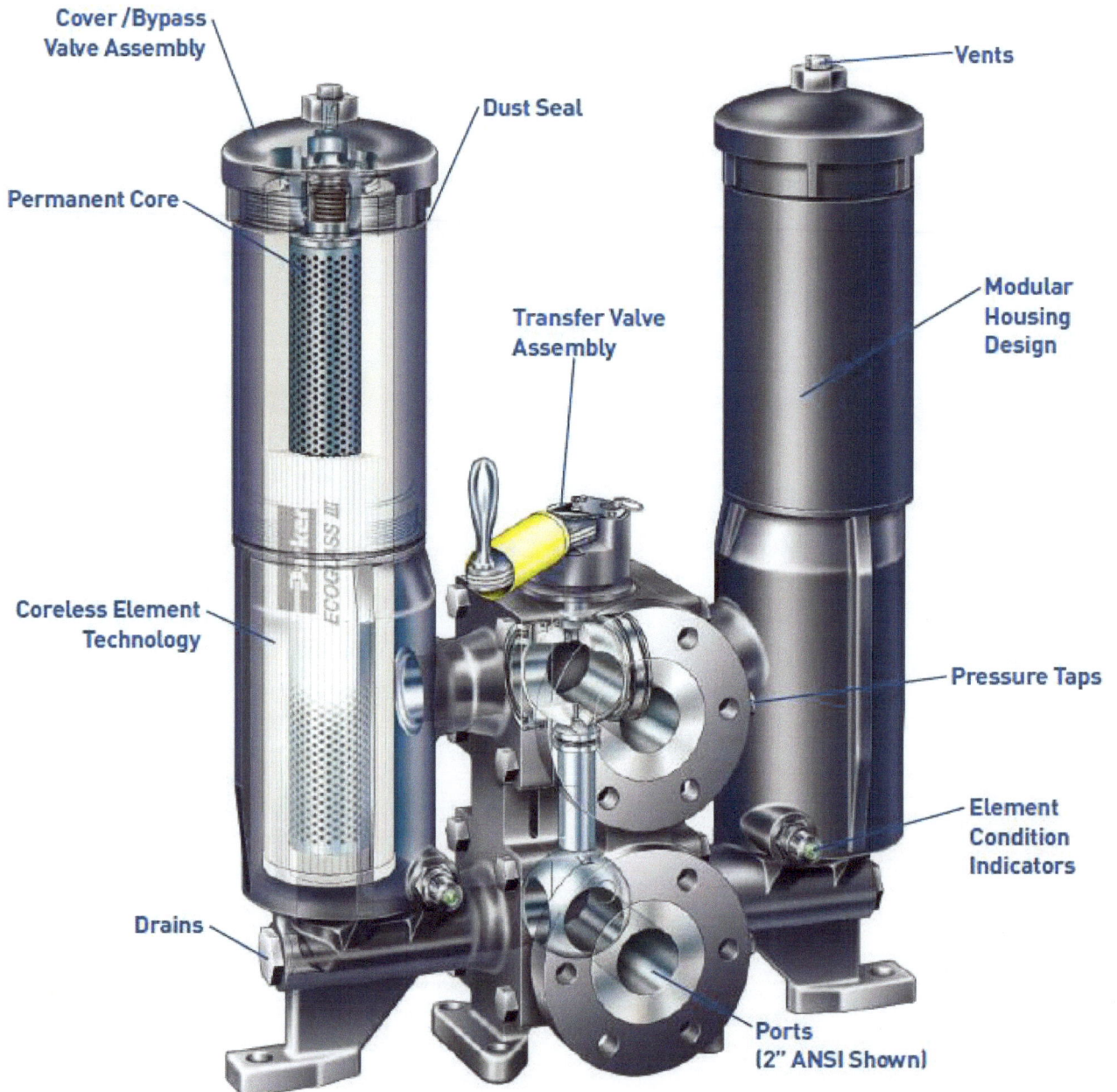

Fig. 1.21- Example of Line-Mounted Base-Ported Duplex Filters - MPD Series (Courtesy of Parker)

1.7.8 – Line-Mounted Custom-Ported Filters

Figure 1.22 shows various models of *Line-Mounted Custom-Ported* filters. In such filters, the housing is made of rolled steel or stainless steel. ANSI flange connections for each filter size provide maximum connection flexibility eliminating additional adapters and intermediate flanges.

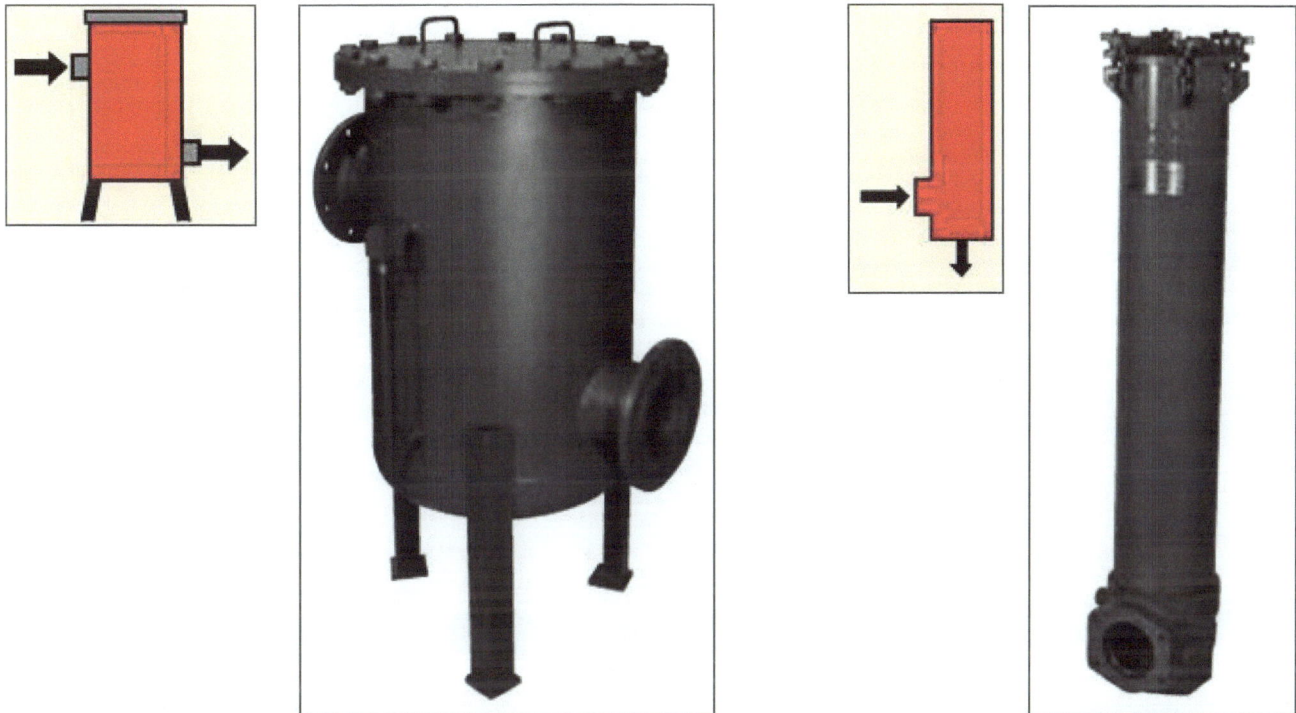

**Fig. 1.22- Examples of Line-Mounted Custom-Ported Filters – RFL Series
(Courtesy of Hydac)**

Figures 1.23 and 1.24, shows detaled construction of *Line-Mounted Custom-Ported* filters.

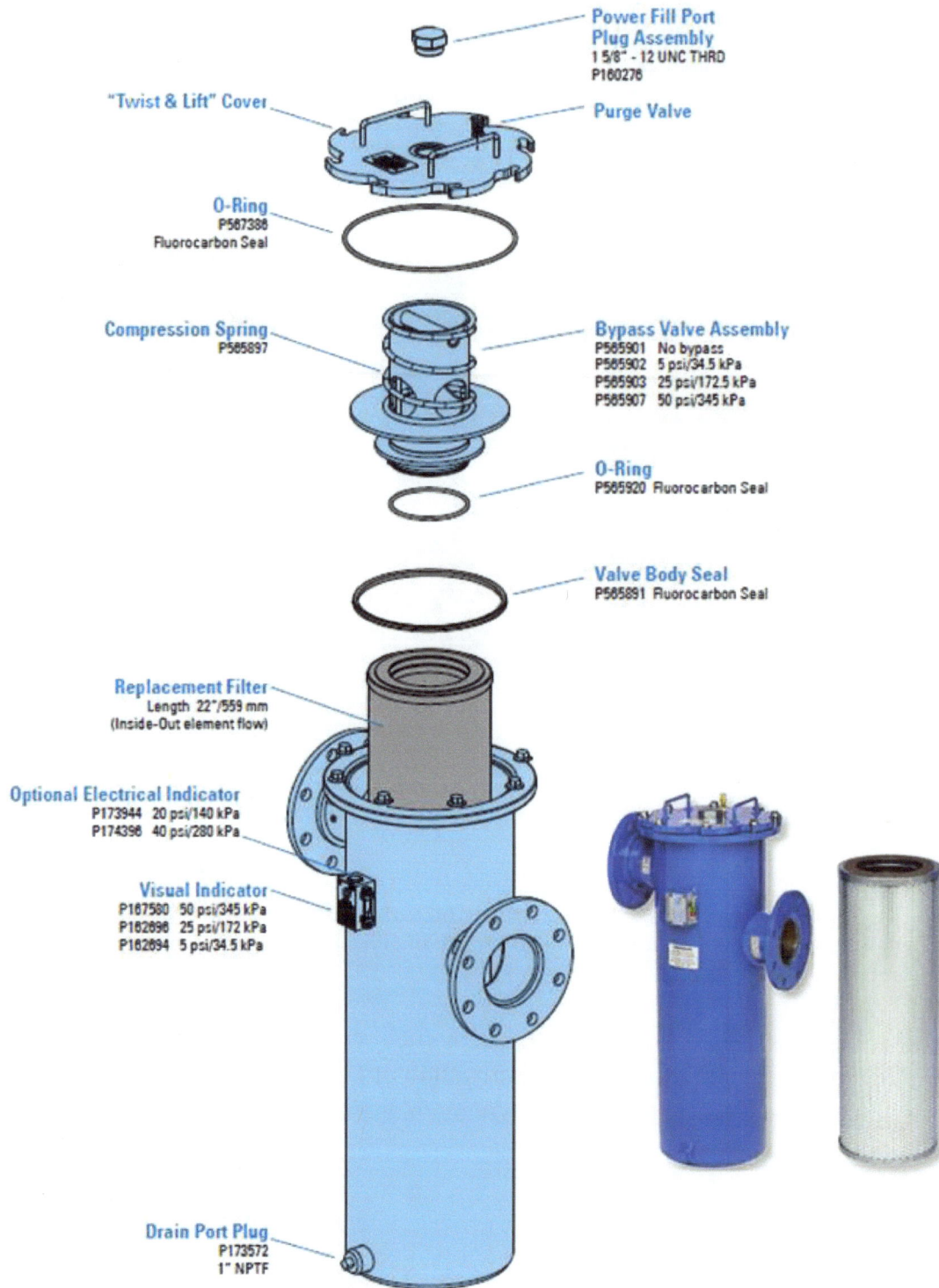

Power Fill Port Plug Assembly
1 5/8" - 12 UNC THRD
P160278

"Twist & Lift" Cover

Purge Valve

O-Ring
P567388
Fluorocarbon Seal

Compression Spring
P565897

Bypass Valve Assembly
P565901 No bypass
P565902 5 psi/34.5 kPa
P565903 25 psi/172.5 kPa
P565907 50 psi/345 kPa

O-Ring
P565920 Fluorocarbon Seal

Valve Body Seal
P565891 Fluorocarbon Seal

Replacement Filter
Length 22"/559 mm
(Inside-Out element flow)

Optional Electrical Indicator
P173944 20 psi/140 kPa
P174398 40 psi/280 kPa

Visual Indicator
P167580 50 psi/345 kPa
P162896 25 psi/172 kPa
P162894 5 psi/34.5 kPa

Drain Port Plug
P173572
1" NPTF

Fig. 1.23- Example of Line-Mounted Custom-Ported Filters - HRK10 Series
(Courtesy of Donaldson)

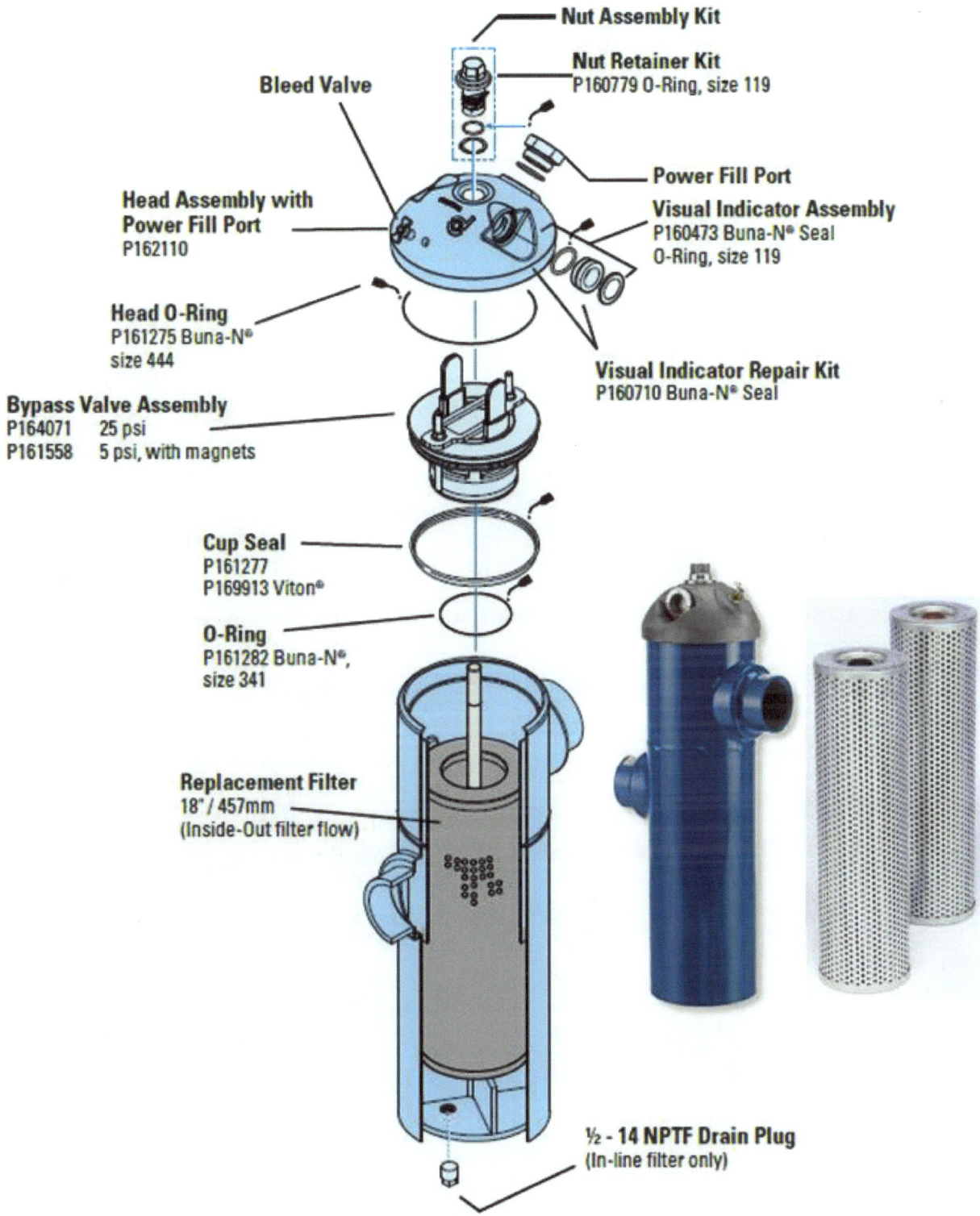

Nut Assembly Kit

Nut Retainer Kit
P160779 O-Ring, size 119

Bleed Valve

Power Fill Port

Head Assembly with Power Fill Port
P162110

Visual Indicator Assembly
P160473 Buna-N® Seal
O-Ring, size 119

Head O-Ring
P161275 Buna-N®
size 444

Visual Indicator Repair Kit
P160710 Buna-N® Seal

Bypass Valve Assembly
P164071   25 psi
P161558   5 psi, with magnets

Cup Seal
P161277
P169913 Viton®

O-Ring
P161282 Buna-N®,
size 341

Replacement Filter
18" / 457mm
(Inside-Out filter flow)

½ - 14 NPTF Drain Plug
(In-line filter only)

**Fig. 1.24- Example of Line-Mounted Custom-Ported Filters - HFK08 Series
(Courtesy of Donaldson)**

1.7.9 – Spin-On Single Filters

Figure 1.25 shows the exterior shape of *Spin-On* filters. Such filters feature non-welded screw-in housing design Figure 1.26 shows detailed construction of a single *Spin-On* filter consists.

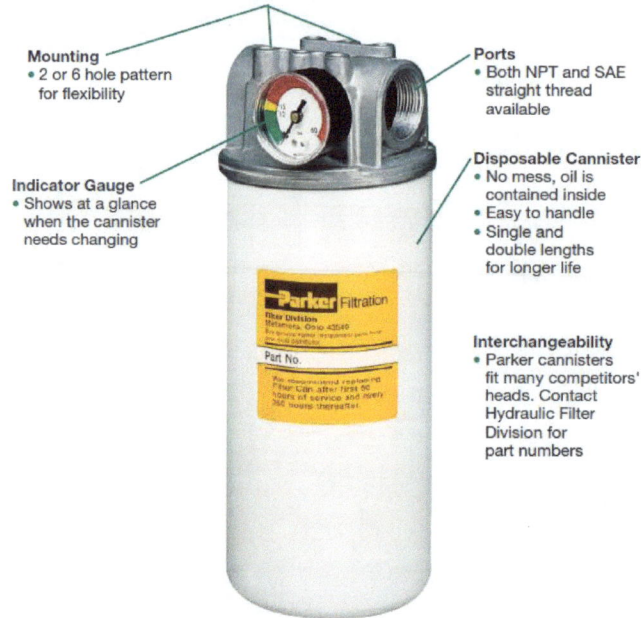

Fig. 1.25- Example of Spin-On Single Filters - 12AT/50AT Series (Courtesy of Parker)

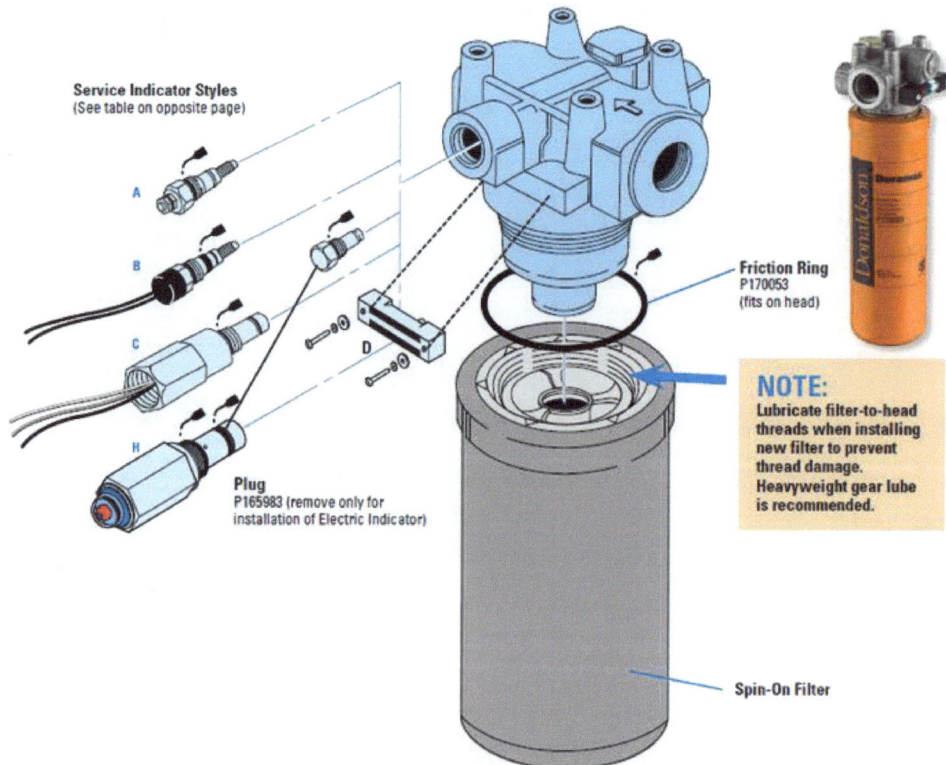

Fig. 1.26- Example of Spin-On Single Filters - HMK03 Series (Courtesy of Parker)

1.7.10 – Spin-On Dual Vertical Filters

Figure 1.27 shows an example of *Spin-On Dual Vertical* filters.

SP100/120 Spin-On Filters

Working Pressures to:	150 psi 1035 kPa 10.3 bar
Rated Static Burst to:	250 psi 1725 kPa 17.2 bar
Flow Range to:	100 gpm 379 lpm

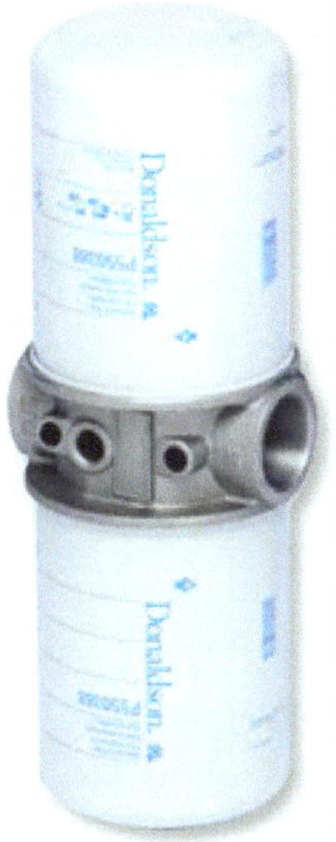

Fig. 1.27- Example of Spin-On Dual Vertical Filters (Courtesy of Donaldson)

1.7.11 – Spin-On Dual Horizontal Filters

Figure 1.28 shows an example of *Spin-On Horizontal Dual* filters.

SP80/90 Spin-On Filters

Working Pressures to:	150 psi 1035 kPa 10.3 bar
Rated Static Burst to:	250 psi 1725 kPa 17.2 bar
Flow Range to:	100 gpm 379 lpm

Fig. 1.28- Example of Spin-On Dual Horizontal Filters (Courtesy of Donaldson)

1.7.12 – Flange-Side-Mounted Filters

Figure 1.29 shows various models of of *Flange-Side-Mouned* filters.

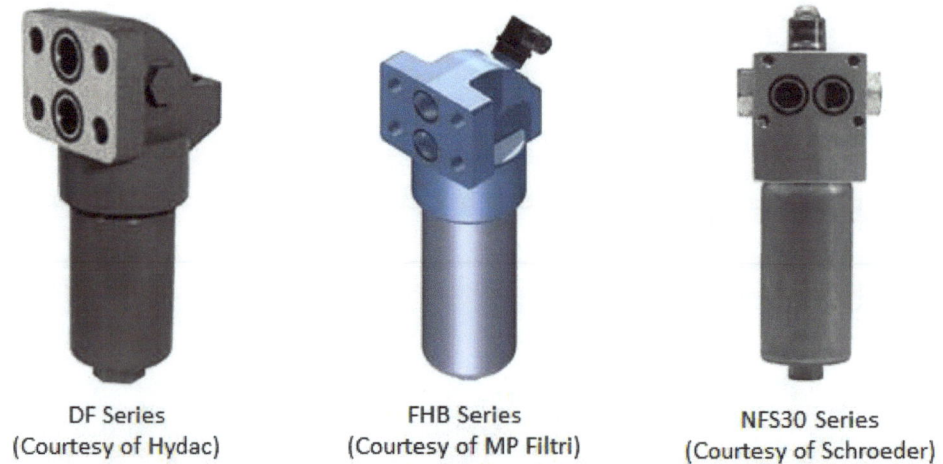

DF Series
(Courtesy of Hydac)

FHB Series
(Courtesy of MP Filtri)

NFS30 Series
(Courtesy of Schroeder)

Fig. 1.29- Examples of Flange-Side-Mounted Filters

1.7.13 – Flange-Top-Mounted Filters

Figure 1.30 shows various models of of *Flange-Top-Mouned* filters.

DFP Series
(Courtesy of Hydac)

FHM Series
(Courtesy of MP Filtri)

Fig. 1.30- Examples of Flange-Top-Mounted Filters

1.7.14 – Flange-Bottom-Mounted Filters

Figure 1.31 shows an example of of *Flange-Bottom-Mouned* filters.

Cover
- Handle protects indicators from damage
- Easy on, easy off, for fast service

Air Bleed
- Helps protect bearings and other sensitive components from trapped air

Fill Port
- Prefilter the fluid, before it gets into the machine's system
- Purge air while filling

Indicators
- You can tell element condition at a glance
- Both visual and electrical available

Bowl
- Rugged cold drawn steel— excellent fatigue resistance
- Three sizes for any application: Single (8"), Double (16"), and Triple (39")

Ports
- SAE straight thread or flange face

Bypass Valve (not visible)
- Soft seat design for zero internal leakage
- Located in cover assembly

Drain Port (not visible)
- Clean and easy servicing
- Lets you drain bowl of fluid- before element changes

Fig. 1.31- Examples of Flange-Bottom-Mounted Filters (Courtesy of Parker)

1.7.15 – Sandwich-Mounted Filters

Figure 1.32 shows various models of of *Sandwich-Mouned* filters.

DFZ Series
(Courtesy of Hydac)

NFS30-05 Series
(Courtesy of Schroeder)

Fig. 1.32- Examples of Sandwich-Mounted Filters

1.7.16 – Screw-In Filters

Figure 1.33 shows an example of *Screw-In* filters. They also refrred to as *Manifold-Mounted* or *Cartridge-Style* filters. As shown in the figure, they are installed in special cavities to protect critical components from oil-born contaminants. They are not intended to replace main filters.

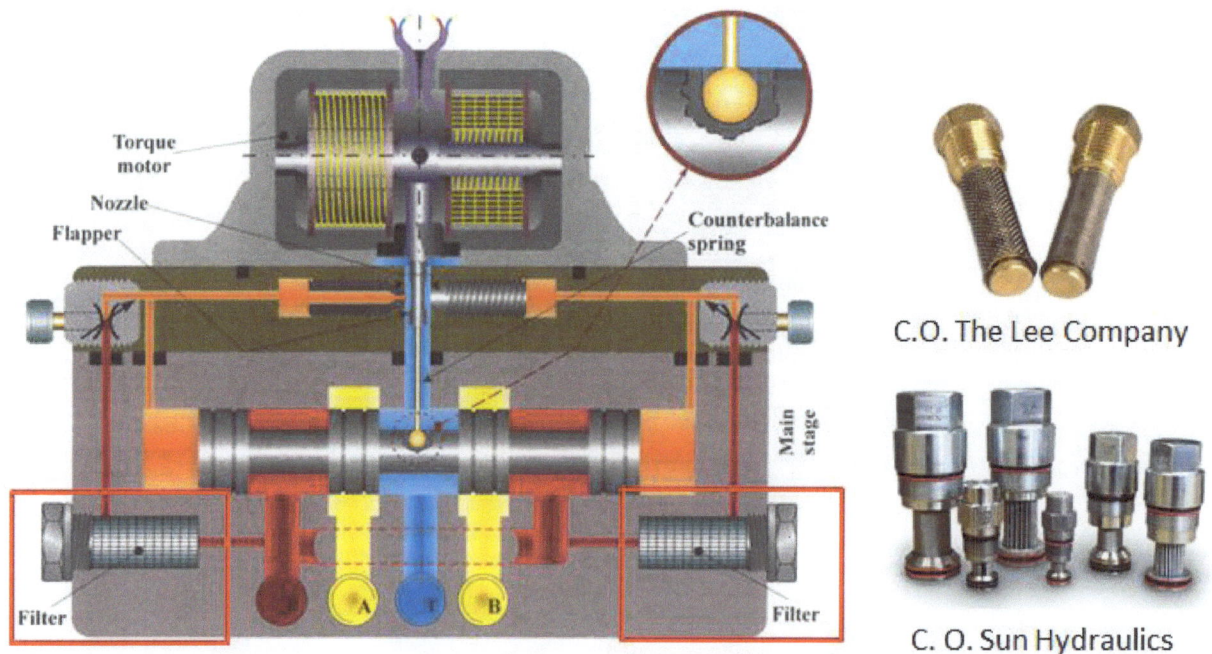

Torque motor

Nozzle

Flapper

Counterbalance spring

Main stage

Filter

Filter

A   T   B

C.O. The Lee Company

C. O. Sun Hydraulics

Fig. 1.33- Example of Screw-In Filters – CP-C16 Series (Courtesy of Hydeck)

1.8 - Filter Clogging Indicators

Purpose of Clogging Indicators: Some users install filters without indicators, preferring instead to change and/or clean elements according to a fixed time schedule or based on number of hours of operation. However, there is some risk in utilizing this approach. It may be difficult to establish a reliable schedule for installing new elements because the rate of dirt ingression is not known and varies from time-to-time and from machine-to-machine. Clogging indicators are warning devices that signal visually and/or electrically that the filter element is filled with contaminants and should be changed or cleaned.

Advantages of using Clogging Indicators:
- Eliminates the need to guess when the element will clog.
- Avoids the unnecessary cost of replacing elements too soon.
- Shutdown a machine that has sensitive components that is intolerant to contamination.

Configurations of Clogging Indicators: As shown in Fig. 1.34, clogging indicators could be visual, electrical, or optoelectrical.

Fig. 1.34- Various Configurations of Clogging Indicators (Courtesy of MP filtri)

Differential vs. Static Pressure Visual Indicator:
- *Differential Pressure Indicators* react to the pressure drop across the filter that is caused by the flow of fluid through the filter housing and element. These devices measure the difference in pressure upstream and downstream of the filter element, regardless of the system pressure. They are utilized in most pressure and inline return filters.

- *Static Pressure Indicators* measure only the build-up of pressure upstream of the filter element *(downstream pressure is ambient – tank vented to atmosphere)*. Consequently, if any components are located downstream of the filter, the indicator will generate a false reading of pressure at the filter entrance. As a result, static indicators are recommended only on filters that discharge directly to vented tanks and have minimal back pressure.

Clogging Indicator Settings: These devices activate *(trip)* when the flow of fluid causes a pressure drop across the filter element that exceeds the indicator setting. The indicator is set to trip well before the element becomes fully clogged and lower than bypass cracking pressure, thereby giving the operator sufficient time to take corrective action. The following are examples excerpted from Hydac literature:

- In a majority of applications, a HYDAC indicator is set to trip at 15 psid (1 bar) below the bypass valve cracking pressure; or, for a non-bypass filter, at 15 psid (1 bar) below the element design changeout pressure.

- Typically, a HYDAC pressure filter bypass valve begins to crack at 87 psid (6 bar), so the indicator is set to trip at 72 psid (5 bar).

- A HYDAC return filter ordinarily begins to bypass at 43 psid (3 bar), so the indicator is set to trip at 29 psid (2 bar).

Manual vs. Automatic Reset:
- *Electrical Clogging Indictors* reset automatically to their original position when the pressure across the filter drops below trip pressure.
- *Visual Clogging Indicator* are reset automatically or manually. The advantage of manual reset is that the indicator shows that the element is dirty even after the system is shut down.

Interchangeability: As shown in Fig. 1.35, a filter is manufactured to be equipped with various types of clogging indicators without the need for special mechanical arrangement.

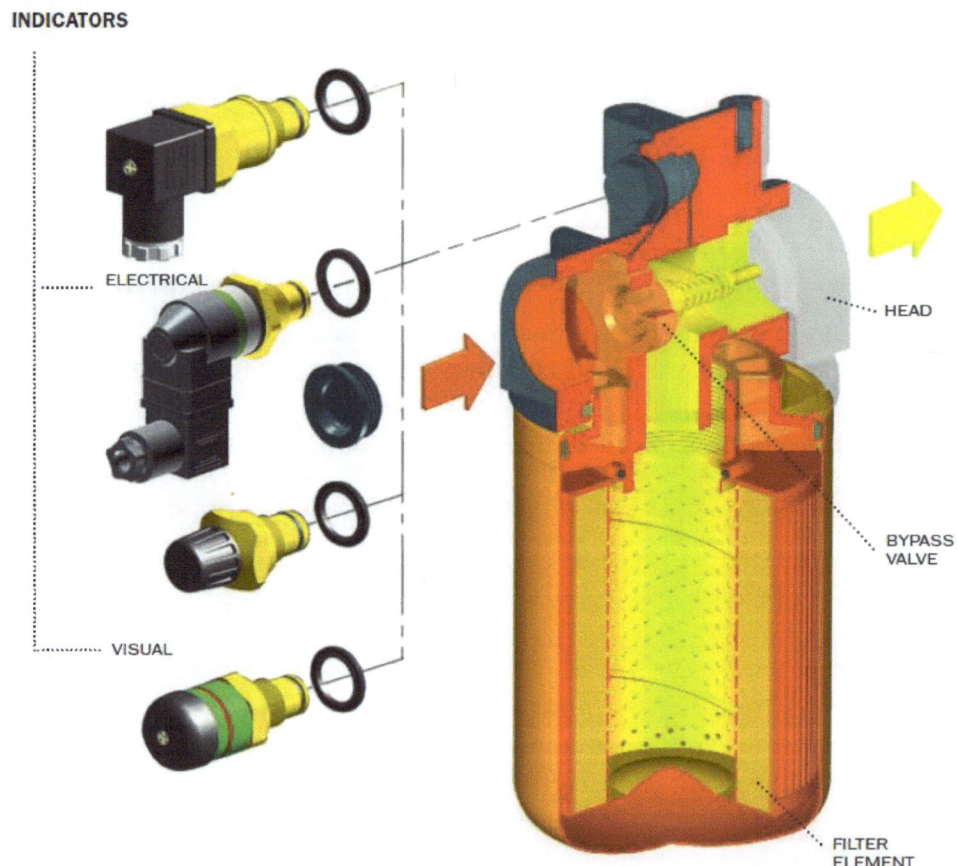

Fig. 1.35- Examples of Clogging Indicators (Courtesy of Assofluid)

1.8.1- Visual Clogging Indicators

Operation of Visual Differential Pressure Clogging Indicators: Figure 1.36 shows a typical *differential pressure* visual clogging indicator. The differential pressure across the filter increases, the piston/magnet assembly is driven down against a spring until the attractive force between the magnet and indicator pin is reduced sufficiently to allow the indicator to trip. Tripping results in the indicator pin rises giving visual indication that the filter must be serviced. Indicator is automatically reset when Δp < trip Δp.

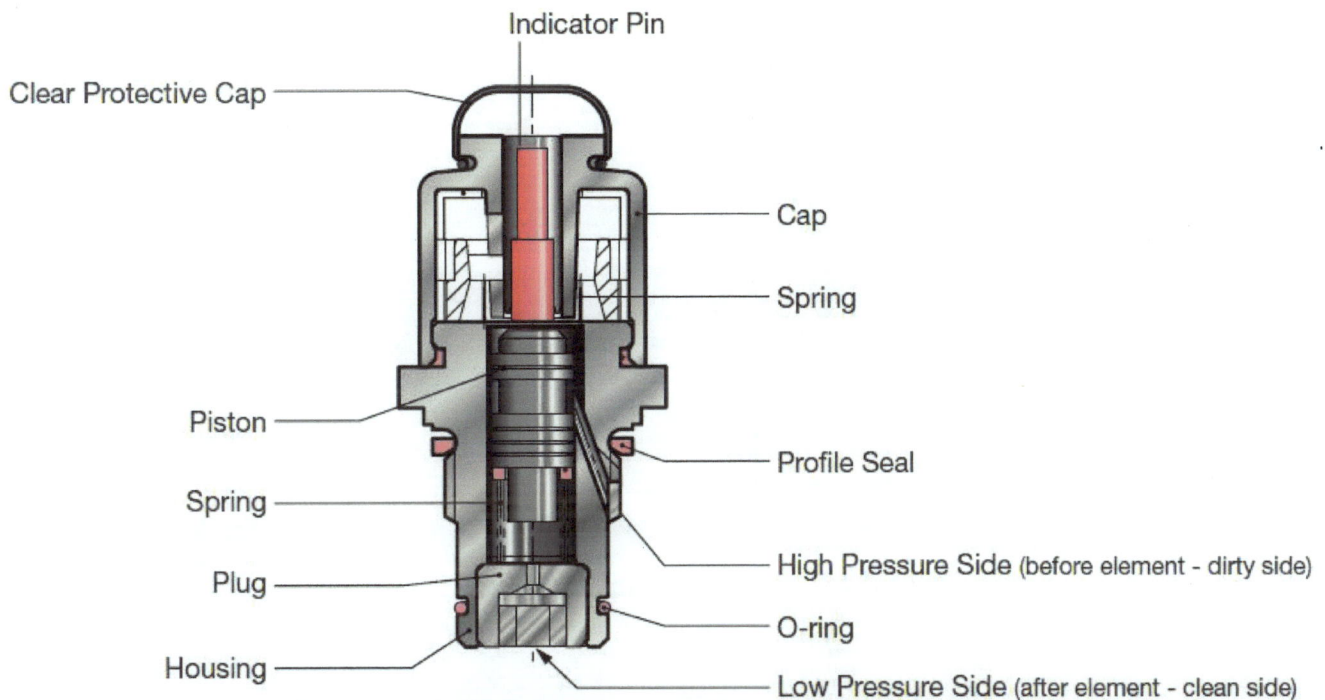

Fig. 1.36- Visual Differential Pressure Indicators (Courtesy of Hydac)

Operation of Static Pressure Visual Indicators: Figure 1.37 shows a typical *static pressure* visual clogging indicator. Increasing pressure upstream of the filter acts upon a diaphragm in the indicator and causes the indicator pin to overcome an opposing spring force until it trips at a pre-set pressure. The indicator pin automatically resets once pressure is reduced below the trip pressure.

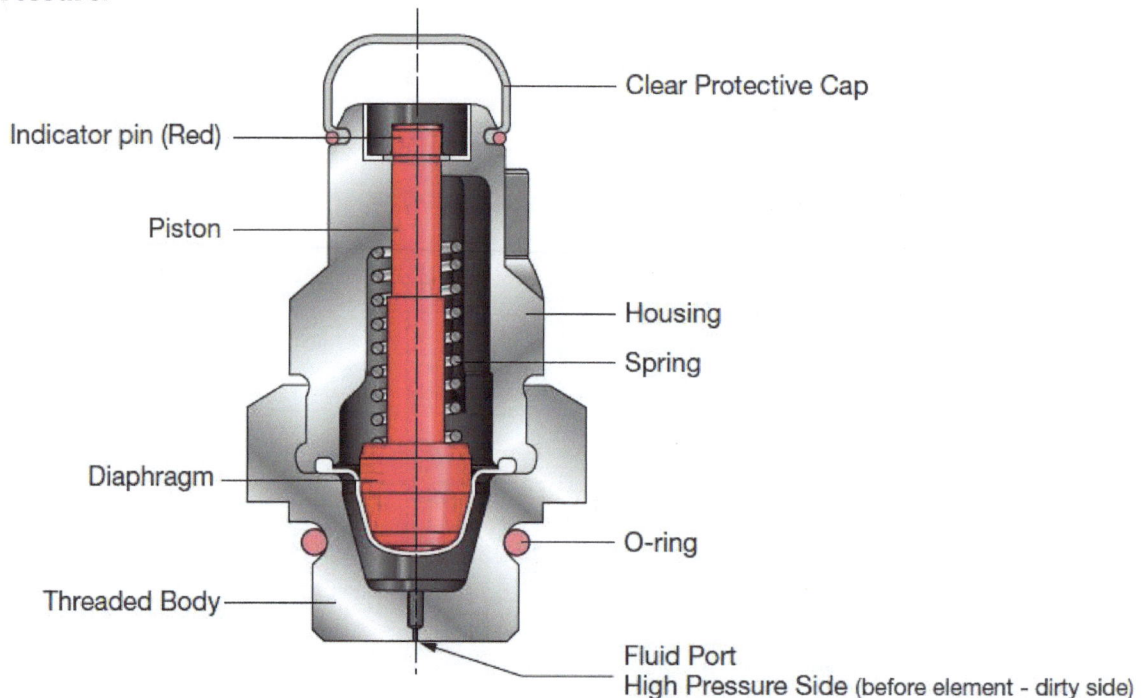

Clear Protective Cap
Indicator pin (Red)
Piston
Housing
Spring
Diaphragm
O-ring
Threaded Body
Fluid Port
High Pressure Side (before element - dirty side)

Fig. 1.37- Static Pressure Visual Indicators (Courtesy of Hydac)

1.8.2- Electrical Clogging Indicators

Concepts: In the *Electrical Clogging Indicator*, the pressure drop across the filter may also be used to actuate an electric switch. The resulting electric signal is used to actuate a remote light, bell or buzzer to indicate the need for service. In critical applications, such as when contamination sensitive expensive components are used, the resulting electric signal may be used to stop the machine to avoid damage that might occur due to a plugged or restricted filter.

Thermal Lockout: When mobile and other equipment is started in the cold, the hydraulic or lube fluid is likely to be highly viscous until it approaches normal operating temperature. The high pressure drop created by a highly viscous fluid can trip the indicator and falsely signify that the element is clogged. An optional thermal lockout device, available on many electric indicators, prevents the indicator from tripping until the fluid reaches a certain specified temperature. The device consists of a switch in series (AND function) in the indicator circuit, which is caused to make or break by a bi-metal strip that alters in shape according to temperature.

Single Pole, Double Throw Switches (SPDT): Differential pressure and most static pressure electrical indicators contain single-pole, double-throw switches. This provides the choice of *normally open (N/O)* or *normally closed (N/C)* contacts when the pressure differential is below trip-point

Operation of Differential Pressure Electrical Clogging Indicator: Figure 1.38 shows a typical differential pressure electrical clogging indicator. The differential pressure across the filter increases, the piston is driven down against a spring. Tripping causes a switch to make or break, permitting a remote indication to warn of the need for servicing.

Electric static pressure indicators, which also operate mechanically, are available.

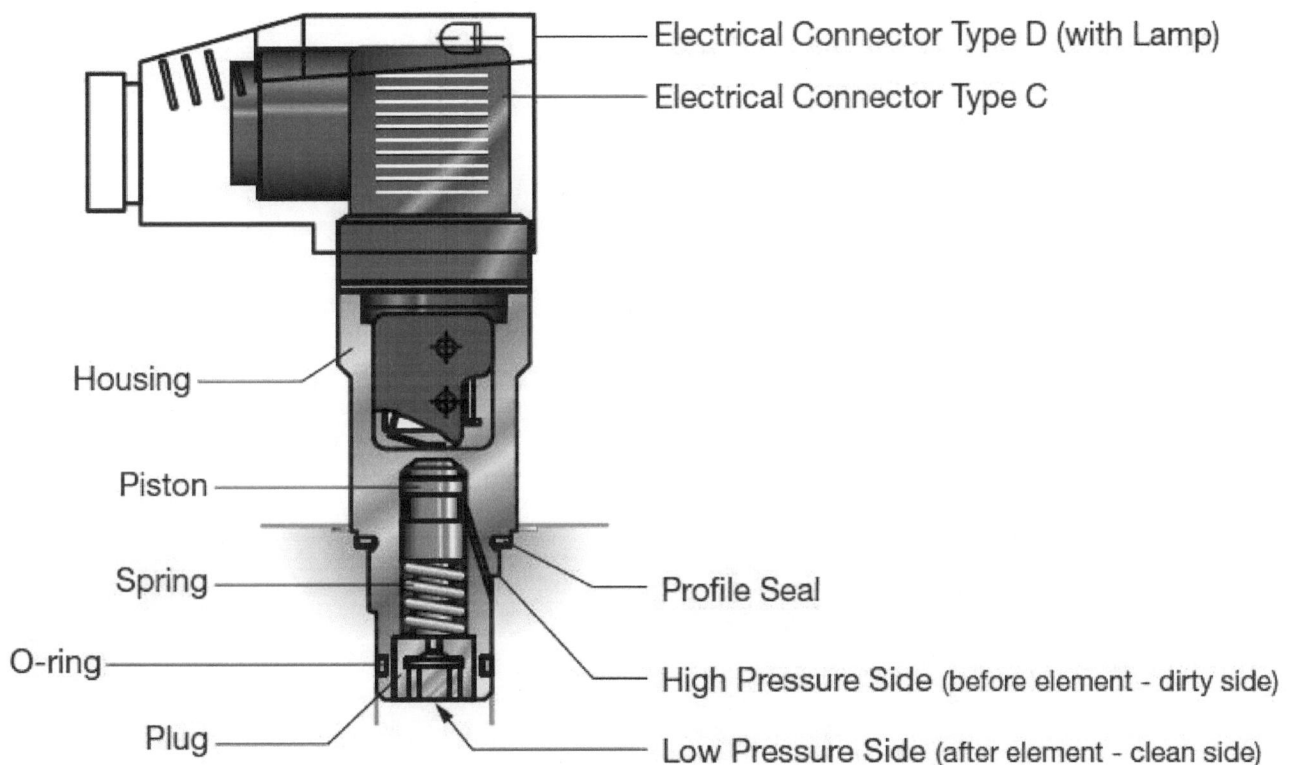

Fig. 1.38- Differential Pressure Electrical Indicators (Courtesy of Hydac)

1.9 – Types of Filters Based on their Placement in the Circuit

Basically, filters shall be installed where they are readily accessible, the space around allows for replacing filter elements, and filter element is replaced without emptying the tank. Ideally, each component in a hydraulic system would be equipped with its own filter, but this is economically impractical in most cases. A more practical solution is to examine the system and identify the most contamination-sensitive components and provide filtration as close as possible to these components. As shown in Fig. 1.39 and Fig. 1.40, filters may be located in a circuit in one or more of the following locations:

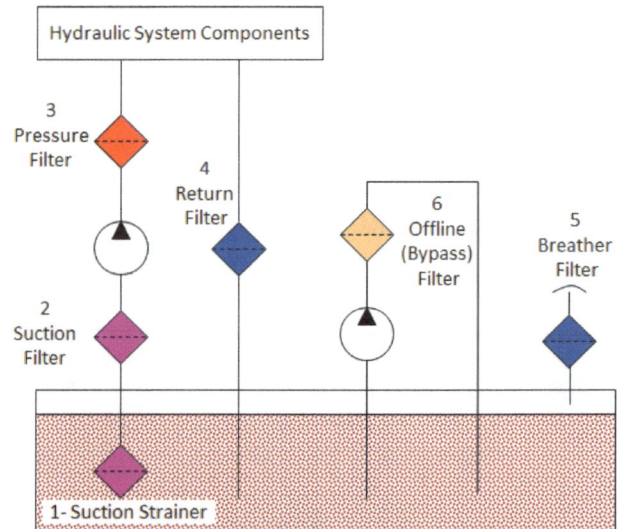

Fig. 1.39- Types of Filters Based on their Placement in the Circuits

- **Online Filters:** These filters are placed in series with the main pump as follows: Suction Strainer (1), Suction Filters (2), Pressure Filters (3), and Return Filters (4).

- **Offline Filters:** These filters are placed in parallel with the main pump as follows: Breather Filters (5), and Offline (Bypass) Filters (6).

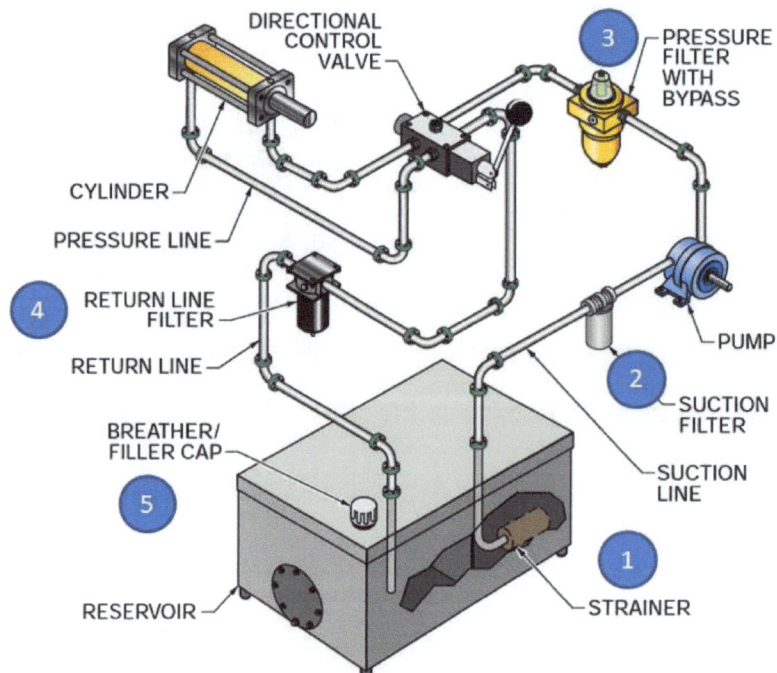

**Fig. 1.40- Types of Filters Based on their Placement in the Circuits
(Courtesy of American Technical Publishers)**

1.9.1- Suction Strainers

Placement: As shown in Fig. 1.41 and 1.42, *Suction Strainers* are connected to the beginning of the suction line before a suction filter (if found). As shown in the figure, a strainer could be aligned with the suction line or a 90° elbow at the bottom of the suction line.

Primary Duty: The primary duty of suction strainers is to capture the relatively large particles, chips or rags before getting into the pump. Generally speaking, because of the cavitation concerns, suction strainers are not recommended for large flow pumps. Installing a suction strainer on the suction line subject to approval from the pump manufacturer.

Cost: Suction strainers are relatively inexpensive as it does not have a complex housing.

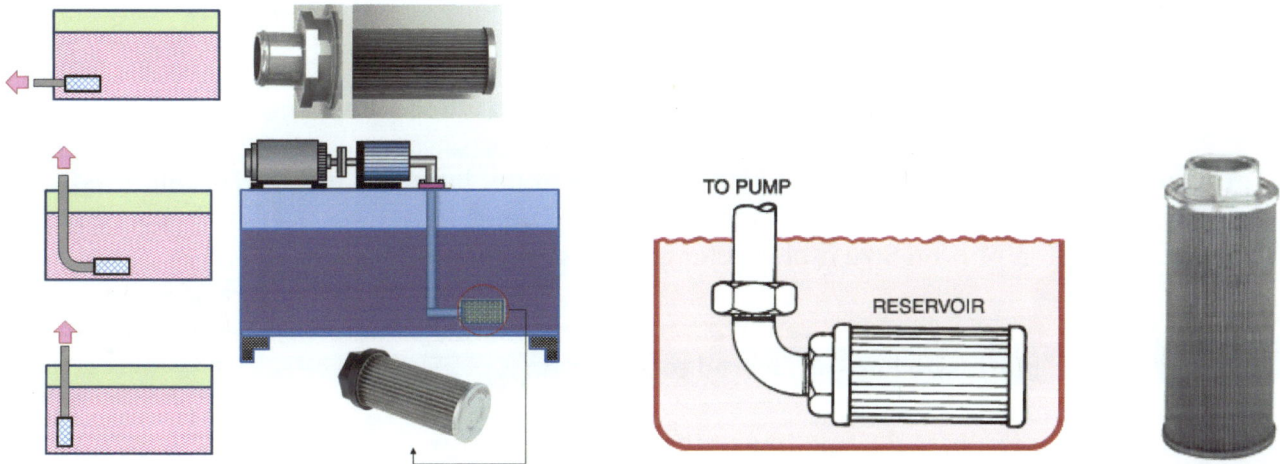

Fig. 1.41 – Suction Strainers   **Fig. 1.42 – Suction Strainers (Courtesy of Schroeder)**

Construction: As shown in Fig. 1.43, suction strainers are furnished with optimized pleat size and screen area for extended life and low pressure drop. Some suction strainers could be magnetic in order to capture metallic contaminants or wear products.

Bypass Valve: Suction strainers mounted at the beginning of the suction line inside the reservoir without a bypass valve or pressure drop indicator are not recommended. They are hidden from sight and there is no easy access to remove and clean them. Overtime, without proper cleaning, they will plug restricting flow and causing pump cavitation leading to pump failure.

One pieced high strength nylon hex cap for reduced cost.

Cap assembly epoxy bonded to body for superior strength.

Pleated stainless wire cloth provides excellent flow with minimum pressure drop. Choice of different mesh sizes.

Sides of end caps are reverse tapered to enhance epoxy bonding.

Inner perforated steel support tube for added strength and rigidity.

Optional 3 or 5 PSI relief valve to prevent failure should the screen become clogged with debris.

Fig. 1.43 – Suction Strainers (ohfab.com)

Micron Size vs. Mesh Size: *Micron size* is the size of the largest particles (in microns) that can pass through the screen. *Mesh Size* is the number of holes in one squared inch. Large mesh size means finer filter. Micron Size is also referred to as *Pore Size*.

Mesh Size: If no *mesh size* is reported by the pump manufacturer, 250-500 mesh size is recommended. Mesh size is also referred to as *Porosity*.

Flow: a strainer receives the full flow of the main pump flow.

Surface Area: A strainer is sized based on the pump flow. Size of the strainer should offer surface area to minimize the pressure drop across the strainer so that pump cavitation is avoided. Review the pump data sheet if found. Otherwise, consult the pump manufacturer in regard to recommended surface area. If no information is found, the following rule of thumb is applicable as a guideline. Surface area shouldn't be less than 2 square inches for every GPM of the pump flow (≈ 3 cm² for every liters/min of the pump flow). Equations 1.1A and 1.1B are used to determine the surface area of the strainer in metric and English system of units, respectively.

$$\text{Suction Strainer Surface Area (cm}^2\text{)} = 3 \times Q_p \text{ (lit/min)} \qquad \text{1.1A}$$

$$\text{Suction Strainer Surface Area (in}^2\text{)} = 2 \times Q_p \text{ (gpm)} \qquad \text{1.1B}$$

Example: A 20 GPM pump. Eq. 1.1B → A minimum strainer surface area of 40 square inches.

Magnetic Suction Strainers: Figure 1.44 shows the specification for magnetic suction strainers. Magnetic suction strainers offer dual protection to the pump inlet without risk of cavitation. Powerful ceramic magnets located parallel to the pleated mesh attract and protect against damaging ferrous particles of all sizes. The pleated stainless-steel screen provides additional filtration protection for larger particles that would result in catastrophic failure. The generous open area of the stainless-steel pleated mesh screen eliminates the possibility of pump cavitation.

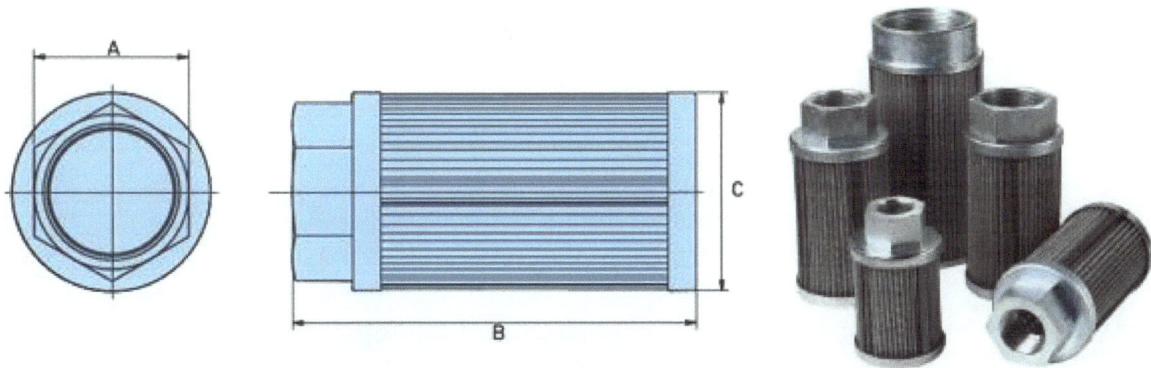

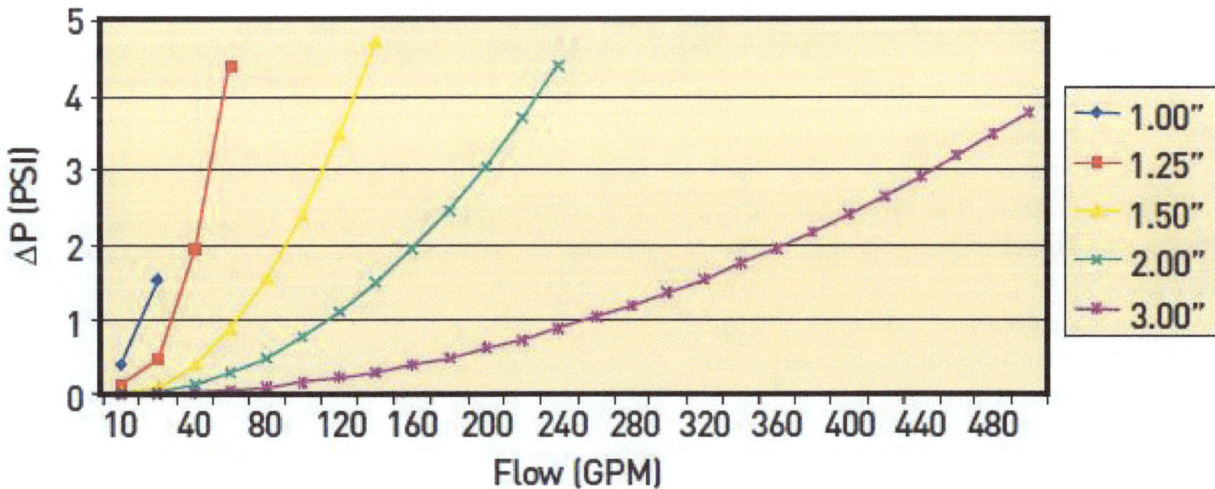

Fig. 1.44 – Magnetic Suction Strainers (Courtesy of Parker)

1.9.2- Suction Filters

Placement: As shown in Fig. 1.45, A *Suction Filter* is placed on the suction line before the inlet port of the main pump, between the pump and a suction strainer (if found). Unlike strainers, suction filters are mounted externally, i.e., outside the reservoir.

Primary Duty: The primary duty of suction filters is to protect the main pump. However, the system components downstream the pump aren't protected from any contamination that may be generated by the pump during operation. Like suction strainers, pressure drop across the filter and pump cavitation issue must be taken into consideration when sizing suction filters. Always consult the pump manufacturer for inlet restrictions. However, if no information is found, the maximum acceptable pressure drops must not be higher than 0.1 bar (1.45 psi).

Cost: Suction filters are more expensive than strainers, but less expensive than pressure filters since no special housing is needed for high pressure.

Construction: A typical suction filter consists of a housing, a filter element, and a filter head that contains filter ports.

Bypass Valve: Bypass valve is highly recommended in suction filters avoiding pump cavitation in case of filter clogging. When expensive pumps are used, adding a vacuum switch to stop the machine when the filter is clogged is an extra protection for the pump.

Mesh Size: Filter media in a suction filter is coarse filter, ranging from ranges from 60 to 250 micron, in order to limit the pressure, drop across the filter.

Flow: Suction filters receive the full flow of the main pump. So, they are sized based on the main pump flow.

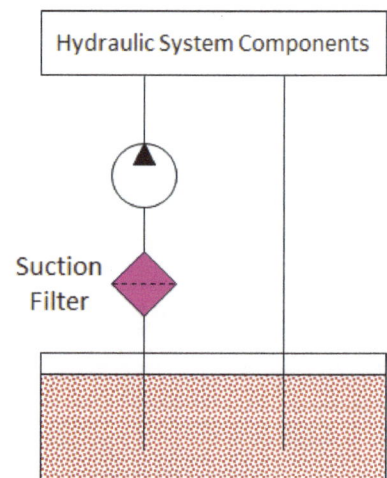

Fig. 1.45 – Placement of Suction Filters

Figure 1.46 shows a typical example of an On-Tank-Top Single suction filter. The filter is equipped with a differential pressure switch to trip the machine if the filter is clogged. The data sheet associated with the filter provides full information about it.

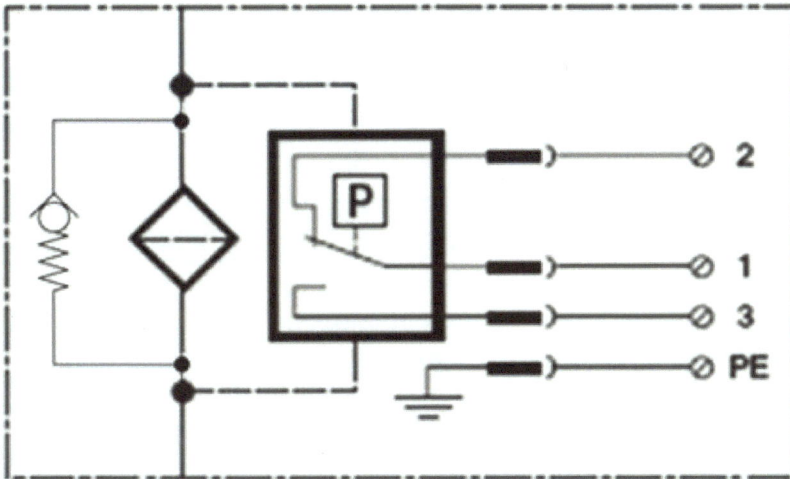

SF Series
In-tank Suction Filters
360 psi • up to 200 gpm

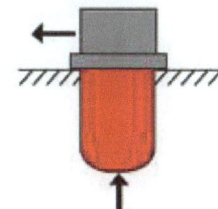

Technical Specifications

Mounting Method	4 mounting holes - filter head	
Flow Direction	Inlet: Bottom	Outlet: Side
Construc. Materials	Housing	Lid
SF 110-330	Aluminum	Aluminum
SF 950-1300	Ductile Iron	Ductile iron

Flow Capacity

110	5 gpm (20 lpm)
240	15 gpm (57 lpm)
330	30 gpm (114 lpm)
950	175 gpm (662 lpm)
1300	200 gpm (757 lpm)

Housing Pressure Rating

Max. allowable working pressure: 360 psi (25 bar)
Fatigue Pressure: 360 psi (25 bar) @ 700,000 cycles

Burst Pressure:
110	1080 psi (75 bar)
240	1230 psi (85 bar)
330	1440 psi (100 bar)
950-1300	>1440 psi (100 bar)

Element Collapse Pressure Rating

W/HC: 290 psid (20 bar)

Fluid Temp. Range 14°F to 212°F (-10°C to 100°C)
Consult HYDAC for applications operating below 14°F (-10°C)

Fluid Compatibility

Compatible with all hydrocarbon based, synthetic, water glycol, oil/water emulsion, and high water based fluids when the appropriate seals are selected

Indicator Trip Pressure

ΔP = 3 psi (0.2 bar) -10% *(standard)*

Bypass Valve Cracking Pressure

ΔP = 3 psi (0.2 bar) +10% *(standard - sizes 60, 950, 1300)*
ΔP = 4.4 psi (0.3 bar) +10% *(standard - sizes 110,160,240,330)*

Fig. 1.46- Example of On-Tank Top Mounted Suction Filter (Courtesy of Hydac)

1.9.3- Pressure Filters

Placement: As shown in Fig. 1.47, In open hydraulic circuits, *Unidirectional Pressure Filters* are placed directly downstream of the main pump. In closed hydraulic circuits, *Bidirectional Pressure Filters* are used to handle reserve flow. This is accomplished using a check valve rectifier built into the filter head.

Primary Duty: The primary duty of pressure filters is to protect the sensitive components in the system such as proportional and servo valves. Since the pump produces wear debris, contamination is captured before it is spread to the rest of the system.

Cost: Pressure filters are the most expensive because it requires a special housing that is sealed and rated work under maximum system pressure.

Mesh Size: Filter media in a pressure filter is selected to maintain the cleanliness level required by the system manufacturer.

Flow: Pressure filters receive the full flow of the main pump. So, they are sized based on the main pump flow.

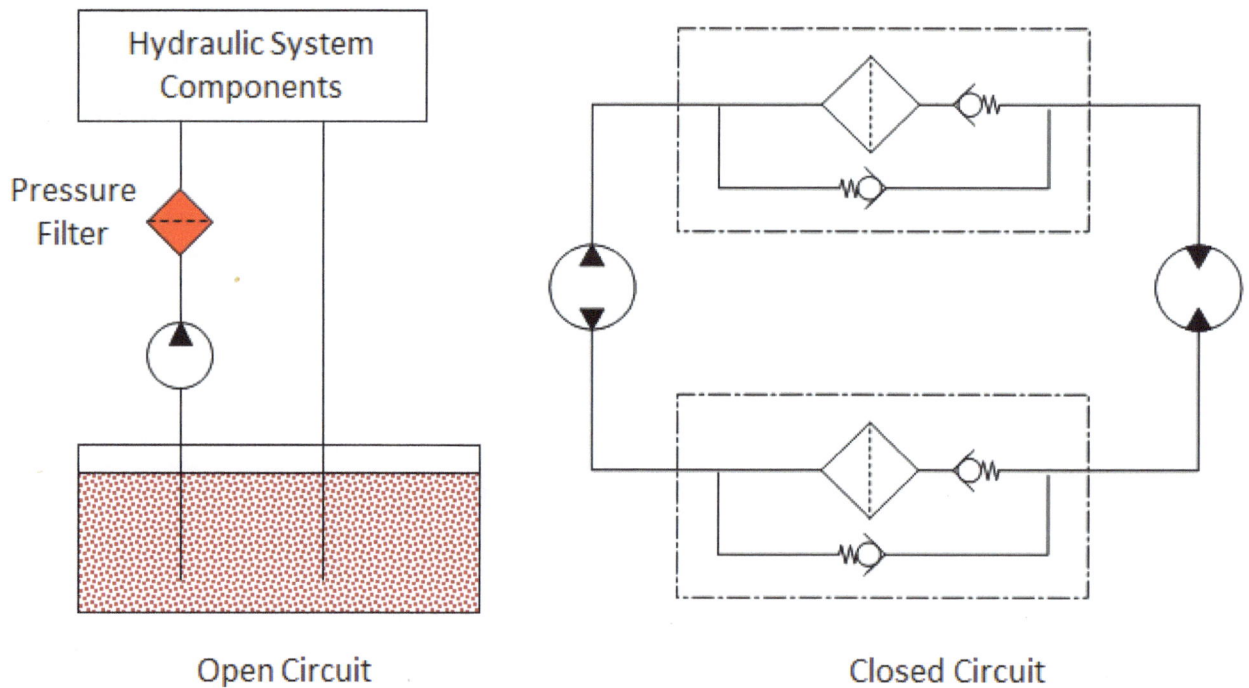

Fig. 1.47 – Placement of Pressure Filters

Construction: As shown in Fig. 1.48, the construction of a pressure filter is typically consisted of same components as suction and return filters. However, housing of such filters must withstand the maximum pressure. Steady state pressure ratings are 100 - 400 bar (1500 psi to 6000 psi). Additionally, pressure distribution analysis must be made to make sure the pressure filter is capable to work under possible pressure spikes and pressure fluctuation without subjecting to collapse of fatigue failure.

Bypass Valve: in pressure filters, bypassing the filter media is an optional feature. Non-Bypass filters are used when sensitive components that are less tolerant to contamination are used. However, if such filters are used, filter media collapse pressure must be at least equal to maximum working pressure. Otherwise, Bypass-to-Tank filters could be used.

1. High strength ductile iron filter head with integral indicator port

2. Steel bowl with standard drain port

3. Proprietary element endcap assembly includes bypass and reverse flow valves

4. Patented deformable tangs secure element in bowl

5. Coreless element assembly

6. Re-usable element support core

Fig. 1.48- Sectional view of Pressure Line Filter WPF Series (Courtesy of Parker)

Figure 1.49 shows a typical example of a Line-Mounted Top Ported pressure filter. The filter is equipped with a differential pressure visual clogging indicator. The specification sheet shows that this filter can work in a pressure up to 420 bar (6000 psi) The data sheet associated with the pump provides full information about the filter.

Technical Specifications

Mounting Method	4 mounting holes	
Port Connection		
30	SAE-8, 1/2" NPT, 1/2" BSPP	
60/110	SAE-12, 3/4" NPT, 3/4" BSPP	
	3/4" SAE, Code 62	
160/240/280	SAE-20, 1 1/4" NPT, 1 1/4" BSPP	
	1 1/4" SAE, Code 62	
330/660/1320	SAE-24, 1 1/2" NPT, 1 1/2" BSPP	
	2" SAE Flange Code 62	
Flow Direction	Inlet: Side	Outlet: Side
Flow Capacity		
30	8 gpm (30 lpm)	
60	16 gpm (60 lpm)	
110	29 gpm (110 lpm)	
160	42 gpm (160 lpm)	
240	63 gpm (240 lpm)	
280	74 gpm (280 lpm)	
330	87 gpm (330 lpm)	
660	174 gpm (660 lpm)	
1320	200 gpm (757 lpm)	
Housing Pressure Rating		
Max. Allowable Working Pressure	6090 psi (420 bar)	
Fatigue Pressure	6090 psi (420 bar) @ 1 million cycles	
Burst Pressure	30	15950 psi (1100 bar)
	60/110	17400 psi (1200 bar)
	160/240/280	17110 psi (1180 bar)
	330/660/1320	15080 psi (1040 bar)
Element Collapse Pressure Rating		
BH4HC, V	3045 psid (210 bar)	
ON, W/HC	290 psid (20 bar)	
Fluid Temp. Range	14°F to 212°F (-10°C to 100°C)	
Consult HYDAC for applications operating below 14°F (-10°C)		
Fluid Compatibility		
Compatible with all hydrocarbon based, synthetic, water glycol, oil/water emulsion, and high water based fluids when the appropriate seals are selected.		
Indicator Trip Pressure		
ΔP = 29 psid (2 bar) -10% *(optional)*		
ΔP = 72 psid (5 bar) -10% *(standard)*		
ΔP = 116 psid (8 bar) -10% *(optional non bypass)*		
Bypass Valve Cracking Pressure		
ΔP = 43 psid (3 bar) +10% *(optional)*		
ΔP = 87 psid (6 bar) +10% *(standard)*		
Non Bypass Available		

DF Series
Inline Filters
6090 psi • up to 200 gpm

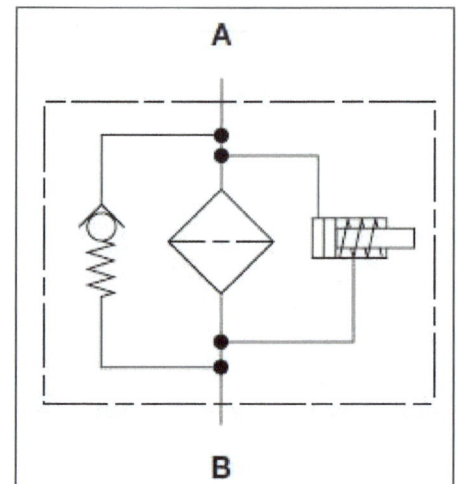

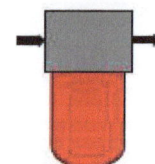

Fig. 1.49- Example of Line-Mounted Top-Ported Pressure Filter (Courtesy of Hydac)

Pressure Drop: As shown in Fig. 1.50, it is to be noted that the overall pressure drop across the filter is the sum of the pressure drop across the housing and the pressure drop across the filter element. As shown in the figure, for the same flow through the filter, the pressure drop across the filter element is inversely proportional to its micron size (i.e. directly proportional to mesh siz).

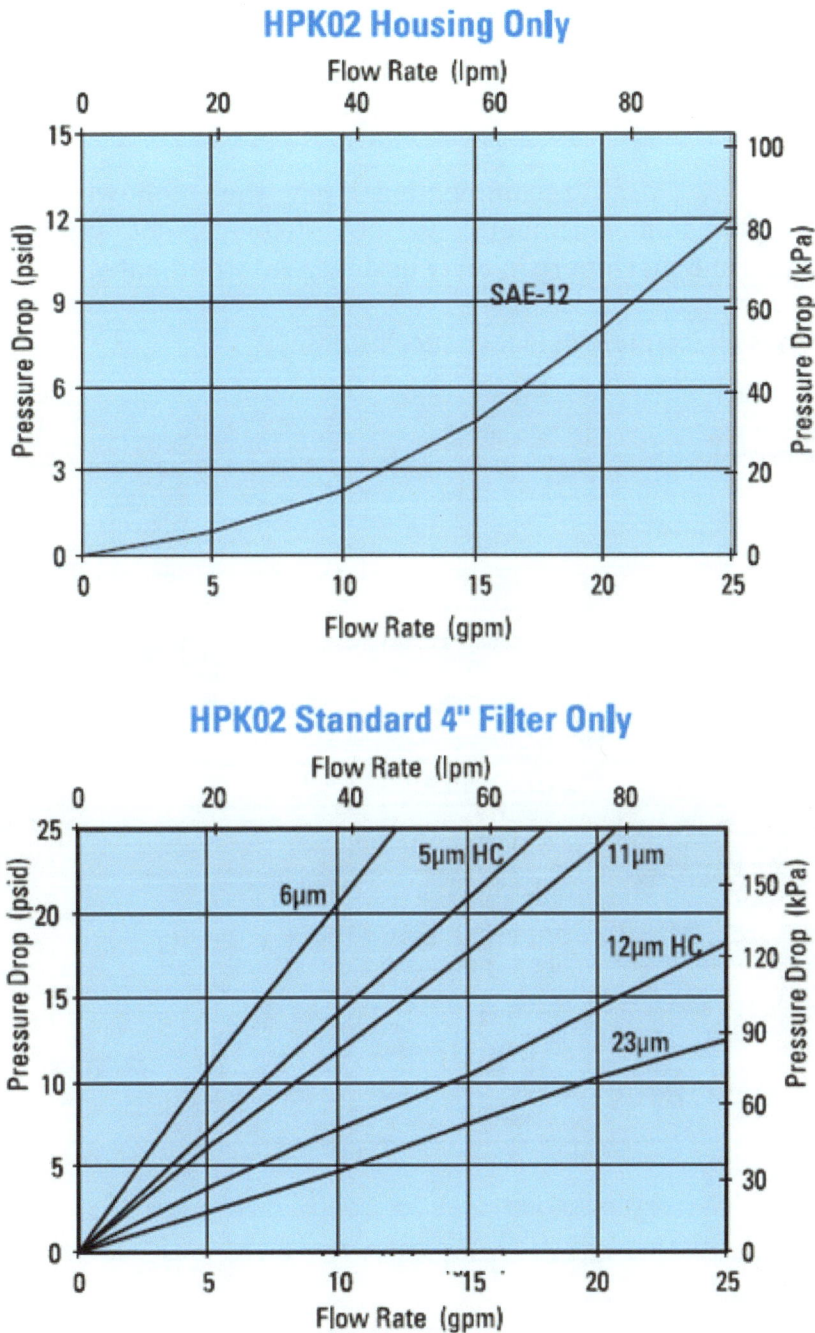

Fig. 1.50- Example of Line-Mounted Top-Ported Pressure Filter (Courtesy of Hydac)

1.9.4- Last Chance Filters

Primary Duty: If abrasive particles pass through main system filters and enter a hydraulic system, they may damage expensive components. These contaminants may prevent hydraulic components from operating properly by causing them to respond slowly or stick open. *Last Chance Filters* are used to protect critical components from catastrophic failure. However, they are not intended to replace the system filter.

Placement: Because they are placed in series and exposed to system pressure, they have a high collapse pressure.

Applications: They are recommended in the following market applications: servo circuits, precision machinery, variable-displacement pump and motor systems, hydrostatic drives, and machinery in dirty or dusty environments.

Figure 1.51 shows an example of last chance filters.

CP-C16 Series
Circuit Protector Manifold Cartridge Filters
3000 psi • up to 12 gpm

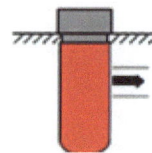

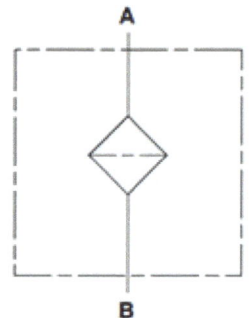

Technical Specifications

Mounting Method	C16-2 Cavity (SAE-16 Threaded Port)	
Flow Direction	Inlet: Bottom	Outlet: Side
Construction Materials	Steel	
Flow Capacity	12 gpm (45 lpm)	
Housing Pressure Rating		
Max. Allowable Working Pressure	3000 psi (210 bar)	
Fatigue Pressure	Contact HYDAC Office	
Burst Pressure	Contact HYDAC Office	
Element Collapse Pressure Rating		
W/HC	250 psid (17 bar)	
Fluid Temperature Range	14°F to 212°F (-10°C to 100°C)	
Consult HYDAC for applications operating below 14°F (-10°C)		
Fluid Compatibility		
Compatible with all petroleum oils rated for use with Nitrile rubber (NBR) seals.		

Fig. 1.51- Example of Last Chance Filters (Courtesy of Hydac)

1.9.5- Return Filters

Placement: As shown in Fig. 1.52, in open hydraulic circuits, *Return Filters* are placed on the main return line that collects return oil from actuators and other components back to the reservoir. As shown in the figure, in closed hydraulic circuits (hydrostatic Transmission), return filters are placed on the case drain line.

Primary Duty: The primary duty of return filters is to capture contaminates generated by all the components in the systems and leave only clean oil to get back to the reservoir. Return line filters are generally designed for lower pressures up to 34 bar (500 psi). Cooling can be integrated by installing an oil cooler downstream of the return filter.

Cost: Return filters are the most cost-effective filtration solution because return line pressure is low pressure, hence no special housing is required. Larger elements with a greater dirt-holding capacity can be used at a fraction of the cost.

Mesh Size: Filter media in return filters are selected to meet the cleanliness level required by system manufacturer. Common range is between 5 and 40 microns is acceptable.

Flow: When sizing a return filter, a through flow distribution analysis must be made to determine the maximum flow in the return line. The maximum flow in return line may be more than the main pump flow if a differential cylinder and/or accumulator are used.

Bypass Valve: Equipping return filters by bypass valve are highly recommended in order to limit the back pressure generated at the filter inlet port when the filter media is clogged. In closed circuits, the filter housing must incorporate a bypass valve with a cracking pressure lower than the maximum allowable case pressure for the pump or the motor, typically 0.5-1 bar (7-15 psi).

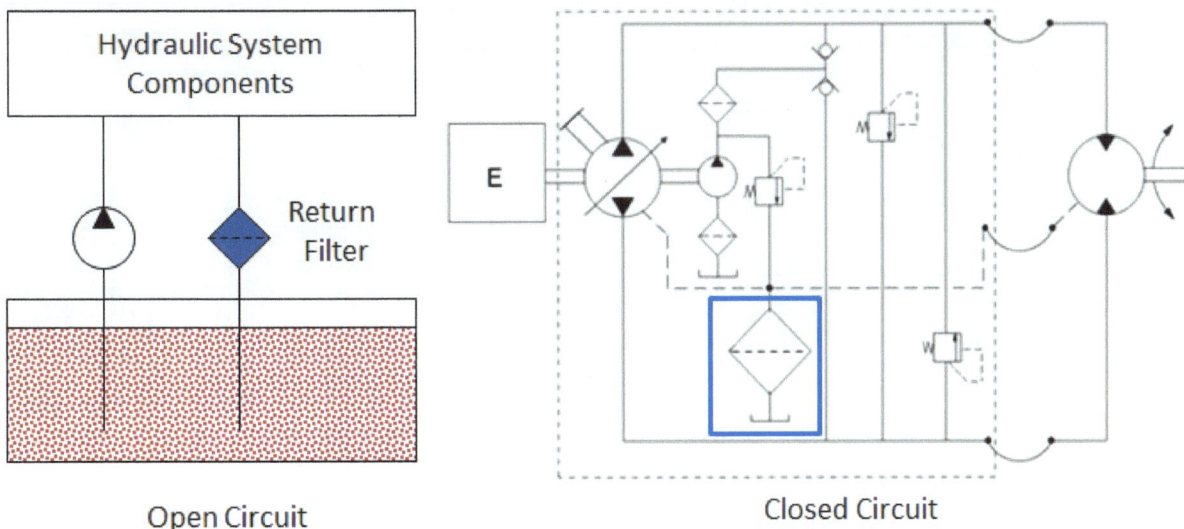

Open Circuit   Closed Circuit

Fig. 1.52 – Placement of Return Filters

Figure 1.53 shows an example of On-Tank Top single filters. As shown in the figure, the filter can be equipped with a clogging indicator and a built-in filter breather. The associated specification sheet shows full details of the filter specifications.

Technical Specifications

Mounting Method	
75/90/150/165/185	2 mounting holes - filter housing
50/75/90/150/165/185/210/270/330/500/661/851/975/1100	4 mounting holes - filter housing

RFM Series

In-Tank Return Line Filters
145 psi • up to 224 gpm

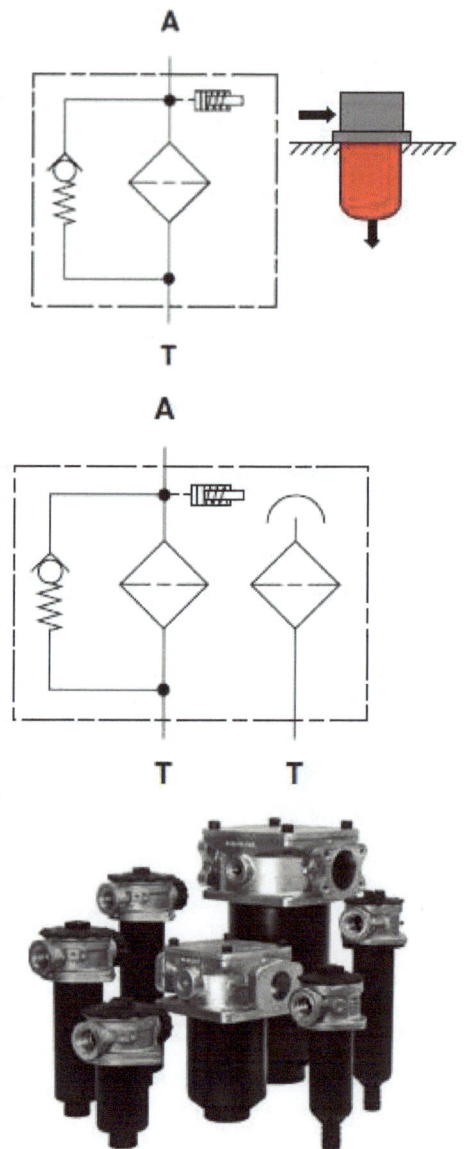

Flow Capacity	
50 - 13 gpm (50 lpm)	270 - 71 gpm (270 lpm)
75 - 20 gpm (75 lpm)	330 - 87 gpm (330 lpm)
90 - 24 gpm (90 lpm)	500 - 132 gpm (500 lpm)
150 - 40 gpm (150 lpm)	661 - 174 gpm (660 lpm)
165 - 43 gpm (165 lpm)	851 - 225 gpm (850 lpm)
185 - 49 gpm (185 lpm)	975 - 258 gpm (950 lpm)
210 - 55 gpm (210 lpm)	1100 - 300 gpm (1100 lpm)

Housing Pressure Rating	
Max. Allowable Working Pressure*	145 psi (10 bar), 101.5 psi (7 bar) *(Sizes 975 & 1100)*
Fatigue Pressure	145 psi (10 bar) @ 1 million cycles
Burst Pressure	75-500 >580 psi (40 bar)
	50, 661/851 536 psi (37 bar)
	975/1100 Consult Factory

Element Collapse Pressure Rating	
BN4HC *(size 50, 975 & 1100 only)*	145 psid (10 bar)
ON *(size 50-851 only)*, W/HC	290 psid (20 bar)
ECON2, BN4AM, AM, P/HC, MM	145 psid (10 bar)
V	435 psid (30 bar)

Fluid Temperature Range	-22°F to 212°F (-30°C to 100°C)

Consult HYDAC for applications below -22°F (-30°C)

Fluid Compatibility

Compatible with all hydrocarbon based, synthetic, water glycol, oil/water emulsion, and high water based fluids when the appropriate seals are selected.

Indicator Trip Pressure

P = 20 psi (1.4 bar) - 10%
P = 29 psi (2 bar) -10% *(standard)*
P = 72 psi (5 bar) -10% *(optional)*

Bypass Valve Cracking Pressure

ΔP = 43 psid (3 bar) +10% *(Standard - All sizes except 50, 975, 1100)*
ΔP = 87 psid (6 bar) +10% *(Optional - Sizes 50, 975 & 1100 not available)*
ΔP = 25 psid (1.7 bar) +10% *(Standard for Sizes 50, 975 & 1100)*

Fig. 1.53- Example of Return Filters (Courtesy of Hydac)

1.9.6- Combined Return and Suction Booster Filter

Description: Many hydraulic circuits may have more than one operating pump. As shown in Fig. 1.54, a return filter is used for the open circuit and a suction filter is used for the boosting pump in the closed circuit.

Advantage: A *Combined Return and Suction Booster Filter* is two filters in one that have the advantages:
- Cost and space saving.
- Easy maintenance.
- Meets automotive standard.
- Offered in pipe, SAE straight thread, flange and ISO 228 porting.
- Available with NPTF inlet and outlet female test ports.
- Available with magnet inserts.
- Various Dirt Alarm options.
- Available with housing drain plug.

Typical Applications: Such a unique design is used in machines with two or more circuits, such as in mobile working machines with hydrostatic traction drives (wheel loaders, forklifts) and automotive engineering.

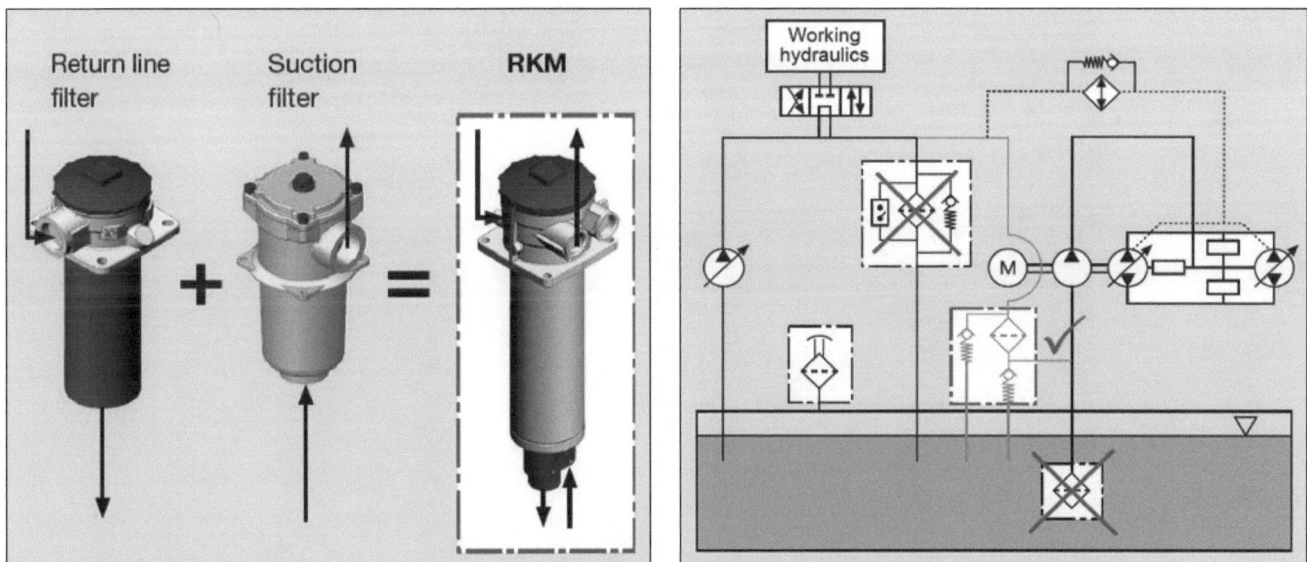

Fig. 1.54- Combined Return and Suction Booster Filter (Courtesy of Hydac)

Function (Refer to Fig. 1.55):

- Inlet to Return Filter:
  - Q_R is supplied via port **A**.

- Inlet to Boosting Pump of a Closed Circuit:
  - Q_R from the outside to the inside the element.
  - Q_s from inside the element to inlet of booting pump.

- Inlet to Suction Filters:
  - Back pressure valve **V1** builds 0.5 bar positive pressure.
  - Filtered oil is supplied to suction ports (**B1, B2**, etc.).
- Bypass-to-Tank **V2**:
  - Backpressure → surplus flow drains (bypassing the element) to port **T**.
- Bypass Valve **V3 (optional)**:
  - Oil can be drawn from the tank for short periods (e.g. for initial filling and for venting).

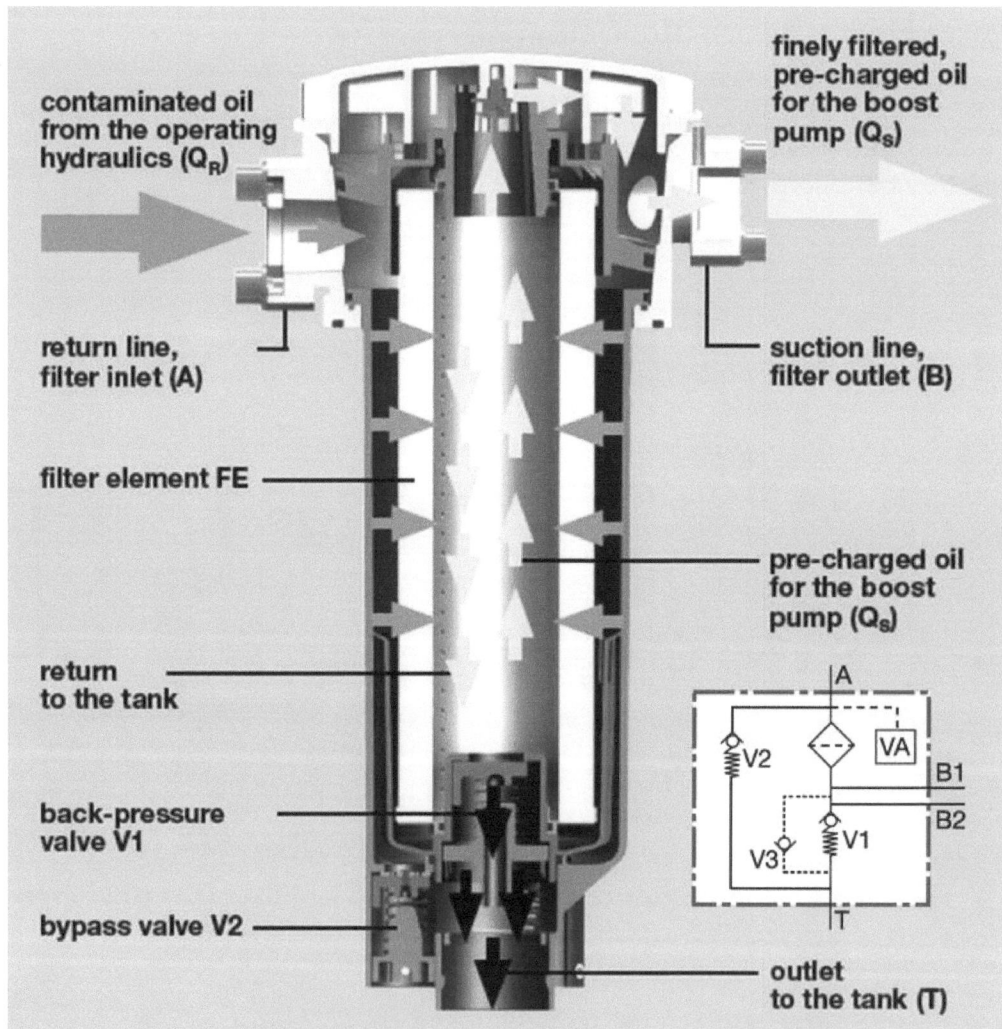

Fig. 1.55- Function of Combined Return and Suction Booster Filter (Courtesy of Hydac)

1.9.7- Diffusers

As shown in Fig. 1.56 installing a *diffuser* in a hydraulic reservoir is a simple addition that makes a big difference in system performance. With special concentric tubes designed with discharge holes 180° opposed, fluid aeration, foaming and reservoir noise are reduced. Pump life is also extended by reducing cavitation to the pump inlet. Figure 1.57 shows the typical stream of flow around the baffle plate between the return line through a diffuser and the suction line through a suction strainer.

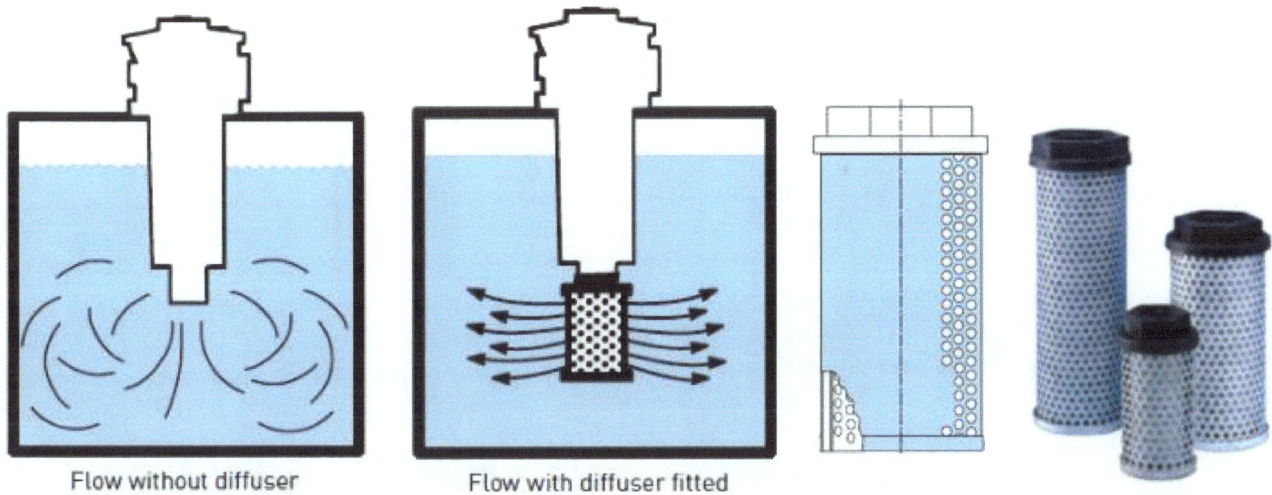

Flow without diffuser   Flow with diffuser fitted

Fig. 1.56- Diffusers on a return line (Courtesy of Parker)

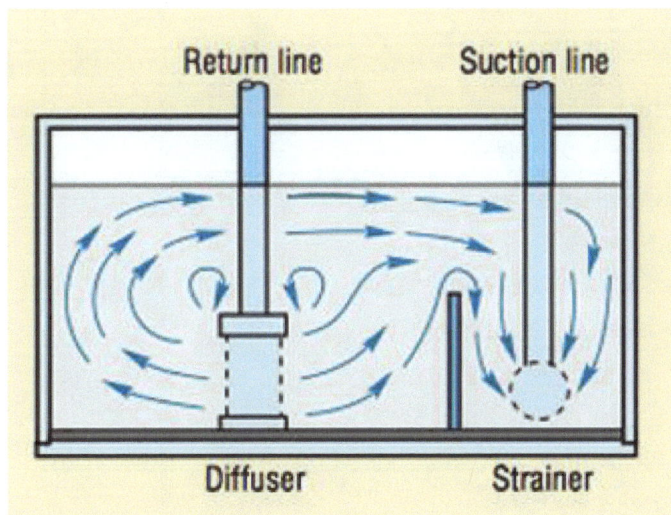

Fig. 1.57- Flow Streams from the Return Line to Suction Line (Courtesy of Parker)

1.9.8- Filler Caps

Figure 1.58 shows the least expensive traditional filler caps. A *filling cap* should be fitted with a sealed cover to prevent the ingress of contaminants when closed. It may contain a filling screen to catch relatively large contaminants during filling. Filler caps should be chained to the reservoir to keep them captive. For more protection, the cover should be lockable. Figure 1.59 shows a typical filler cap that allows reservoir breathing (but without filter breather) as follows:

1- Air intake to reservoir through vacuum breaker when pressure decreases (0.435 psi)
2- Venting to atmosphere through relief valve to maintain a 5 or 10 psi.

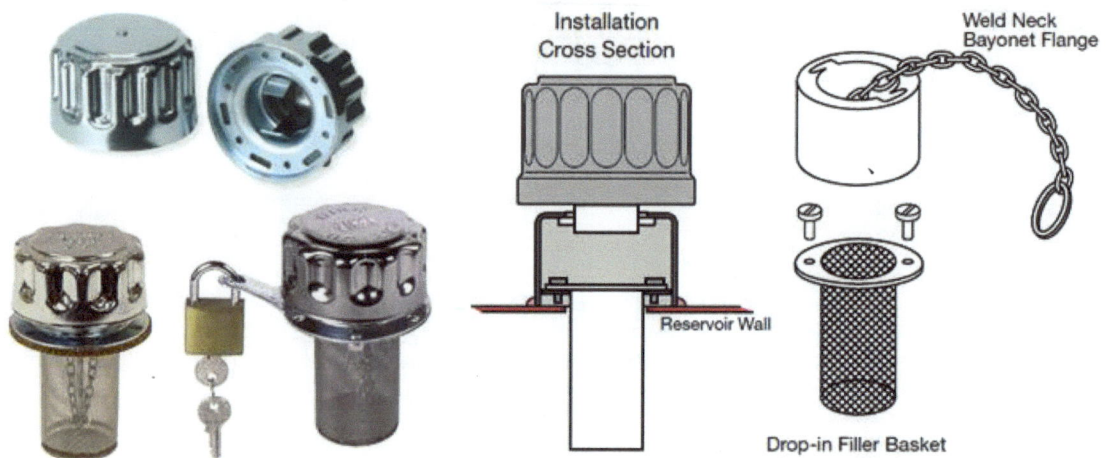

Fig. 1.58- Filler Caps

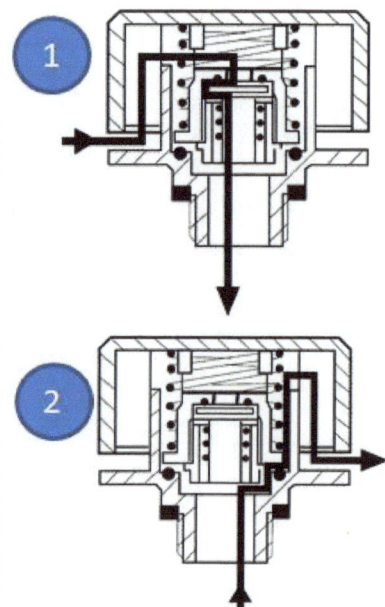

Specifications
- Chrome plated, epoxy coated or zinc plated steel cap
- Airflow to 30 cfm/850 lpm
- Compatible with petroleum based fluids
- Temperature to 212°F / 100°C
- 1/2", 3/4" and 1" NPT on ABS
- 1/4" and 3/8" NPT on MBS

Options
- 3, 10 and 40 micron (ABS), 10 and 40 micron (MBS)
- Zinc and epoxy coated weather-proof cap versions

Fig. 1.59- Example of Filler Caps (Courtesy of Donaldson)

1.9.9- Filter Breathers

Concept: In all systems using accumulators, single-acting cylinders or double-acting differential cylinders, the reservoir oil level falls as the cylinders extend and raises up as they retract. So, an open reservoir breathes during the machine operation like a human and hence a very large air volume passes in and out to the air space in the reservoir. If the air is allowed to freely move in and out of the reservoir, everything contained within that air is exchanged with the oil. This can include air-born contaminants and moisture, that both have severe effect on the oil lifetime and the hydraulic system reliability. Therefore, to prevent the dust from getting into the tank, *air breathers* are used. Breathers are available in various configurations, sizes, and working features. Filtration rating of an air breather must be equal to or better than main system filter.

Standard Filler Breathers: As shown in Fig. 1.60, *Standard Filter Breathers* are very similar in shape to the filler caps. They are usually screwed onto a threaded pipe that provides air exchange through the top of the reservoir. Other styles can look like a spin-on oil filter. They are available in various forms such as metallic or non-metallic, flange-mount or thread-mount, and with 10 microns size of a conventional or telescopic strainer. As shown in the figure, manufacturers report the differential pressure across the breather versus the air flow.

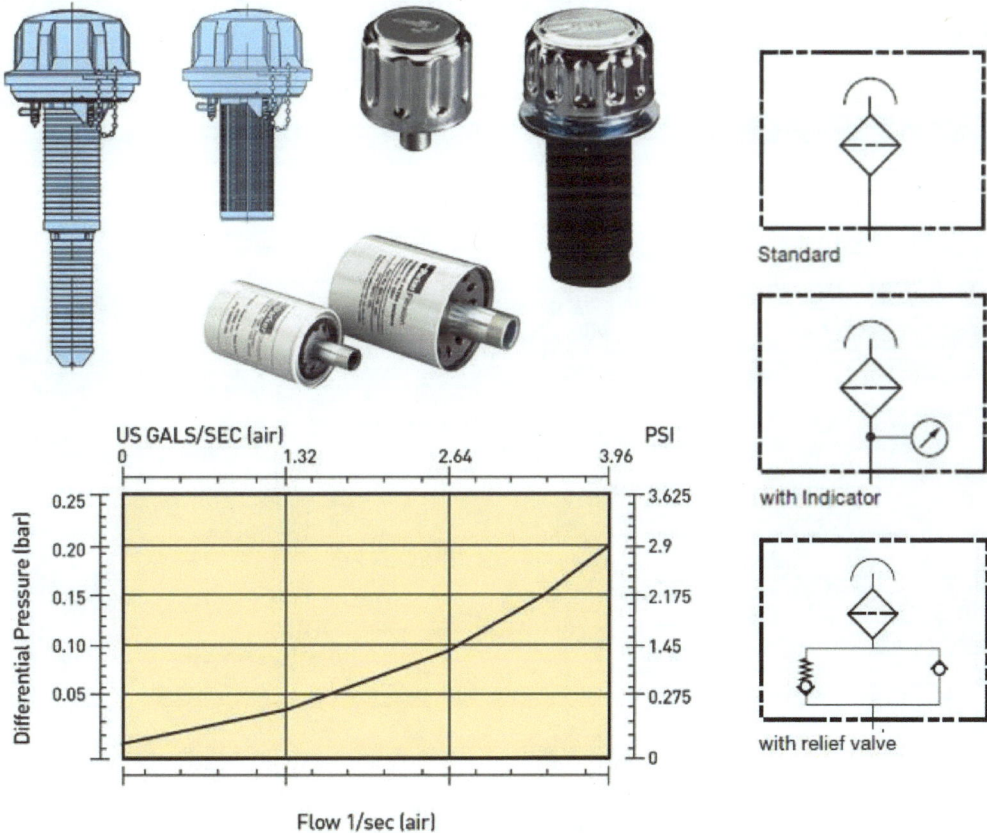

Fig. 1.60- Standard Filter Breathers (Courtesy of Parker)

Desiccant Breathers: In hydraulic systems that use petroleum-based hydraulic fluids, water ingress into the tank is a familiar problem. In such systems, if water content in oil increases above the allowable limit, system faces frequent breakdowns and high maintenance costs. For detailed information about contamination by water, review Volume 3 of this textbook series. Therefore, using *Desiccant Breathers* is a must for applications that work in very humid environments such as marine and offshore applications.

Figure 1.61 shows a typical example from industry for a desiccant breather and its hydraulic symbol. As shown in the figure, desiccant breathers are designed with a transparent body filled with silica-gel that are designed to absorb as much as 40% of its weight. The gel's color changes from a blue to light pink color when saturated. The unit also contains regular filter element to capture contaminants as small as 3 microns. As an option, an inlet check valve can be assembled on the air inlet to prevent the saturation of the desiccants during the system shutdown. An outlet check valve can be assembled on the air outlet to prevent exhaust air from the tank from flowing back through the desiccant so that the desiccant is protected from oil mist. Replace breather when desiccant color changes or when a built-in clogging indicator shows expiration of the drying material.

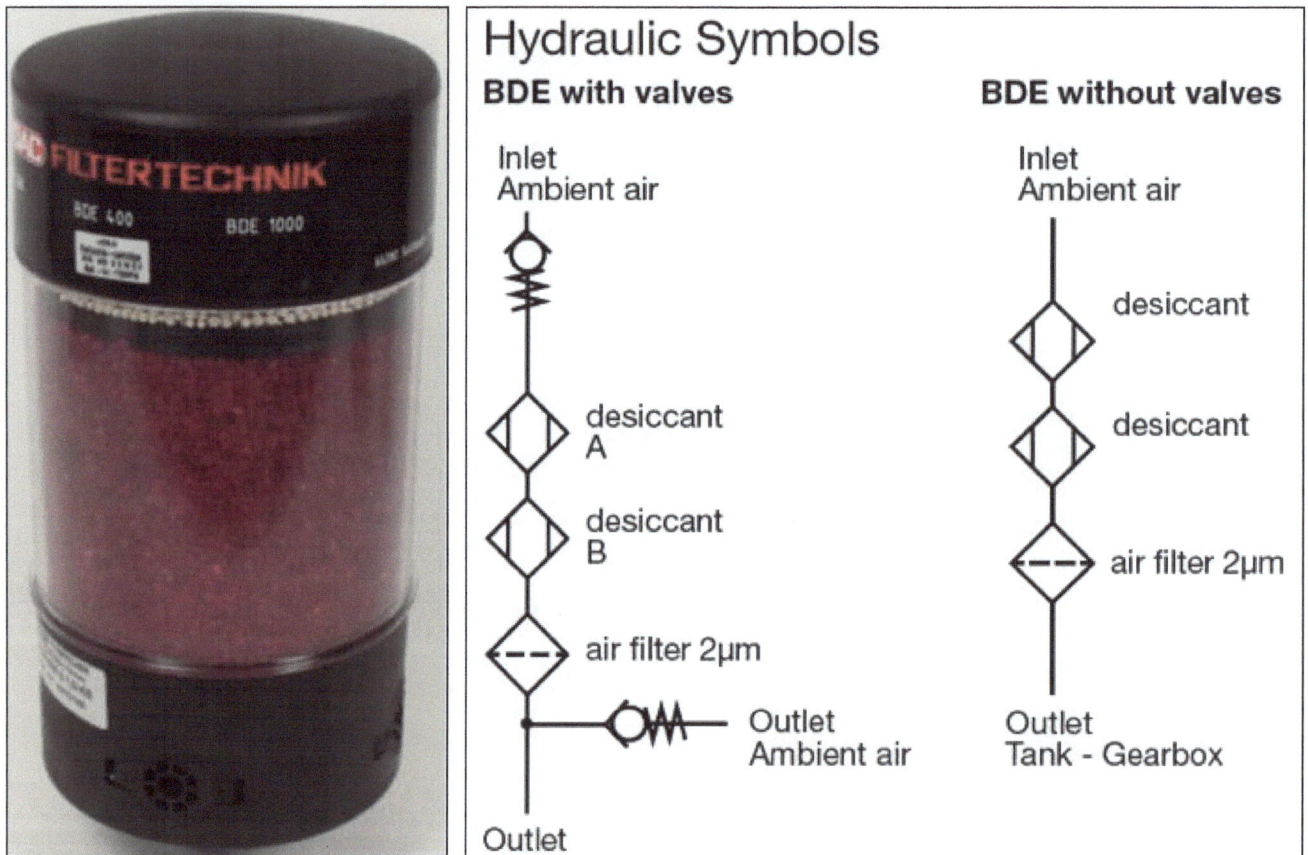

Fig. 1.61 – Desiccant Breather Dryer (Courtesy of HYDAC)

Construction and Operation of Desiccant Breather: As shown in Fig. 1.62, the desiccant breather consists of two separate chambers which can be filled with two desiccants, which in combination increase total water retention because of two-stage dewatering. The figure shows a built-in pleated air filter element (absolute filtration of particles > 2 µm) provides the filter with a very high contamination retention capacity (26 g). Such breather dryers can work in temperatures range -30 °C to 100 °C (-22 °F to 212 °F).

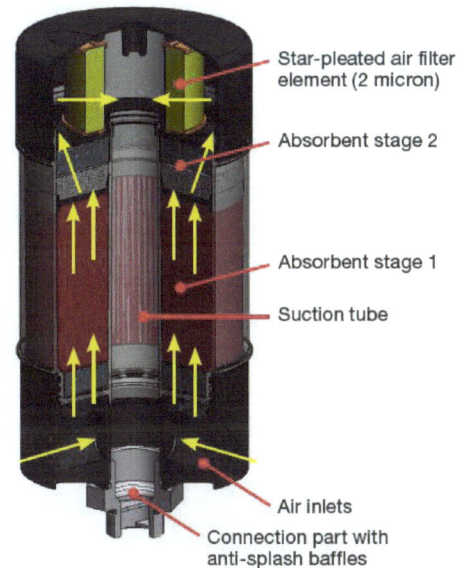

Star-pleated air filter element (2 micron)

Absorbent stage 2

Absorbent stage 1

Suction tube

Air inlets

Connection part with anti-splash baffles

Fig. 1.62 – Construction and Operation of the Desiccant Breather Dryer (Courtesy of Hydac)

Breathers Dryers: Alternative to using desiccant breathers, *Breather Dryer* can be used. They are breathers with water absorption cartridges. Breather dryers collect and expel moisture out of reservoirs. This means that, unlike desiccant filters, breather dryers will not be changed due to water saturation. Figure 1.63 shows how breather dryers work. An indicator shows when maintenance is required and a new cartridge shall be installed.

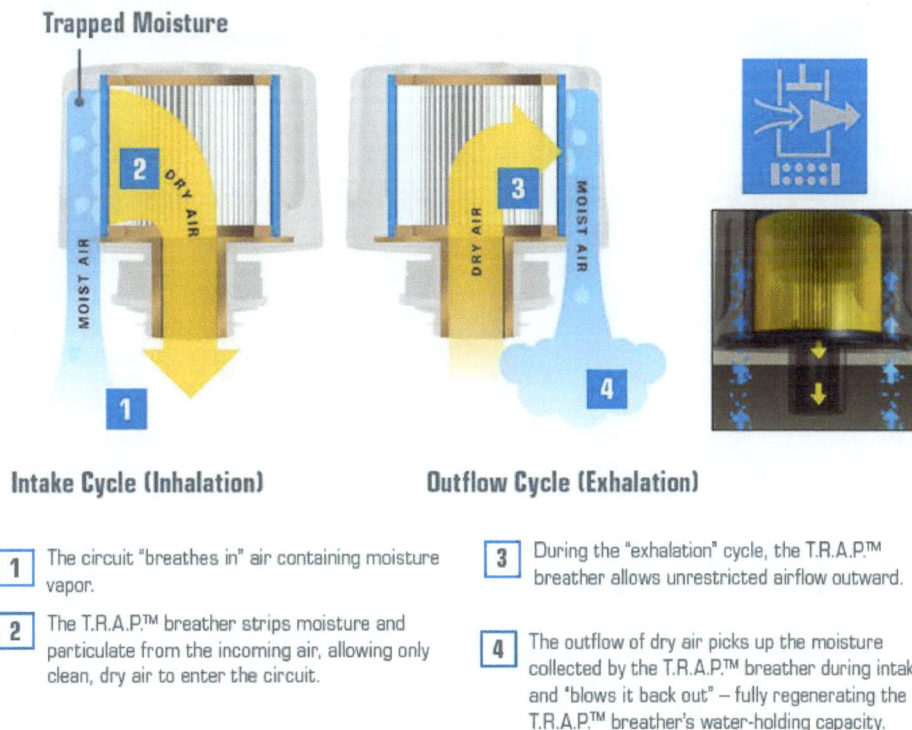

Trapped Moisture

Intake Cycle (Inhalation)

Outflow Cycle (Exhalation)

1 The circuit "breathes in" air containing moisture vapor.

2 The T.R.A.P.™ breather strips moisture and particulate from the incoming air, allowing only clean, dry air to enter the circuit.

3 During the "exhalation" cycle, the T.R.A.P.™ breather allows unrestricted airflow outward.

4 The outflow of dry air picks up the moisture collected by the T.R.A.P.™ breather during intake, and "blows it back out" – fully regenerating the T.R.A.P.™ breather's water-holding capacity.

Fig. 1.63 – Construction and Operation of Breather Dryers (Courtesy of Donaldson)

Sizing of Breather Dryer: Breather dryers are provided in different sizes. Undersized tank breather filters can place additional strain on the system and reduce the service life of breather. As shown in Fig. 1.64, larger size has a better water retention, but a relatively larger pressure drops.

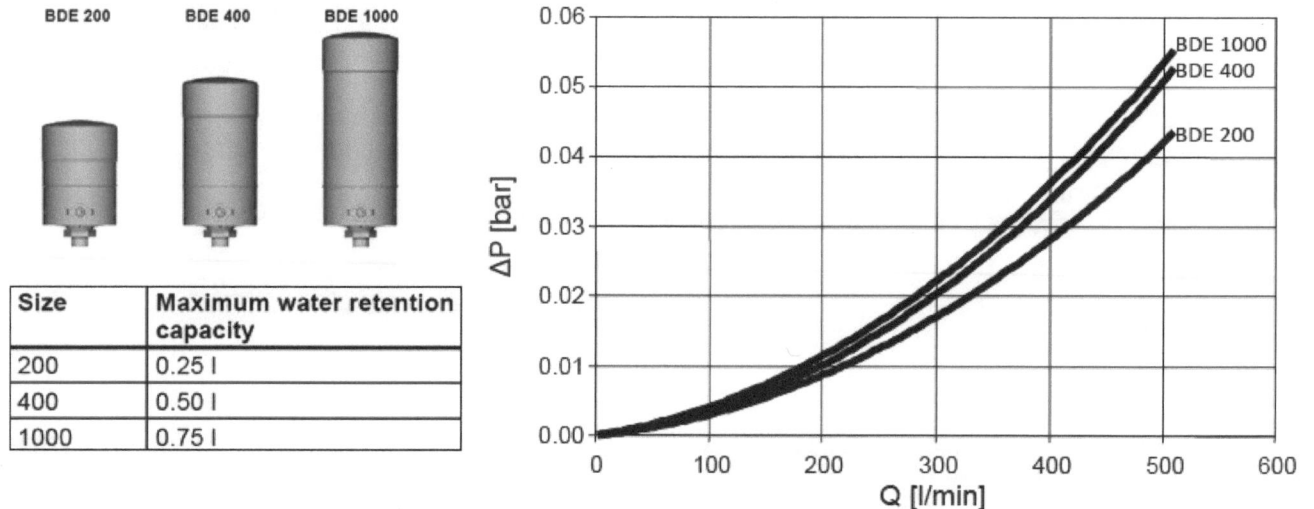

Size	Maximum water retention capacity
200	0.25 l
400	0.50 l
1000	0.75 l

Fig. 1.64 – Sizing of the Breather Dryer (Courtesy of HYDAC)

Air Breathers for Closed Tanks: In some highly contaminated applications, such as mills and foundries, closed reservoirs (pressure-sealed) are recommended. In such reservoirs, the inside pressure is increased above atmospheric by gas bladders and no conventional air breather can be used because the pressure inside the tank is always higher than the atmospheric. As shown in Fig. 1.65, in some cases a tank may be completely sealed from the atmosphere. Any air returns with the oil back to the tank will be accumulated on top of the oil surface. An air check valve is used to protect the tank against accidental over-pressure. Such a valve should be connected to any point on the tank above oil level. The valve allows free one air flow direction toward the atmosphere. Cracking pressure of 1 to 3 PSI is common.

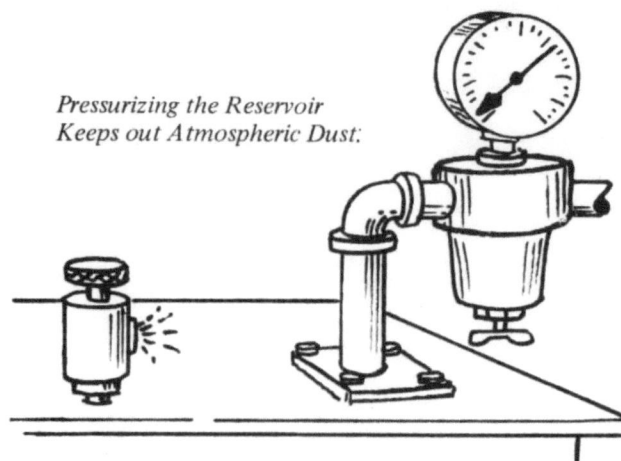

Fig. 1.65 – Sizing of the Breather Dryer (Courtesy of Womack)

1.9.10- Offline (Bypass) Filtration Units

Placement: *Offline Filters* are placed in a separate circuit. They are also referred to as *Bypass Filters.* Offline filtration can be done by a portable unit or by a permanently installed unit.

Primary Duty: The primary duty of offline filters is to filter the full volume of oil in the reservoir apart from the oil circulation in the main circuit. It has the advantage of the filter isn't subjected to surge flows, filter can be replaced without interrupting the system, and fine filtration is possible.

Cost: Offline filtration has high initial cost. However, the cost is justified through extended system life..

Sizing: Flow rate is controlled based on the oil volume in the reservoir and how many fluid circulations is required per hour. However, commonly the pump in such a loop is rated at 5%-10% of the oil volume in the reservoir. For example, if a reservoir has 100 liters of oil, pump is rated for 5-10 liter/min so that the whole reservoir will be filtered in 10-20 minutes. In other words, the whole reservoir is filtered 3-6 times per hour.

Filtration Rating: Filtration rates of 2 microns or less are possible, and polymeric (water-absorbent) filters and heat exchangers can be included in the circuit for total fluid conditioning.

Construction: Offline filtration units are available in various styles and sizes. Figure 1.66 shows an example of <u>hand-portable</u> offline filtration unit. The unit consists of a pump/motor unit, a filter, and the supporting frame.

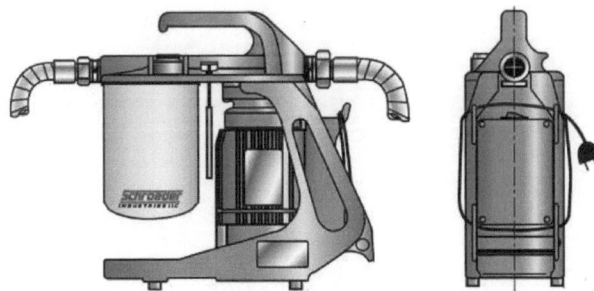

Fig. 1.66 – Example of Hand-Portable Offline Filtration Unit (Courtesy of Schroeder)

Figure 1.67 shows an example of <u>portable-cart</u> offline filtration unit. This provides a convenient portable mode of kidney loop filtration, flushing and fluid transfer. It can be used for in-plant machinery and hydraulic equipment to achieve and maintain proper ISO cleanliness levels. The *Filter Cart* includes a pump/motor unit, a filter, and the wheeled supporting frame. The cart is used in many cases such as *kidney loop filtration*, transferring new oil, cleaning stored oil, system filling/draining, line flushing, and flushing equipment after commissioning or rebuild.

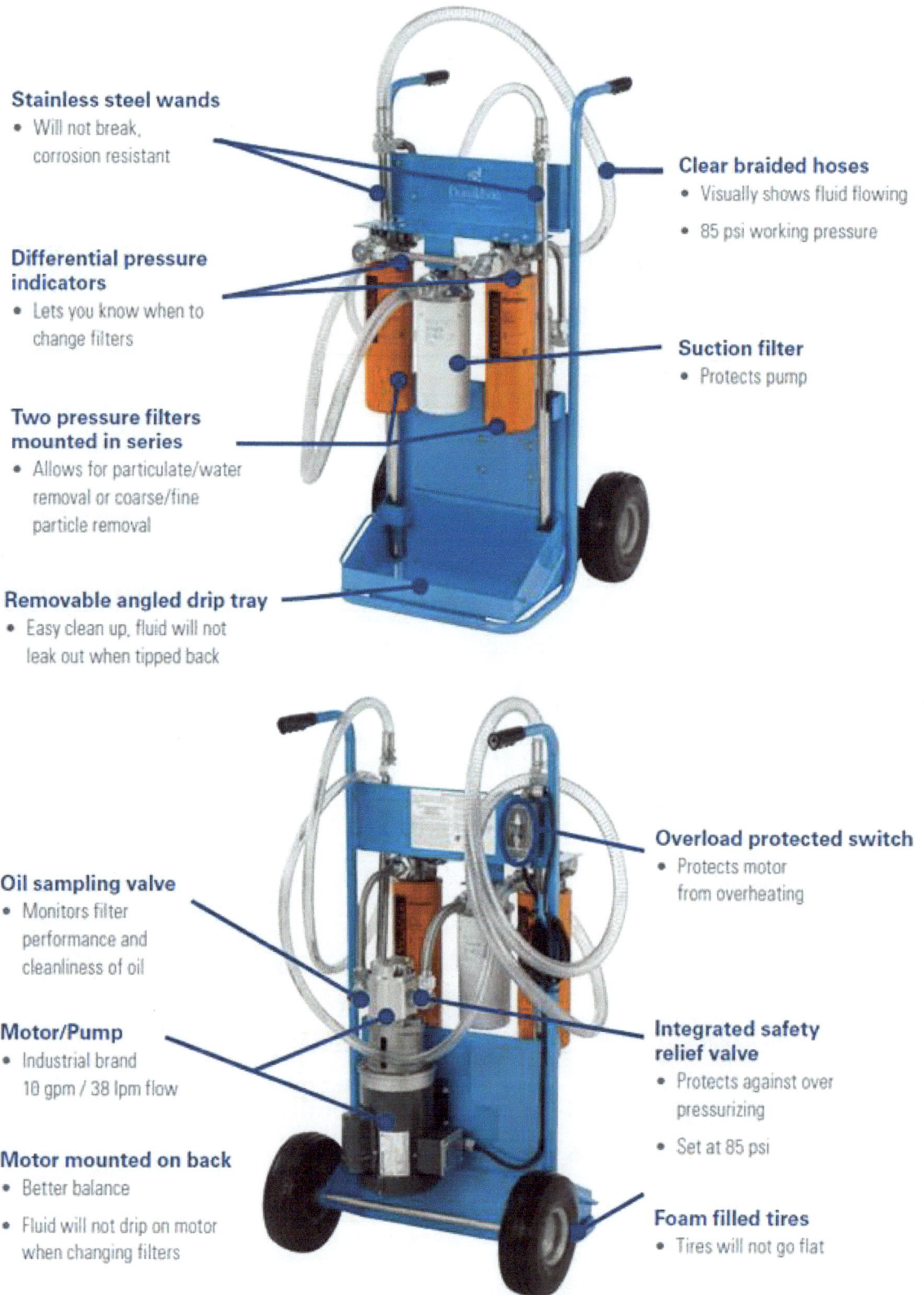

Stainless steel wands
- Will not break, corrosion resistant

Differential pressure indicators
- Lets you know when to change filters

Two pressure filters mounted in series
- Allows for particulate/water removal or coarse/fine particle removal

Removable angled drip tray
- Easy clean up, fluid will not leak out when tipped back

Clear braided hoses
- Visually shows fluid flowing
- 85 psi working pressure

Suction filter
- Protects pump

Oil sampling valve
- Monitors filter performance and cleanliness of oil

Motor/Pump
- Industrial brand 10 gpm / 38 lpm flow

Motor mounted on back
- Better balance
- Fluid will not drip on motor when changing filters

Overload protected switch
- Protects motor from overheating

Integrated safety relief valve
- Protects against over pressurizing
- Set at 85 psi

Foam filled tires
- Tires will not go flat

Fig. 1.67 – Example of Cart-Portable Offline Filtration Unit (Courtesy of Donaldson)

Figure 1.68 shows an example of <u>Fixed-Mounted</u> offline filtration unit. This unit is permanently mounted to offer supplemental filtration for in-plant machinery and hydraulic equipment helping to reduce costs and achieve and maintain proper ISO cleanliness levels. Figure 1.69 shows an example of a hydraulic system with offline filtration.

Fig. 1.68 – Example of Fixed-Mounted Offline Filtration Unit (Courtesy of Donaldson)

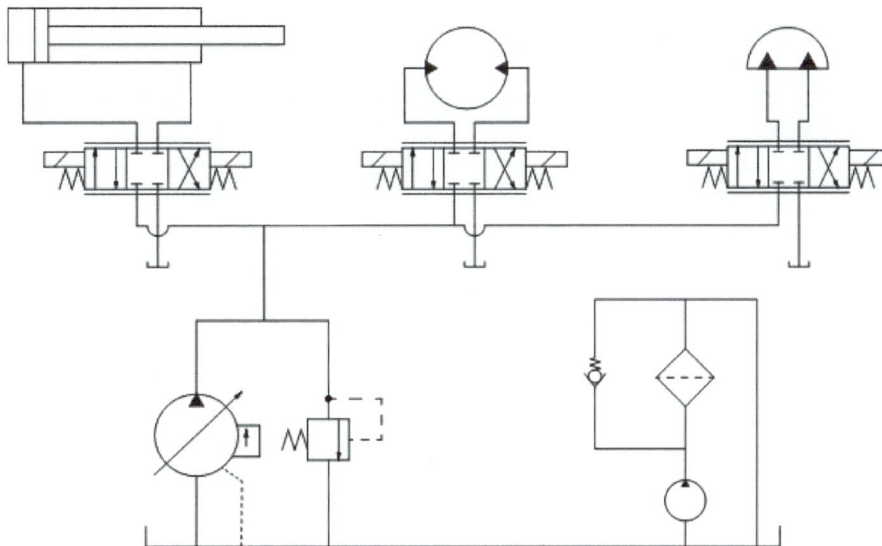

Fig. 1.69 – Example of Hydraulic Systems with Offline Filtration

Chapter 2

Filter Media and Filtration Mechanisms

Objectives

This chapter presents an overview of filter elements including the construction and material of the filter media. This chapter discusses surface filters versus depth filters. The chapter discusses also the principles of various filtration mechanisms that are applicable in hydraulic filters such as direct interception, absorption, adsorption, and magnetic separation.

Brief Contents

2.1- Filtration Mechanisms

2.2- Materials for Filter Media

2.3- Filter Media Structure

Chapter 2 – Filter Media and Filtration Mechanisms

2.1- Filtration Mechanisms

Capturing and retaining the particulate contaminants depends on one of the following mechanisms:

Retaining Large Size Particles by Inertia: As shown in Fig. 2.1, when the fluid is accelerated between fibers and when the fluid changes direction to enter the fiber space, large and heavy particles suspended in the flow stream are slower than the fluid surrounding them because of particles *Inertia*. As a result, the particle continues in a straight line and is trapped by the media fibers where it is held and retained.

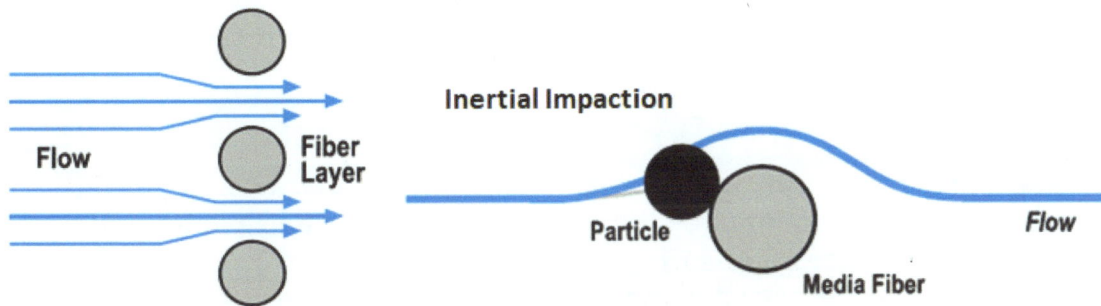

Fig. 2.1- Retaining Large Size Particles by Inertia (Courtesy of Donaldson)

Retaining Medium Size Particles by Direct Interception: As shown in Fig. 2.2, particles are retained by *Direct Interception*. The mid-range size particles that are neither quite large enough to have inertia nor small enough to diffuse within the flow stream. These mid-sized particles are mechanically captured and retained just because they are larger than the micron size in the filter media. Direct interception s also referred as *"Sieving"*.

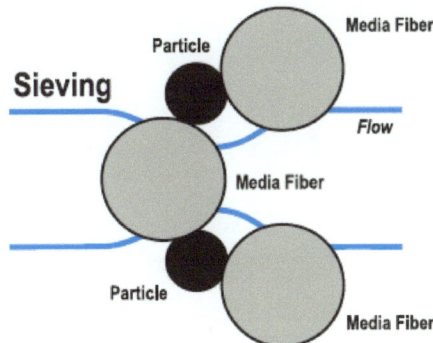

Fig. 2.2- Retaining Medium Size Particles by Direct Interception (Courtesy of Donaldson)

Retaining Small Size Particles by Absorption: As shown in Fig. 2.3, *Absorption* filtration mechanism works on the smallest particles. Absorption is also referred to as *Diffusion*. Small particles are not held in place by the viscous fluid and diffuse within the flow stream. As the particles traverse the flow stream, they are captured and collected by the fiber. Best interpretation of diffusion is that filter media attracts and retain particles by *electrostatic forces* or molecular attraction. Obviously special material is used in developing such filter media.

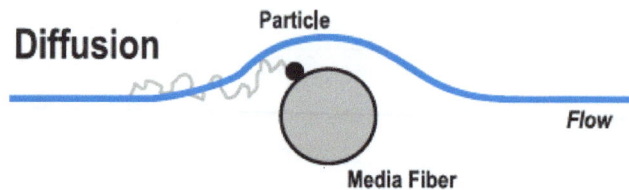

Fig. 2.3- Retaining Small Size Particles by Diffusion (Courtesy of Donaldson)

Retaining Particles by Adsorption: *Adsorbent filter media* use chemically treated filter media to remove contaminants. Charcoal, chemically treated paper, and other materials are used in this process. Adsorbent filter media are not typically used in hydraulic systems as they may remove desirable additives from the system fluid.

Oil Cleaning by Centrifugal Separators (explained in Volume 3): Water and solid particles with a density higher than that of oil can be removed by *Centrifugal Separators*. However, this method can't guarantee 100@ water removal and it can remove oil additives too.

Oil Cleaning by Vacuum Dehydration (explained in Volume 3): Air and water in oil can also be removes by *Vacuum Filters*.

Oil Cleaning by Magnetic Separation: In the control of contamination in hydraulic fluids, *Magnetic Separation* useful in separation of ferrous solids from fluid streams.

2.2- Materials for Filter Media

Cellulose Fibers Filter Media (Traditional): As shown in Fig. 2.4, *Cellulose Fibers* filter media has the following characteristics:
- **Material:** Wooden fibers held together by resin.
- **Shape:** Irregular shape.
- **Size:** Irregular small (microscopic) pores size.
- **Flow and ΔP:** Has more flow resistance, resulting higher pressure drop.
- **Filtration:** Good in catching contamination through the <u>depth</u> of the media. Poor filtration performance as compared to synthetic media.
- **Fluid:** Provides effective filtration for a wide variety of petroleum-based fluids.

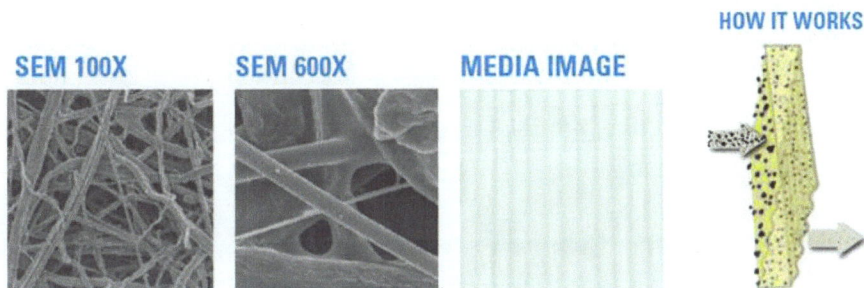

Fig. 2.4- Cellulose Fibers Filter Media (Courtesy of Donaldson)

Synthetic Fibers Filter Media (Fully Synthetic): As shown in Fig. 2.5, *Synthetic Fibers* filter media has the following characteristics:
- **Material:** Man-made, smooth, rounded fibers.
- **Shape:** Consistent shape.
- **Size:** Controlled size and distribution pattern through the media.
- **Flow and ΔP:** Provides low flow resistance, and consequently low pressure drop.
- **Filtration:** Consistency of fiber shape improves contaminant-catching ability on the <u>surface</u> and increases dirt holding capacity.
- **Fluid:** Ideal for use with synthetic fluids, water glycols, water/oil emulsions, HWCF and petroleum-based fluids.

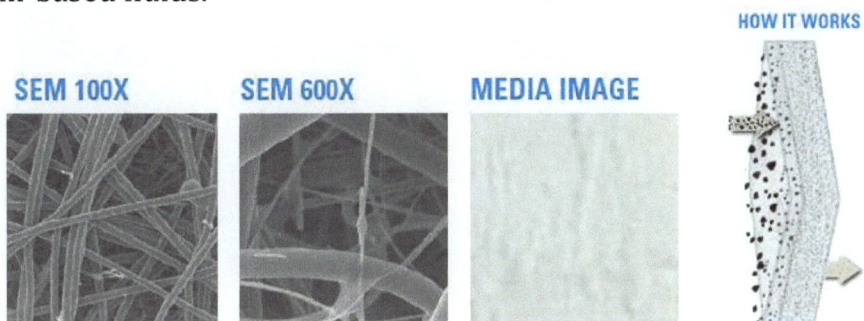

Fig. 2.5- Synthetic Fibers Filter Media (Courtesy of Donaldson)

Combined Fibers Filter Media (Cellulose & Synthetic): As shown in Fig. 2.6, *Combined Fibers filter media* is developed to provide effective fuel filtration performance for optimal protection.

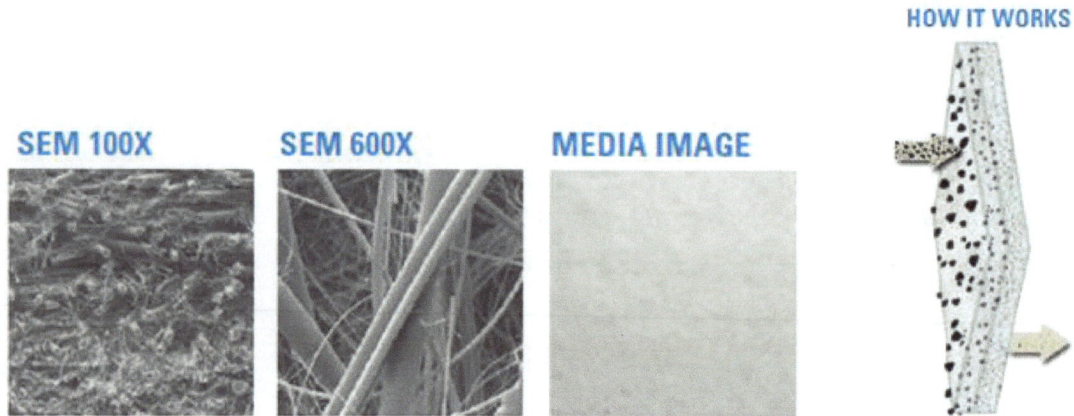

Fig. 2.6- Combined Fibers Filter Media (Courtesy of Donaldson)

High Performance Synthetic Fibers Filter Media: Today's fluid systems are often tailored towards the special needs of fire resistance, biodegradability, chemical and thermal resistances, and electrical insulating ability. As shown in Fig. 2.7, *High Performance Synthetic Fibers* filter media has the following characteristics:

- **Material:** A blend of borosilicate *Glass Fiber* whose matrix is bonded together with an epoxy-based resin system.
- **Flow and ΔP:** Provides high flow resistance, and consequently high pressure drop.
- **Fluid:** They provide the best chemical resistance for the broadest array of hydraulic fluid. Ideal for use with phosphate ester and water glycol fluids.
- **Filtration:** Ideal for fine filtration and precision components.

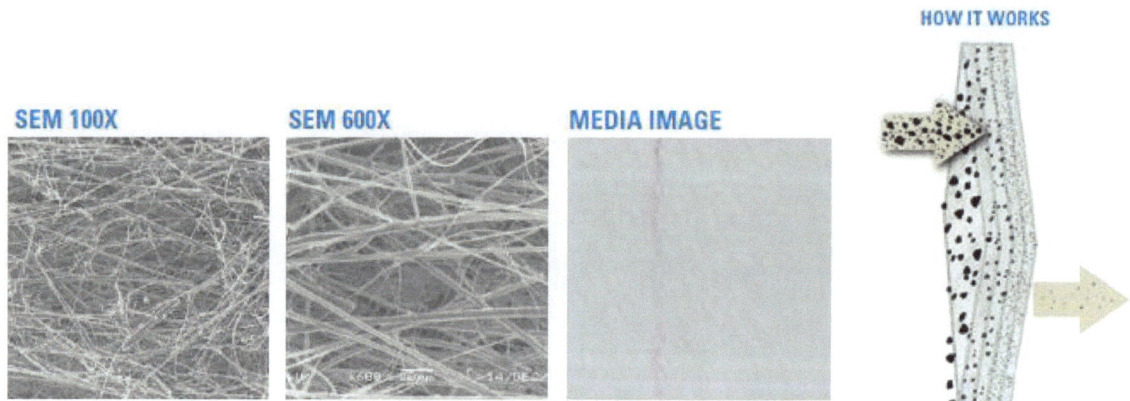

Fig. 2.7- High Performance Synthetic Fibers Filter Media (Courtesy of Donaldson)

Wire Mesh Filter Media: As shown in Fig. 2.8, *Wire Mesh* filter media has the following characteristics:

- **Material:** Stainless steel, epoxy-coated wire mesh.
- **Shape:** Consistent shape.
- **Size:** Available in different sizes.
- **Flow and ΔP:** Provide the least flow resistance, and consequently lowest pressure drop.
- **Filtration:** Available in various micron sizes and ranging from 100 to 500 microns. Typically wire-mesh filters will be applied to catch very large and harsh particles that would plug up a normal filter. Generally used in suction strainers.

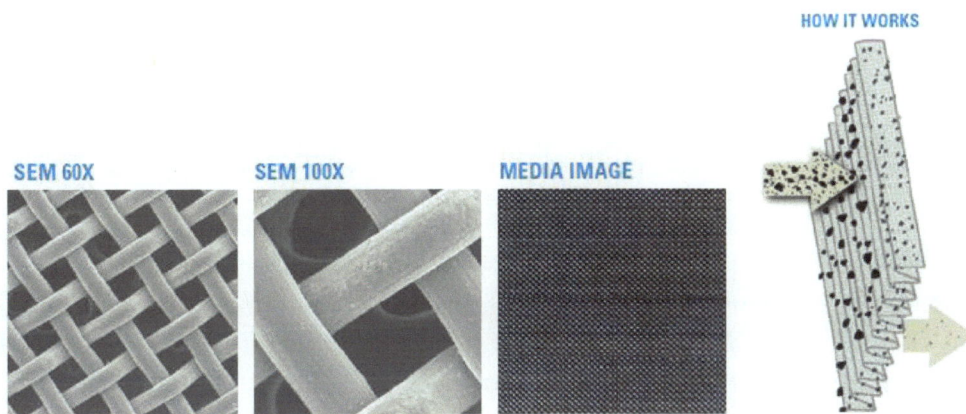

Fig. 2.8- Wire Mesh Filter Media (Courtesy of Donaldson)

Water Absorption Filter Media: As shown in Fig. 2.9, *Water Absorption* filter media quickly and effectively removes free water from hydraulic systems. Using super-absorbent polymer technology, with a high affinity for water absorption, prevents many of the problems associated with water contamination found in petroleum-based fluids.

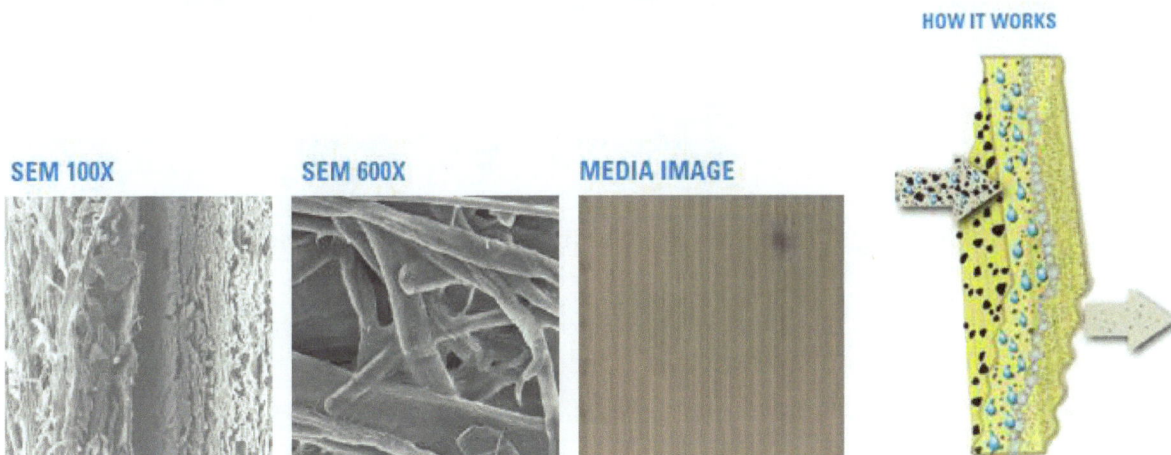

Fig. 2.9- Water Absorption Filter Media (Courtesy of Donaldson)

2.3- Filter Media Structure

Fiber Structure (Fig. 2.10 - 1): For fluid to pass through, the media must have pores or channels to direct the fluid flow and allow it to pass. That's why filter media is a porous material made of fibers that structured to twist, turn, and accelerate during passage.

Uniform vs. Graded Pore Size (Fig. 2.10 - 2): Based on the pore size along the depth of the filter media, it can be constructed to form uniform or graded pore size. Graded pore size with larger pore size on the surface. Graded pore size allows holding more dirt, but it causes higher pressure-drop across the filter media.

Fixed vs. Non-Fixed Pore Size (Fig. 2.10 - 3): Based on the method of bonding the fibers together on each layer, filter media can be constructed to form fixed or non-fixed pore size. In fixed pore media, fibers are bonded with specifically formulated resin to resist deterioration from pressure and flow fluctuations, temperature and aging conditions. Fibers in non-fixed pore media are inconsistently or poorly bonded. This facilitates movement of fibers under pressure and flow surges allowing media migration.

Surface vs. Depth Filters (Fig. 2.11): Hence, filter media are structured as *Surface Filter Media* and *Depth Filters Media*. As it works its way through the depths of the layers of fibers, the fluid becomes cleaner and cleaner. Generally, the thicker the media, the greater the dirt-holding capacity it has.

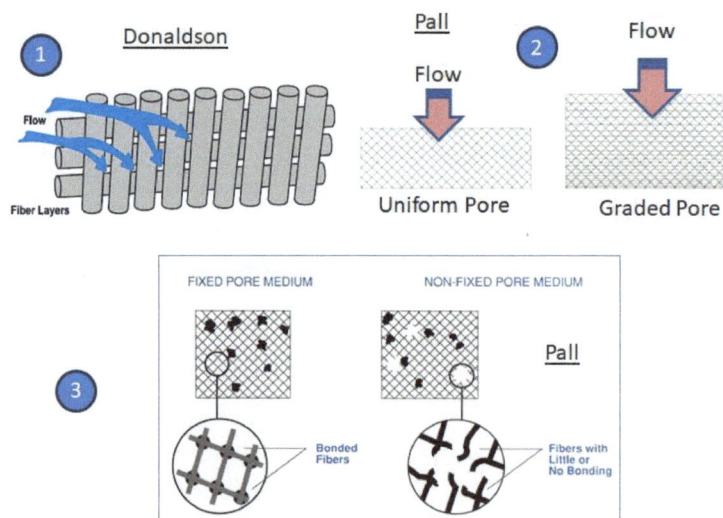

Fig. 2.10- Filter Media Fibers (Courtesy of Donaldson)

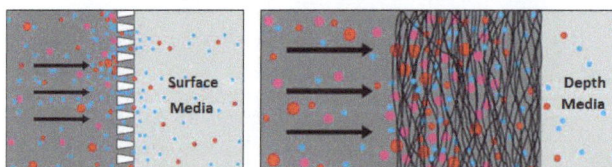

Fig. 2.11- Surface versus Depth Filter Media (Courtesy of Bosch Rexroth)

2.3.1- Surface Filter Media

Filtration Process: As shown in Fig. 7.12, primary filtration mechanism of a *Surface Filters* is direct interception. Most surface-type filters are exposed to the flow of contaminated fluid. Pore size of surface filter media is gradually reduced due to intrusion of soft and deformable particles. Over the time, the filter is completely clogged.

Applications: *Surface Filter* media is used commonly for strainers or suction filters.

Advantages of Surface Filter Media:
- Are washable and cleanable.
- No media migration with the oil.
- Low flow resistance and pressure drop.
- High fatigue and corrosion resistant.
- Work at high temperature.

Disadvantages of Surface Filter Media:
- Catch only relatively large contaminants.
- Can't be used to maintain high cleanliness Level.
- Needle-shaped contaminants that have less diameter than the pore size, even if its length is larger than the pore size, can pass through these filters.

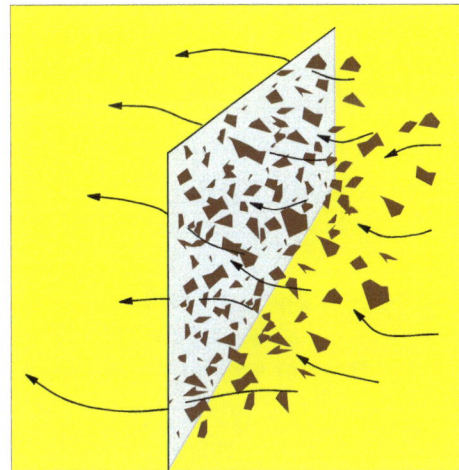

Fig. 2.12- Filtration using Surface Filters

Material: Various materials are used for surface filters such as stainless-steel wires, galvanized iron, or phosphor bronze, accordion-pleated paper, ribbon-shaped metal, and stacked metal disks.

Structure of Metallic Surface Filters: As shown in Fig. 7.13, *Square Wire Mesh* and *Braided Wire Mesh* are most commonly used materials for surface filters. Both are made from stainless steel, Both are washable, and provide low pressure drop. They are commonly used for coarse filters, lubricating systems, and suction filters. Braided mesh wire has better filter rating.

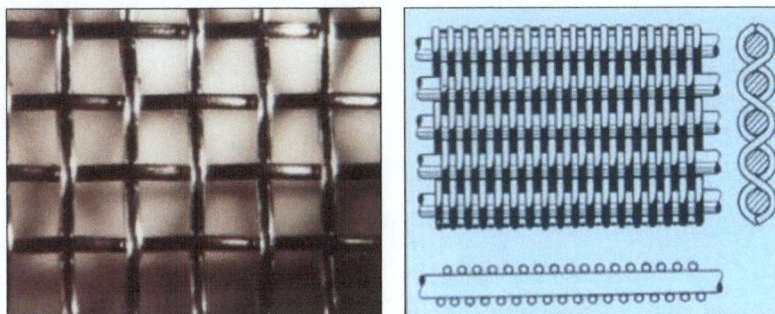

Fig. 2.13- Square versus Braided Wire Mesh Surface Filters Media (Courtesy of Bosch Rexroth)

2.3.2- Depth Filter Media

Filtration Process: As shown in Fig. 2.14, *Depth Filter* media particles are removed by *Direct Interception* and *Absorption.* Fine filtration is done by trapping solid dirt particles within the depth material. Also, water and water-soluble contaminants suspended in the hydraulic fluid in one of the many flow routes in the porous material. Such depth filters are classified as *Absorbent* filters.

Applications: *Depth Filter* media is used commonly for pressure, return, and offline filters.

Advantages of Surface Filter Media:

- Effective filtration for small contaminants.
- Used to maintain high cleanliness Level.
- Has large dirt holding capacity.

Disadvantages of Surface Filter Media:

- Are not washable or cleanable.
- Possible media migration with the oil.
- High flow resistance and pressure drop.

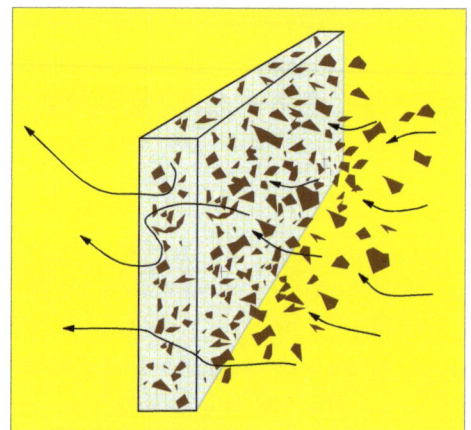

Fig. 2.14- Filtration using Depth Filters

Material: Various materials are used for depth filters such as organic (such as Cellulose, or Cotton) or synthetic fibers. Some depth filters are made from randomly oriented steel wires. However, the materials from which the depth media is constructed must be compatible with the hydraulic fluid and operating temperature of the system.

Structure: As shown in Fig. 2.15, depth filter media is composed of layers of porous material such fibers. It does not have consistent pore size that is why it is rated based on average pore size. Fiber diameter is from 0.5 to 30 microns. Fibers are wounded in layers on top of each other. Each layer depth is 0.25 – 2 mm (0.01-0.08 in). The quality of the elements varies considerably between manufacturers depending on:

- Fiber bonding and the ability to prevent media migration.
- Central support of the filter element pleats.
- Sealing of the filter media to the end caps.
- Consistency of pore size through the media

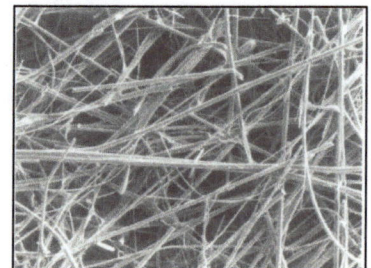

Fig. 2.15- Structure of Depth Filters Media

Example 1– Conventional Glass Fiber Pressure Filter: Conventional inline pressure filters are typically *glass fiber* based, because they need to operate under high pressure and high flow conditions, while creating as little restriction as possible.

As shown in Fig. 2.16, the filter element is pleated, conventionally using Fan-Pleating method, in order to increase the surface area and reduce the pressure drop. Since they are installed after the main system, they often live a tough life with cyclic flows and many stops and starts, which is very harmful for the efficiency of any filter. Capturing and retaining fine silt particles is therefore very difficult, which is why most of these inline filters have a rating of 5 – 50 microns.

However, many captured particles will be released again when the filter is exposed to pressure shocks at stop/start. The glass fiber-based pressure filter is capable of removing solid particles only, and due to the relatively small filter depth and volume, it has a limited dirt holding capacity (1 – 100 grams).

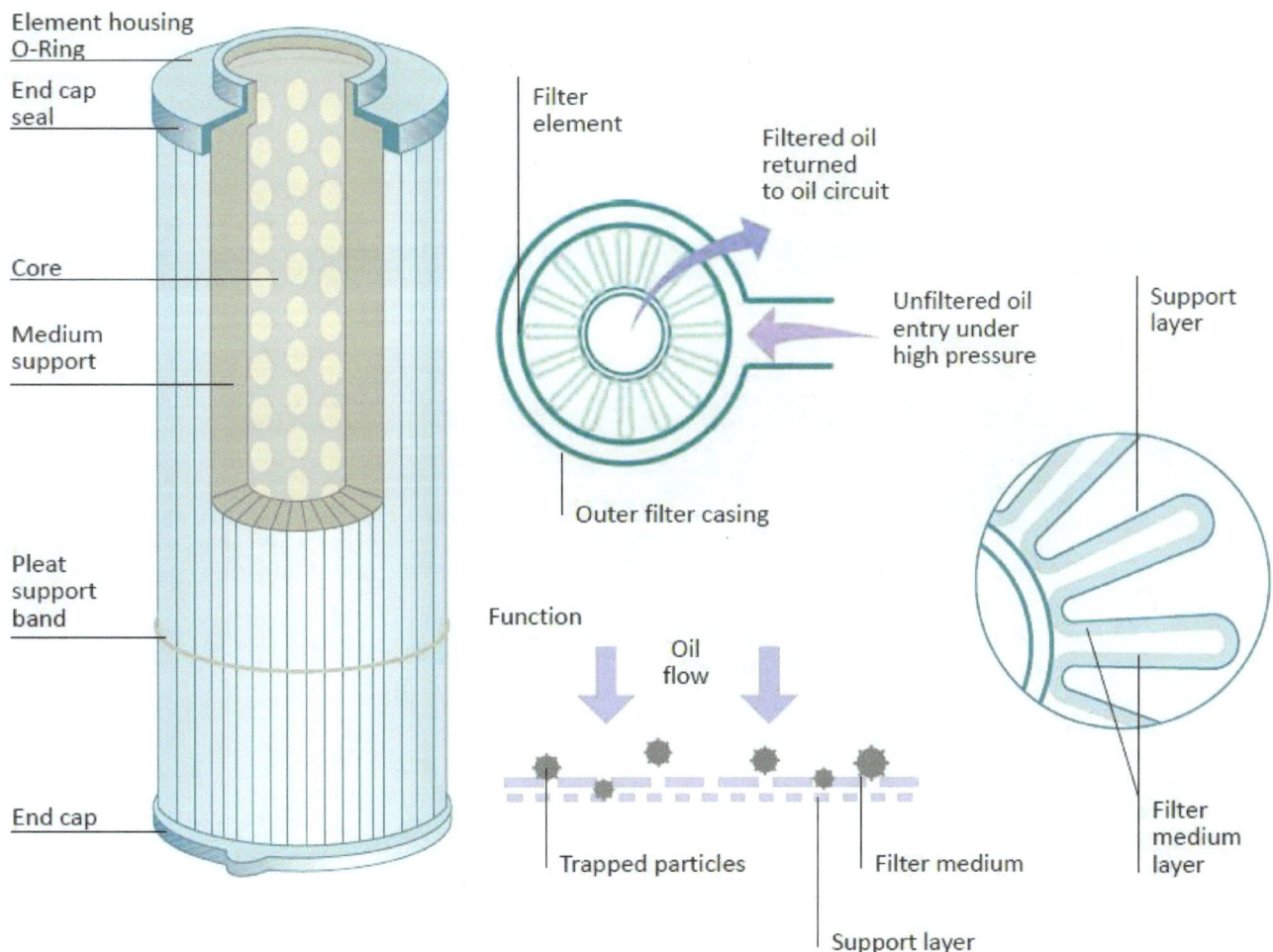

Fig. 2.16- Example of Glass Fiber Pressure Filter (Courtesy of C.C. Jensen Inc.)

Example 2– Conventional Micro Glass Fiber Pressure Filter: Figure 2.17 shows construction of a special class of micro-glass and other fibers depth filter Z-Media®. As shown in the figure, the filter media constructed from multiple layers, each successive layer performs a distinct and necessary function. Filter manufacturer reported the following features:

- Manufactured with utmost precision, to specific thicknesses and densities.
- Layers are bonded with select resins to create material with extra fine passages.
- Maximum dirt-holding capacity and superior particle capture.
- Excellent beta ratio (filter efficiency) stability.
- Minimum pressure drop.
- High flow rate and low operating cost.

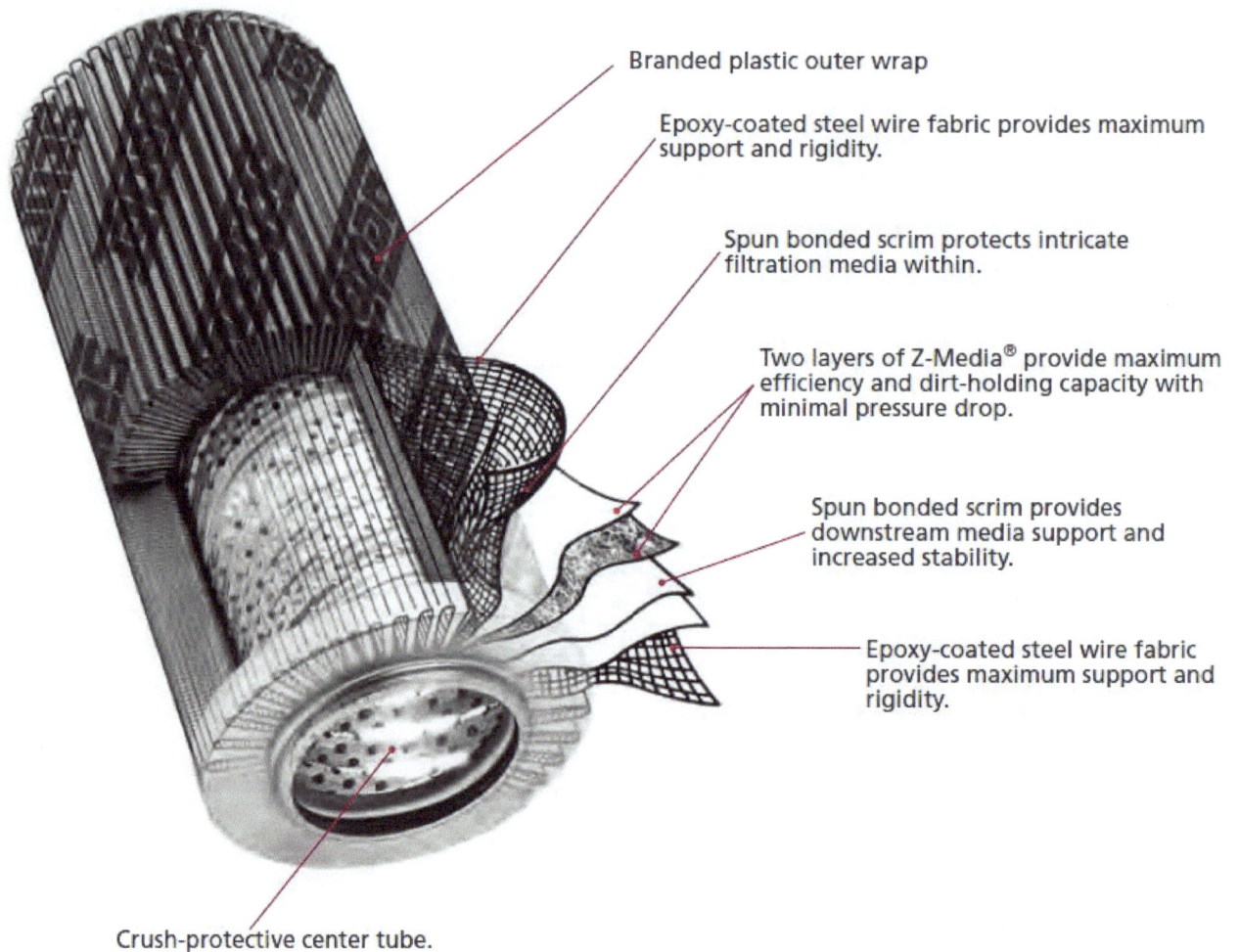

Branded plastic outer wrap

Epoxy-coated steel wire fabric provides maximum support and rigidity.

Spun bonded scrim protects intricate filtration media within.

Two layers of Z-Media® provide maximum efficiency and dirt-holding capacity with minimal pressure drop.

Spun bonded scrim provides downstream media support and increased stability.

Epoxy-coated steel wire fabric provides maximum support and rigidity.

Crush-protective center tube.

Fig. 2.17- Example of Glass Fiber Pressure Filter "Z-Media®" (Courtesy of Schroeder)

Examples 3– Laid-Over Pleating in Depth Filter Element Technology: Figure 2.18 shows description of new depth filter technology (Ultipleat). As shown in the figure, the conventional *fan-pleating* method results in nonuniform volumes between the pleats, and consequently nonuniform flow distribution. The laid-over pleating new pleating technology (Ultipleat) has the following features:

- Allows more filtration area to be packed into a given filter element envelope.
- Creates uniform flow distribution through the filter element.
- Protects pleat against collapse and bunching.
- Anti-static construction minimizes static discharges.
- Resistance to cyclic flow and pressure.

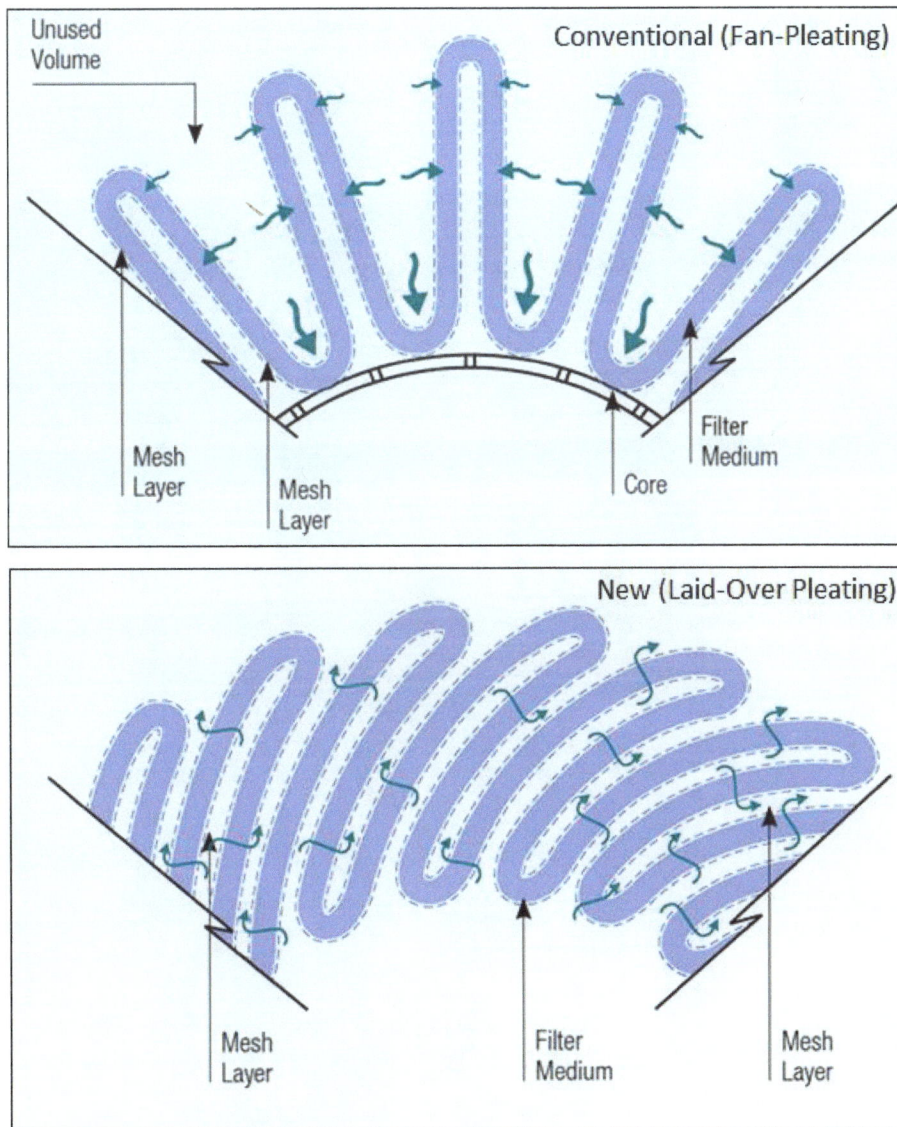

Fig. 2.18- Example of New Depth Filter Element Technology "Ultipleat" (Courtesy of Pall)

Figure 2.19 shows a typical depth filter using laid-over pleating technology. Athalon™ Filter is the next generation in *Anti-Static*, stress-resistant filters. This type of filter has enhanced performance that ensures equipment protection and extends component and fluid life.

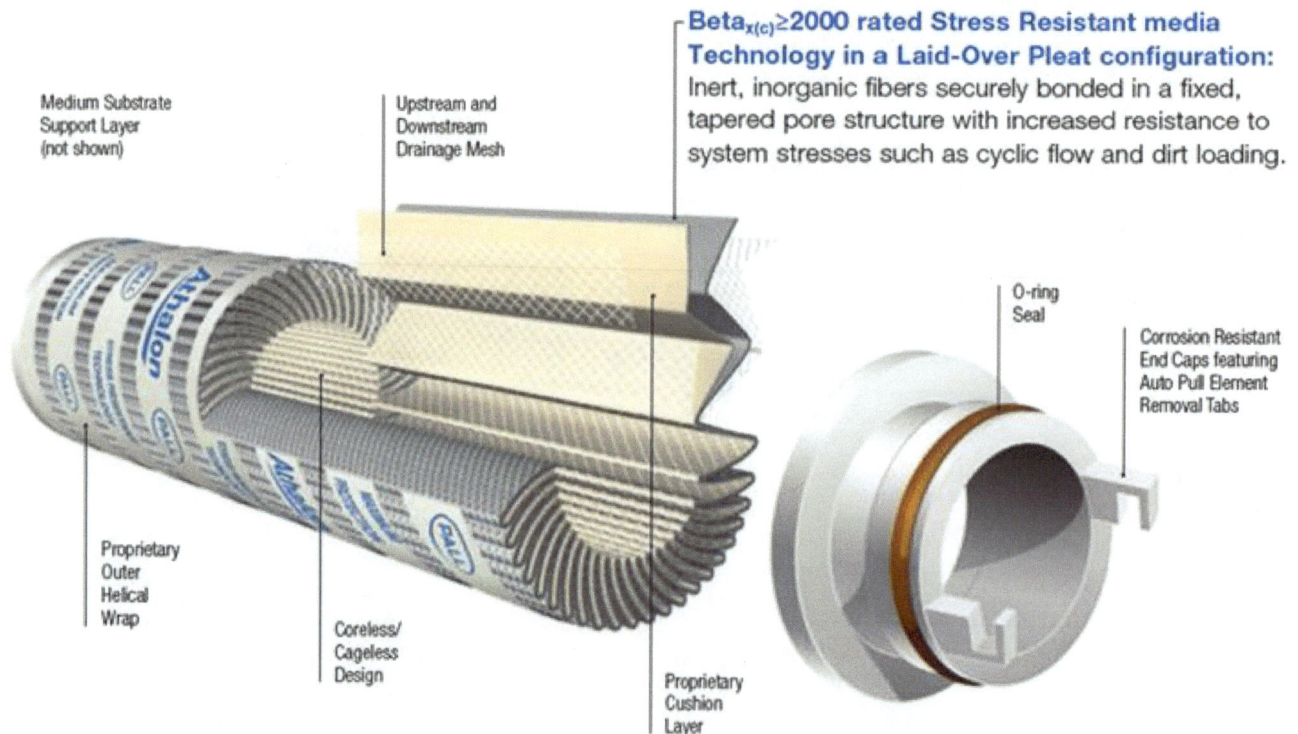

Beta$_{x(c)}$≥2000 rated Stress Resistant media Technology in a Laid-Over Pleat configuration: Inert, inorganic fibers securely bonded in a fixed, tapered pore structure with increased resistance to system stresses such as cyclic flow and dirt loading.

Medium Substrate Support Layer (not shown)

Upstream and Downstream Drainage Mesh

O-ring Seal

Corrosion Resistant End Caps featuring Auto Pull Element Removal Tabs

Proprietary Outer Helical Wrap

Coreless/ Cageless Design

Proprietary Cushion Layer

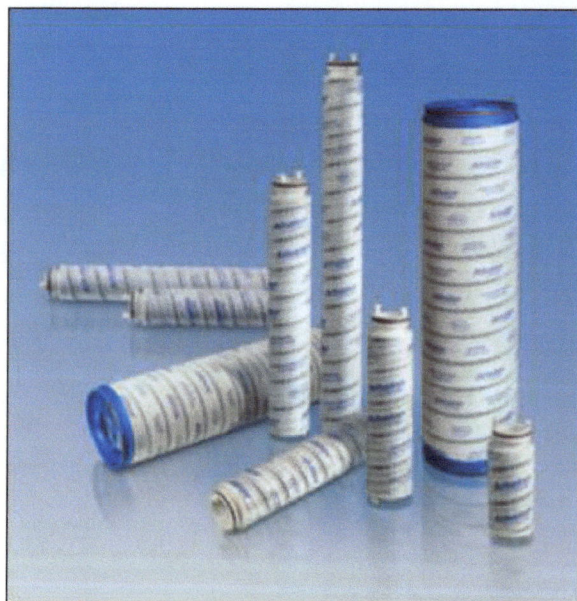

Fig. 2.19- Example of New Depth Filter Element Technology "Athalon" (Courtesy of Pall)

Example 4– New Depth Filter Element Technology: Figure 2.20 shows the of new depth filter technology (Optimicron). The filter has the following features:

- The filter element has an outer rap (1) around the outer surface to protect the sensitive filter media from fluid flow and increase the filter media robustness.
- Optimized crosssection (2) with new pleating shape. This new shape doubles the flow surface, lowers the flow velocity and ensures lower pressure drop across the media.
- Filter media consists of 7 consecutive layers (3) to increase the effectiveness of filtration.

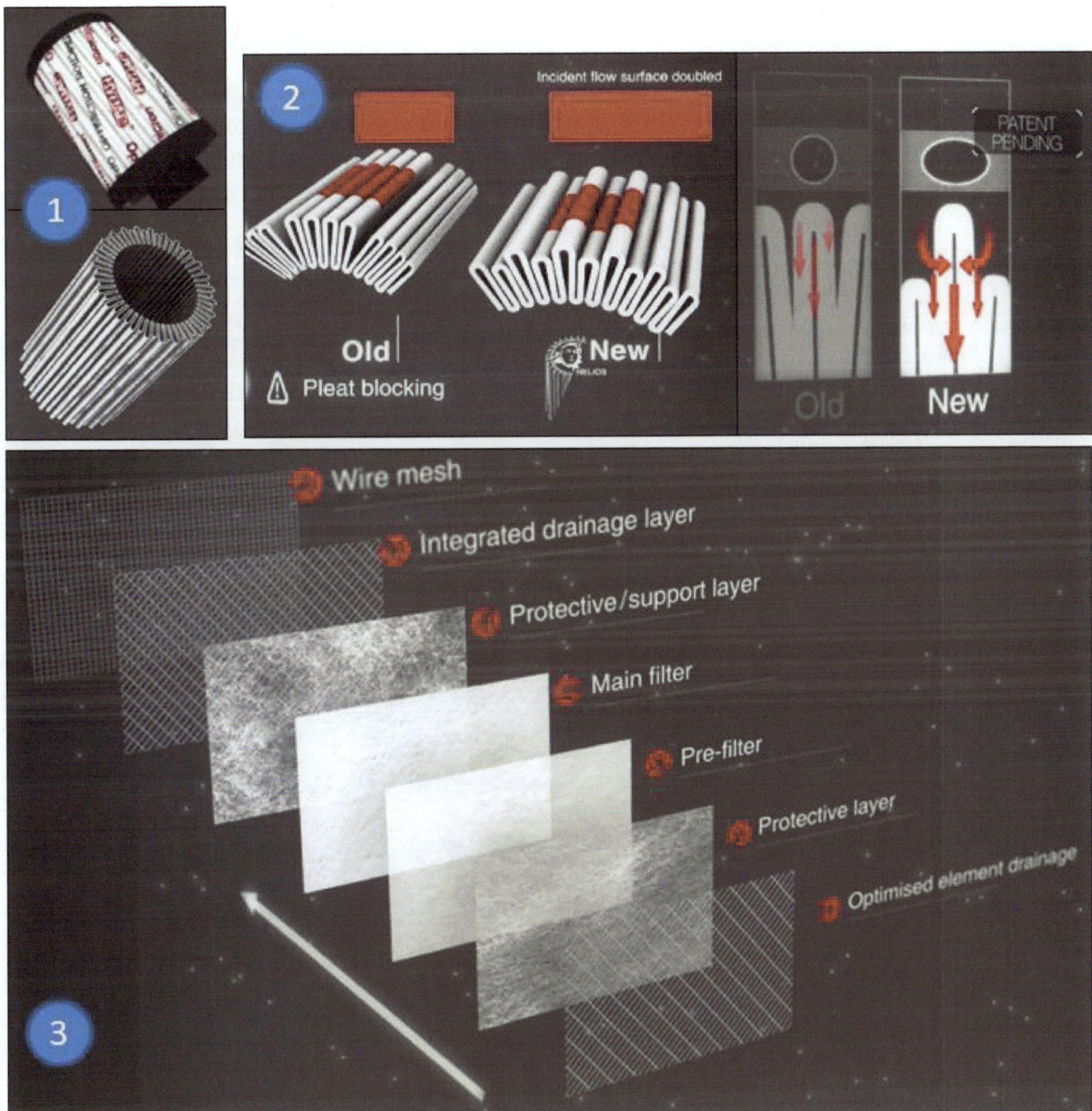

Fig. 2.20- Example of New Depth Filter Element Technology "Optimicron" (Courtesy of Hydac)

Example 5– Synthetic Depth Filter: Figure 2.21 shows a synthetic depth filter that has the following features:

- High-efficiency filtration rating.
- Exceptionally low flow resistance
- Consistent performance throughout filter life.
- Excellent fluid compatibility.
- Ideally suited for a variety of demanding applications, including heavy-duty mobile equipment, in-plant hydraulics, transmissions, and bearing lube oil systems.

The filter consists of the following elements:

Epoxy-Coated Steel Support vs. (1): These two layers at the upstream and downstream sides provide excellent pleat support and spacing, which allows for maximum effective media area. They protect the media against damage during handling and installation.

Media Support Layers (2): These two layers at the upstream and downstream sides protect media during pressure surges

Synteq™ Media Technology (3): Synthetic filter media has smooth, rounded fibers for low resistance to fluid flow. This media is ideal for filtering synthetic fluids, water glycols, water oil emulsions, HWCF (high water content fluids), and petroleum-based fluids.

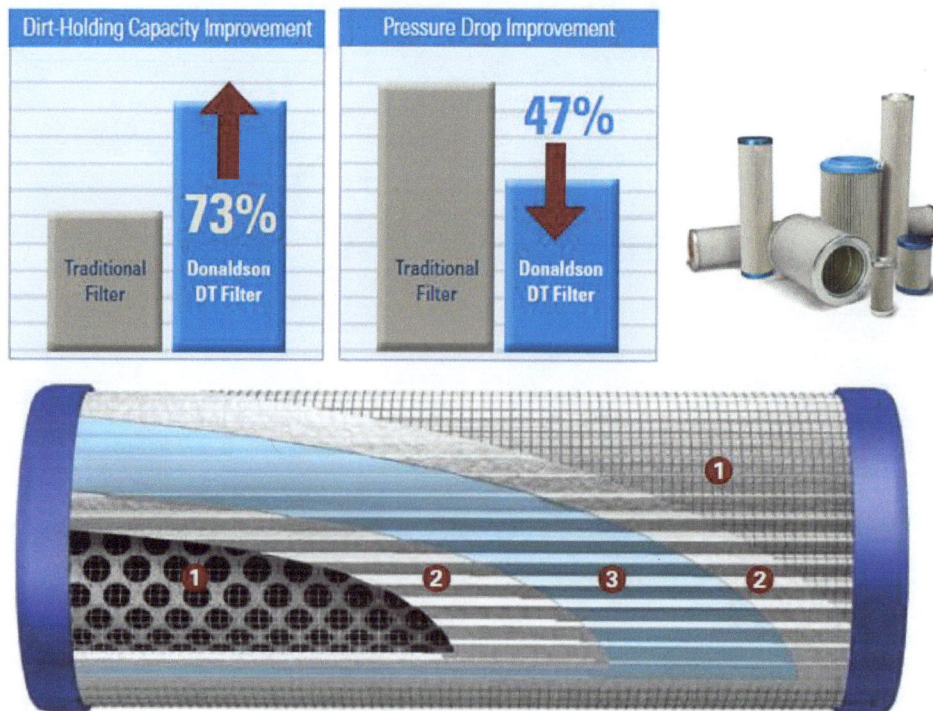

Fig. 2.21- Example of Synthetic Filter Element "DT Filters" (Courtesy of Donaldson)

Examples 6– Offline Filter for High Dirt Holding Capacity: Figure 2.22 shows a typical example of offline filters. Offline filters generally have a large dirt holding capacity of approximately 4 liters solid, 2 liters water, and 4 liters of oil degradation products (*Varnish, Sludge, and Oxidation*). They typically are replaced only on annual bases. Such filters are good to filter particles as small as 3 microns.

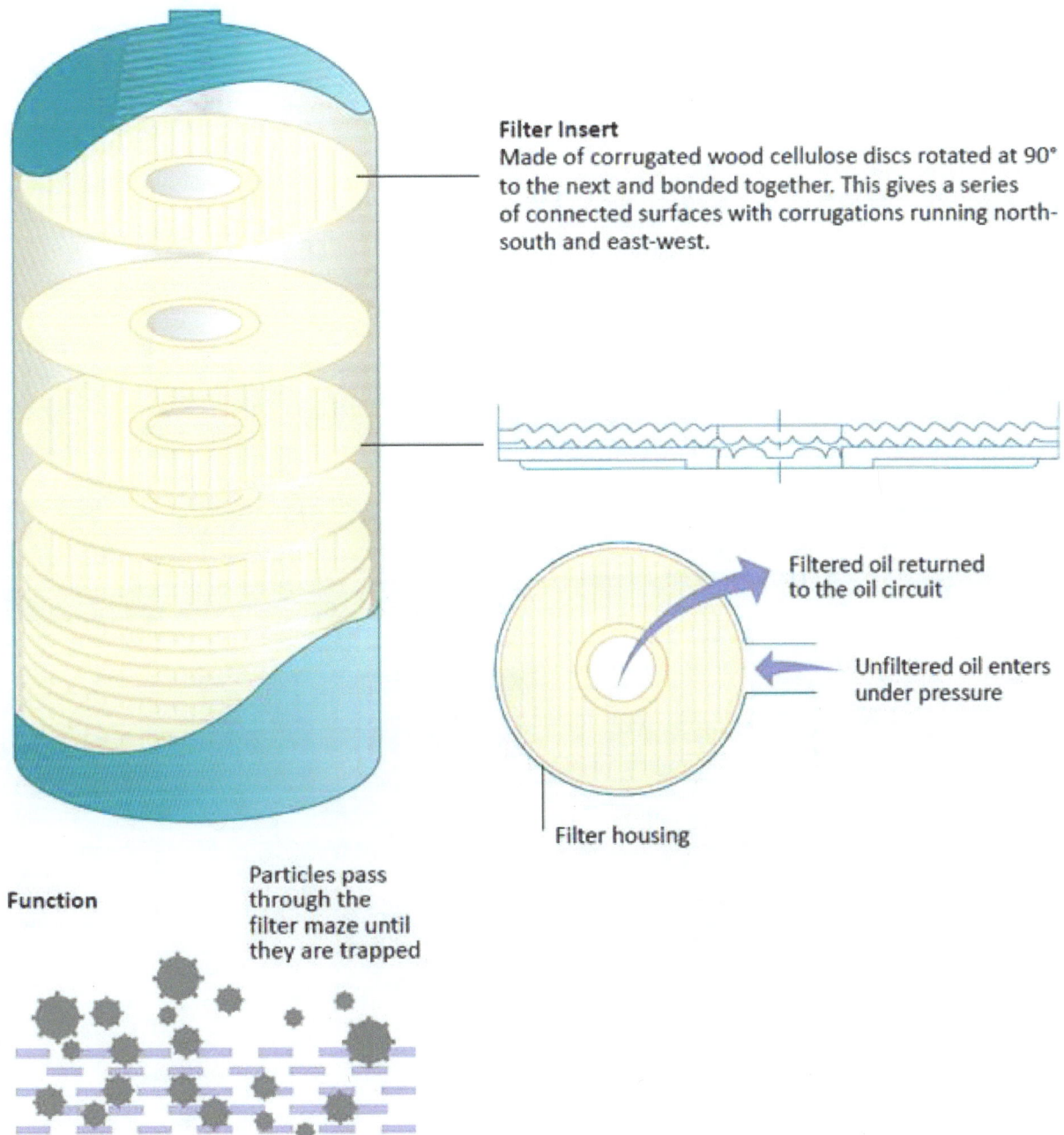

Filter Insert
Made of corrugated wood cellulose discs rotated at 90°
to the next and bonded together. This gives a series
of connected surfaces with corrugations running north-
south and east-west.

Filtered oil returned
to the oil circuit

Unfiltered oil enters
under pressure

Filter housing

Function

Particles pass
through the
filter maze until
they are trapped

Fig. 2.22- Example of Offline Filter for High Dirt Holding Capacity
(Courtesy of C.C. Jensen Inc.)

Example 7- Water Removal Filer Element: Figure 2.23 shows specification for a water removal filter element. Aquamicron® filter elements are specially designed to separate free water from mineral oils. Pressure drop is monitored by clogging indicator. A bypass valve is used to limit the pressure drop across the filter element. The figure shows the pressure drop versus the flow rate for mineral oils with a specific gravity of 0.86. Correction must be applied for fluid with different specific gravity.

Technical Specifications

Collapse Rating	145 psid (10 bar)
Temperature range	32°F to 212°F (0°C to 100°C)
Compatibility with hydraulic media	Mineral oils: Test criteria to ISO 2943 Lubricating oils: Test criteria to ISO 2943 Other media available on request
Opening pressure of by-pass valves	$\Delta P0 = 43$ psid ±7 psi (3 bar ±0.5 bar)
Bypass valve curves	The bypass valve curves apply to mineral oils with a specific gravity of 0.86. The differential pressure of the valve changes proportionally with the specific gravity.

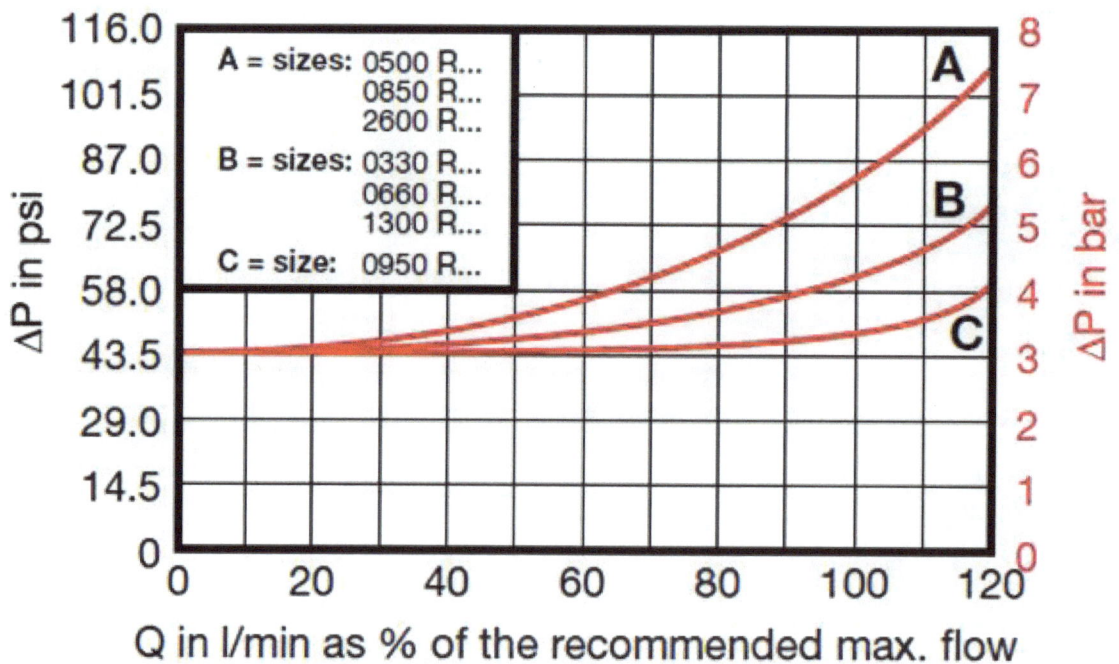

A = sizes: 0500 R...
          0850 R...
          2600 R...
B = sizes: 0330 R...
          0660 R...
          1300 R...
C = size: 0950 R...

Q in l/min as % of the recommended max. flow

Fig. 2.23- Example of Water Removal Filter Element "Aquamicron®" (Courtesy of Hydac)

Example 8- Water Removal Filer Elements: Figure 2.24 shows dry water removal filter media. When it becomes wet (swollen) with absorbed water. The shown water removal filter media is an effective way of removing free water contamination from hydraulic systems. It is highly effective at removing free water from mineral-base and synthetic fluids. This filter media is a highly absorbent copolymer laminate with an affinity for water. The water is bonded to the filter media and forever removed from the system. It cannot even be squeezed out.

The figure also shows a conversion factor table to calculate the water content in a specific volume of oil. For example, assuming a reservoir stores 200 gallons of oil that is highly water contaminated (1000 ppm), it has water content equal 0.0001x1000 = 0.1% of the oil volume water contents.

If the acceptable water content in the oil is 300 ppm, then the water that should be removed is 700 ppm. This is equivalent to 0.0001x700 = 0.07 % of the oil volume. For 200 gallons oil volume stored in the reservoir, 0.07x200 = 14 gallons total water volume is to be removed.

If each filter element capacity is 2 gallons/elements, 7 elements are required.

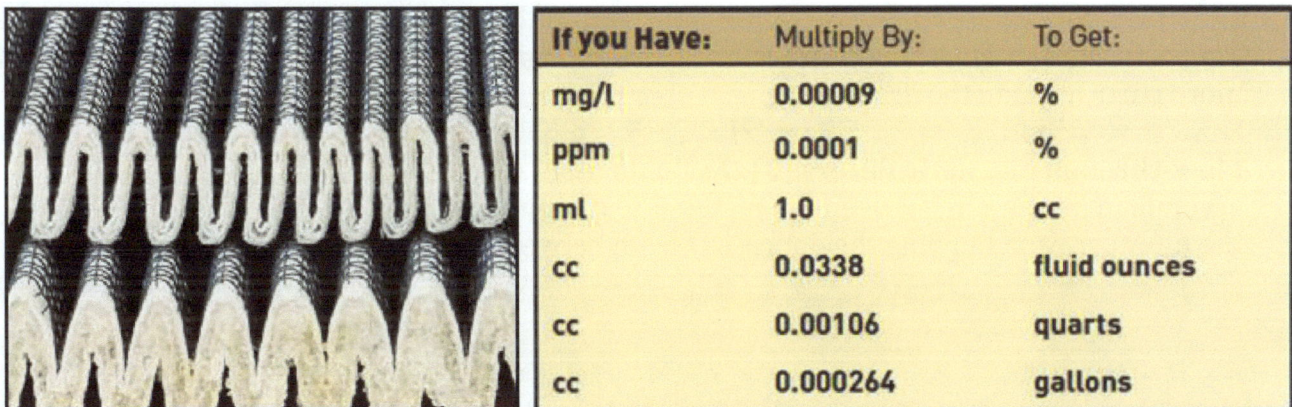

If you Have:	Multiply By:	To Get:
mg/l	0.00009	%
ppm	0.0001	%
ml	1.0	cc
cc	0.0338	fluid ounces
cc	0.00106	quarts
cc	0.000264	gallons

Fig. 2.24- Example of Water Removal Filter Element "Par-Gel" (Courtesy of Parker)

Example 9– Oil Cleaning by Magnetic Separation: Hydraulic fluid analysis shows that in some machines as much as 90% of all particles suspended in the oil can be ferromagnetic (iron or steel particles). Therefore, currently, there are a number of conventional and advanced products on the market that employ the use of magnets in various configurations and geometry.

While it is true that conventional mechanical filters can remove particles in the same size range as magnetic filters, these conventional filters have the following challenges:
- Cost of filter element disposal.
- Cost of Energy wasted due to pressure drop across the filter.
- Possibilities of filter burst under high pressure.
- Possibility of filter media collapse and migration when it becomes clogged.

Figure 2.25 shows the common magnetic products used in lubricating oil and hydraulic fluid applications.

Magnetic Drain Plug (1): The most basic type of magnetic filter is a *Magnetic Drain Plug*. It should be periodically removed and inspected for ferromagnetic particles, which are then wiped from the plug. such plugs are commonly used in engine oil pans, gearboxes and occasionally in hydraulic reservoirs.

Magnetic Rods (2): *Magnetic Rods* are placed inside the reservoir. Magnetic rods can hold more particles than the drain plugs.

Flow-through Magnetic Filters (3): As fluid passes through the slots, ferromagnetic particles accumulate in the gap between the plates. The cleaning process typically involves removing the filter core and blowing the debris out from between the collection plates with an air hose.

Fig. 2.25- Example of Magnetic Filters (Courtesy of Noria)

Example 10– Combo Mechanical and Magnetic Filter: Figure 2.26 shows a Spin-on mechanical filter with steel housing (Bowl). *Magnetic Wraps* are held on the exterior wall of the housing. These wraps transmit a magnetic field through the steel filter housing (bowl). A high-power magnet is installed at the bottom of the housing. The filter operates normally while the ferromagnetic debris are held tightly against the internal surface and at the bottom of the housing. The magnetic filter wraps can be used repeatedly.

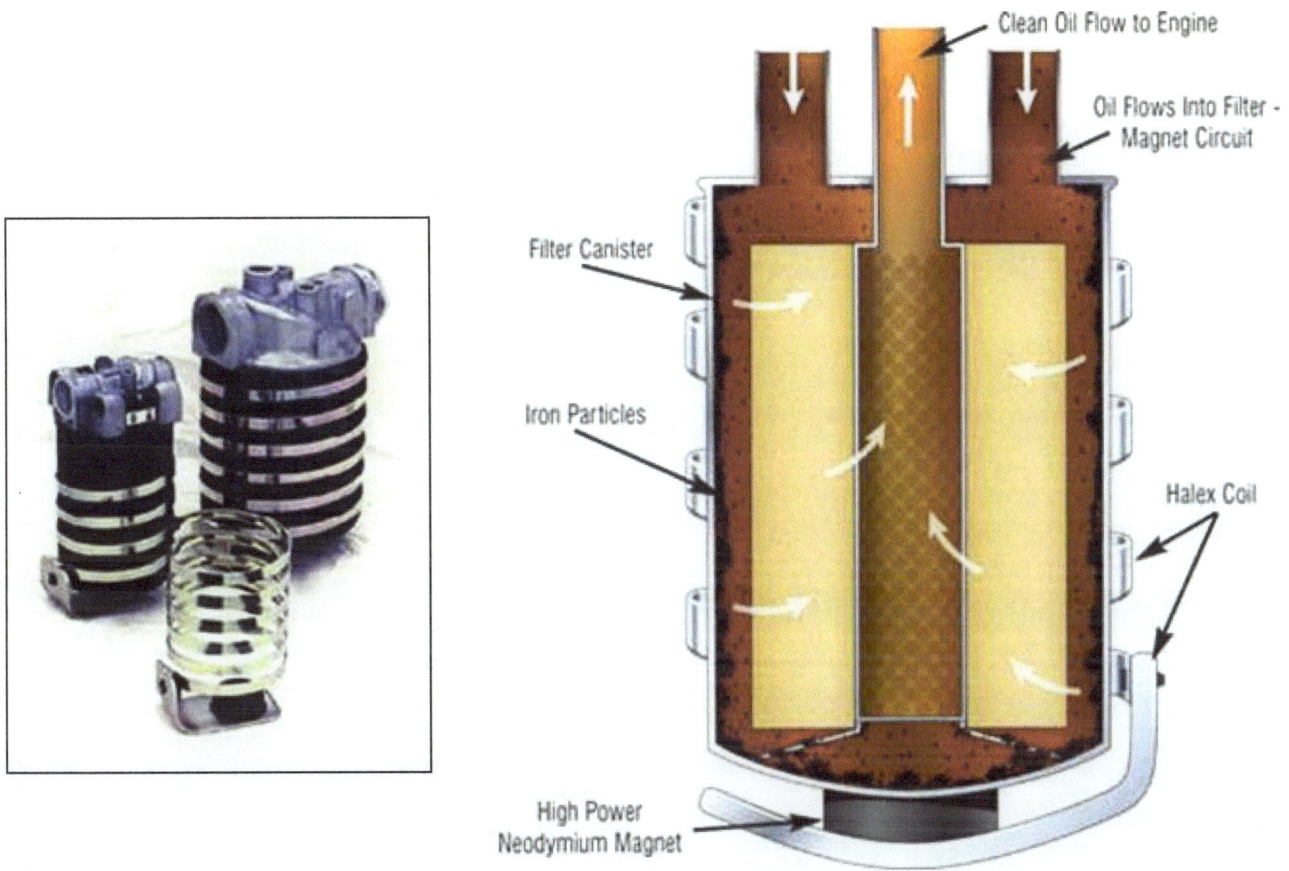

Fig. 2.26- Example of Mechanical and Magnetic Filters (Courtesy of Noria)

Chapter 3

Hydraulic Fluid Analysis

Objectives

This chapter discusses standard methods for hydraulic fluid analysis including methods for particle and material analysis. The chapter covers the various standard cleanliness classes used to evaluate the contamination level in hydraulic fluids. The chapter also provides examples for interpretation of hydraulic fluid analysis reports.

Brief Contents

3.1- Introduction to Hydraulic Fluid Analysis
3.2- Hydraulic Fluid Sampling
3.3- Hydraulic Fluid Material Analysis
3.4- Hydraulic Fluid Cleanliness Standards
3.5- Hydraulic Fluid Particle Analysis
3.6- Interpretation of Fluid Analysis Report

Chapter 3 – Hydraulic Fluid Analysis

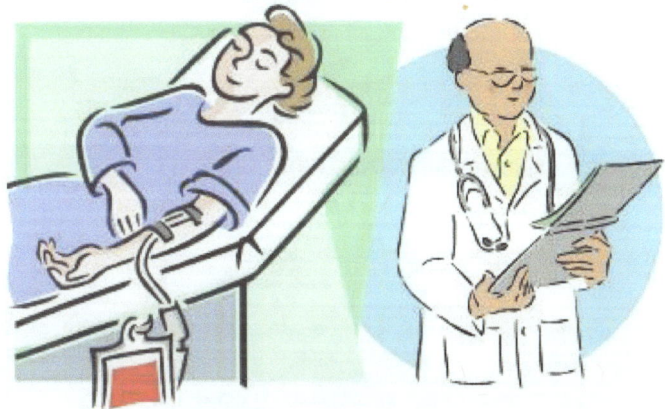

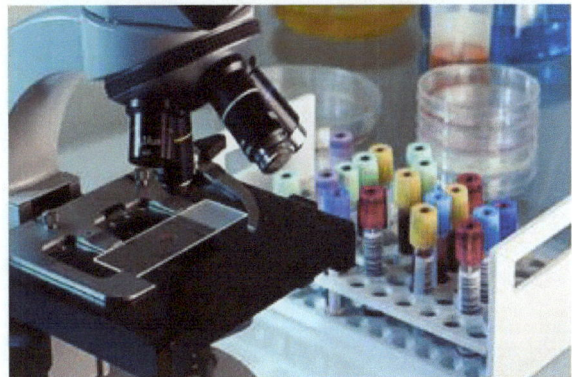

3.1- Introduction to Hydraulic Fluid Analysis

The first question is why is hydraulic *fluid analysis* important? The following set of bullets answers this question:

- As shown in Fig. 3.1, like blood analysis, hydraulic fluid analysis is a snapshot of what is happening inside the equipment and summarizes its condition.
- Equipment warranty support programs require routine hydraulic fluid analysis to maintain coverage just like medical service providers require periodic checkup to maintain one's health.
- It identifies contamination level, type of contaminants, and potential component wear.
- It identifies opportunities for optimizing filtration performance.
- It minimizes downtime by identifying minor problems before they become major.
- It maximizes asset reliability and extends equipment life.

Fig. 3.1- Hydraulic Fluid Analysis and System Reliability

Figure 3.2 shows the various types of hydraulic fluid analysis. As shown in the figure, there two types of fluid analysis, *Material Analysis* and *Particle Analysis*.

The **Material Analysis** concerns with the type of contaminants. It investigates the sources and the material distribution of contaminants. The outcome of this analysis is used for purposes of troubleshooting and predictive maintenance. For example, if there is a content of bronze or brass, this means there is wear of a bearing inside a component. If there is silica content, this means that dust somehow found its way to the system. Material analysis also reports air and water contents, as well as the acidity in a fluid sample.

The **Particle Analysis** (Cleanliness Level Analysis) reports the number, the size, and the shape of particulate contaminants in a sample of fluid. This information, based on specified standards, indicates the cleanliness level of the fluid according to the approved standards.

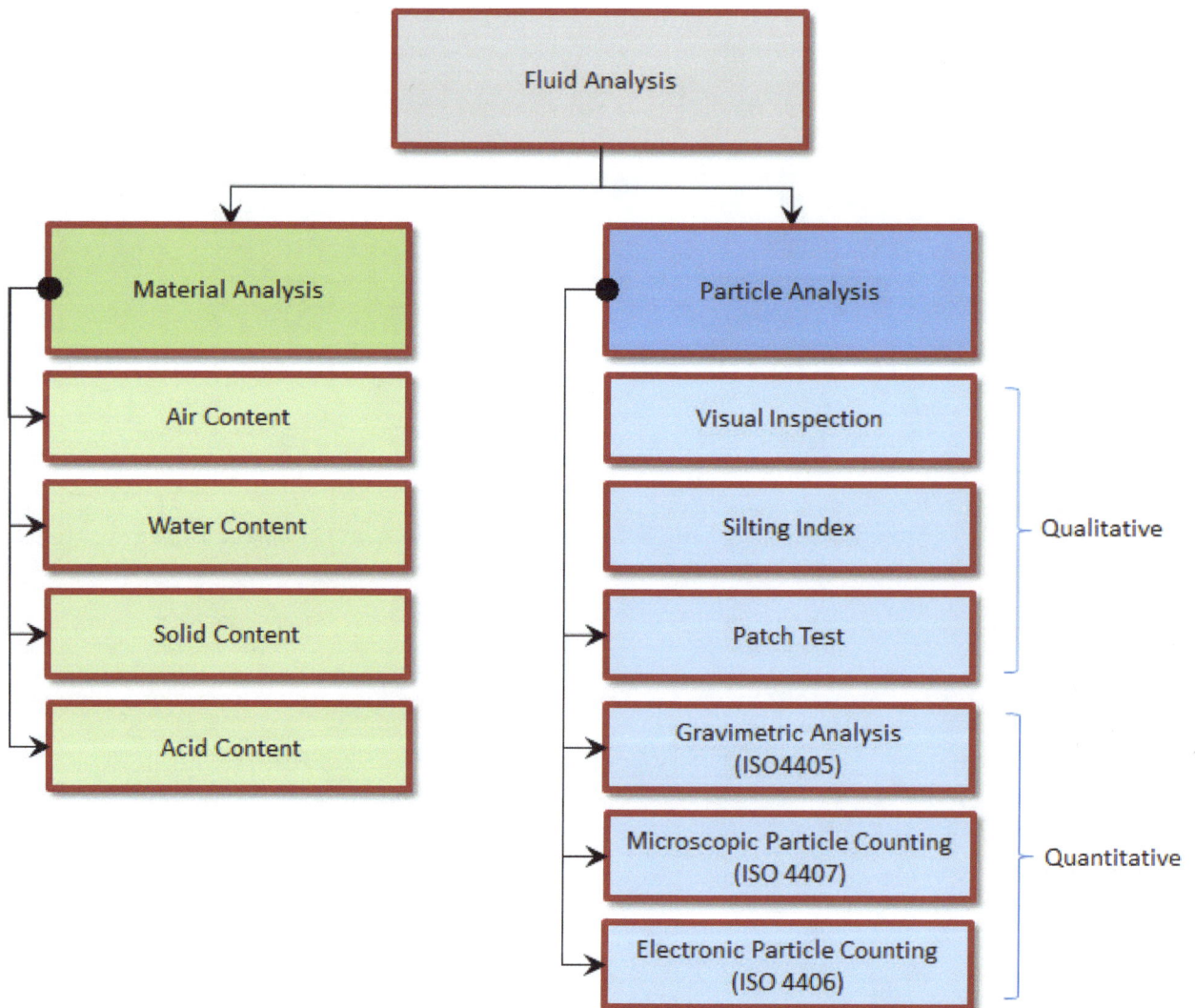

Fig. 3.2- Common Types of Hydraulic Fluid Analysis

The quality of analysis results depends first on correct sampling and handling of the sample, secondly on the quality of the laboratory performing the analysis. Figure 3.3 shows the essential steps for hydraulic fluid analysis.

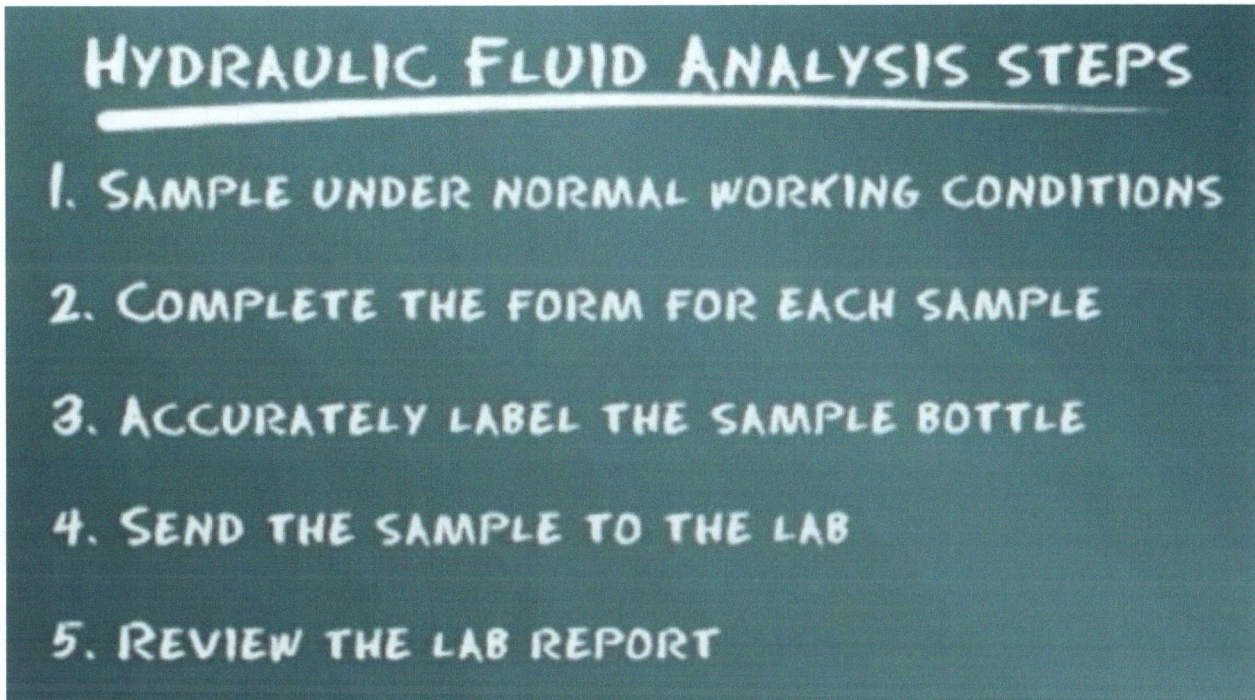

HYDRAULIC FLUID ANALYSIS STEPS

1. SAMPLE UNDER NORMAL WORKING CONDITIONS

2. COMPLETE THE FORM FOR EACH SAMPLE

3. ACCURATELY LABEL THE SAMPLE BOTTLE

4. SEND THE SAMPLE TO THE LAB

5. REVIEW THE LAB REPORT

Fig. 3.3- Hydraulic Fluid Analysis Steps (Courtesy of Donaldson)

3.2- Hydraulic Fluid Sampling

For a representative hydraulic fluid *sample*, the following should be considered:
- **Sampling Interval:** Always take samples at regularly scheduled intervals.
- **Sampling Location:** Points of withdrawing the fluid should be defined.
- **Sampling Kit:** Standard sampling kit should be used.
- **Sampling Procedure:** Fluid sampling should follow prescribed procedure.

3.2.1- Hydraulic Fluid Sampling Intervals

Machinery manufacturers will often suggest a sampling interval. In general, as shown in Table 3.1, a quarterly or monthly sampling interval is appropriate for most important industrial machinery.

Industrial and Marine			
Equipment Type	*Normal Use Sampling Frequency (Hours)	(Calender)	Occasional Use Sampling Frequency (Calendar)
Steam Turbines	500	Monthly	Quarterly
Hydro turbines	500	Monthly	Quarterly
Gas Turbines	500	Monthly	Quarterly
Diesel Engines-Stationary	500	Monthly	Quarterly
Natural Gas Engines	500	Monthly	Quarterly
Air/Gas Compressors	500	Monthly	Quarterly
Refrigeration Compressors	500	Monthly	Quarterly
Gearboxes-Heavy Duty	500	Monthly	Quarterly
Gearboxes-Medium Duty		Quarterly	Semi-Annually
Gearboxes-Low Duty		Semi-Annually	Annually
Motors-2500 hp and higher	500	Monthly	Quarterly
Motors-200 to 2500 hp		Quarterly	Semi-Annually
Hydraulics		Quarterly	Semi-Annually
Diesel Engines-On and Off Highway	150 hours/10,000 miles	Monthly	Quarterly

**Table 3.1- Fluid Analysis Intervals for Common Industrial Machines
(Courtesy of Spectro Scientific)**

Table 3.2 shows recommended sampling intervals for mobile machines. Because mobile machines work outdoor where the contamination is more than industrial applications, sampling intervals are reduced to 300 hours instead of 500.

Off-Highway/Mobile Equipment	
Equipment Type	Normal Use Sampling Frequency (Hours/Miles)
Gasoline Engines	5,000 miles
Differentials	300 hours/20,000 miles
Fina Drives	300 hours/2,000 miles
Transmissions	300 hours/20,000 miles
Hydraulic Systems	1,000 hours/Annually

**Table 3.2- Fluid Analysis Intervals for Mobile Equipment
(Courtesy of Spectro Scientific)**

Table 3.3 shows the recommended sampling intervals for aerospace industry. For safety of human life, sampling intervals are reduced to 50-100 hours.

Aviation	
Equipment Type	*Normal Use Sampling Frequency in hours
Reciprocating Engines	50 hours
Gas Turbines	100 hours
Gearboxes	100 hours
Hydraulics	100 hours

**Table 3.3- Fluid Analysis Intervals for Aerospace Industry
(Courtesy of Spectro Scientific)**

3.2.2- Hydraulic Fluid Sampling Locations

When taking a sample of hydraulic fluid, there are prohibited places because the sample will not be representative. Hence, Do NOT take samples from places where:
- Oil flow is restricted.
- Contaminants or component wear products tend to settle.
- Oil is cold after oil coolers.
- Bottom of reservoirs.

However, sampling location should be:
- Low pressure lines with turbulent flow such as elbows and tees.
- Easily accessible for operators to quickly and easily take the sample.
- Does not require disassembly of other parts.
- Equipped with sampling valve.
- Labeled sampling location as shown in Fig. 3.4.

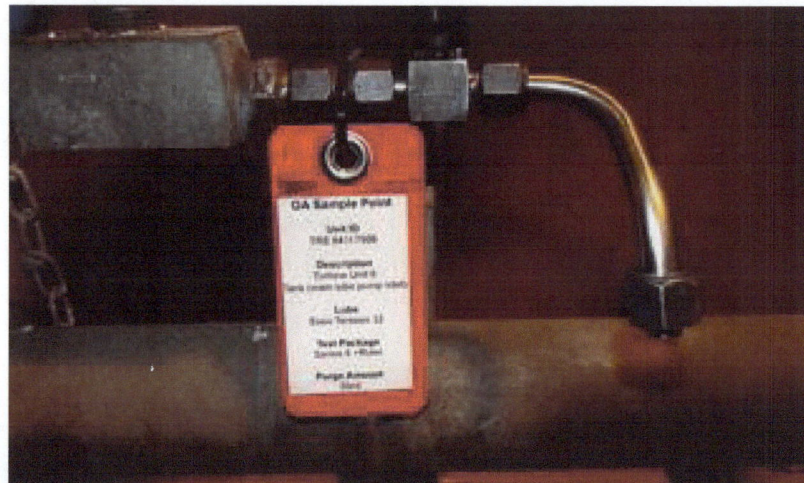

Fig. 3.4- Labeling of Sampling Points (Spectro Scientific)

3.2.2.1- Sampling from Low Pressure Return Line **(ISO 4021)**

As shown in Fig. 3.5, to acquire a representative fluid sample, withdraw the sample on the downstream side of the system before any filtration and before the oil is returned to the system tank.

Per **ISO 4021**, preferably withdraw the oil from an upwards pointing sampling point at an elbow with turbulent flow. Sampling points fitted on the lower or side perimeter of a pipe tend to allow depositing of particles in the sampling valve.

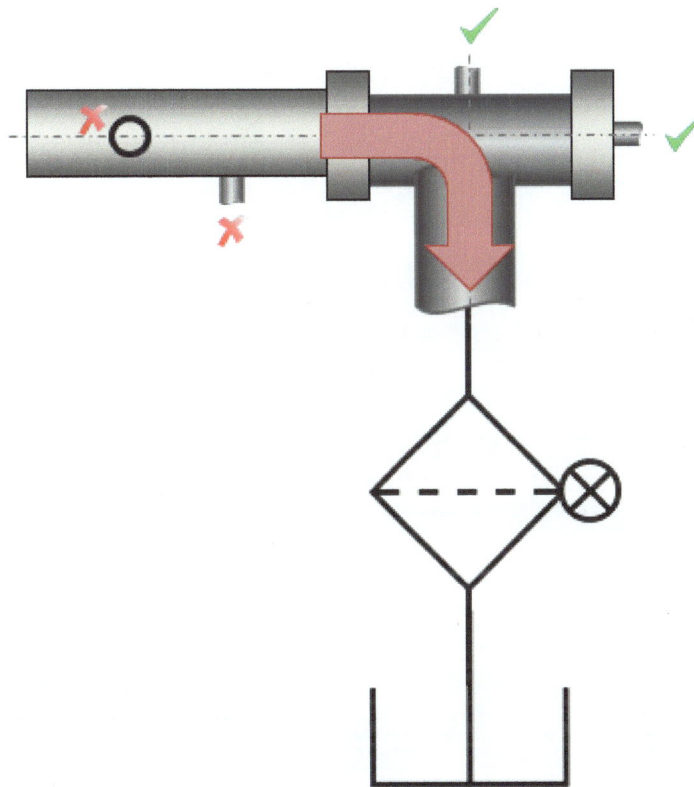

Fig. 3.5- Sampling from a Return Line

3.2.2.2- Sampling from High Pressure Line

If a sampling valve is provided from a high-pressure line, the following precautions must be considered:
- A warning label of a high-pressure jet hazard shall be posted at the sampling location.
- Sampling valve shall be shielded and equipped by a check valve.

3.2.2.3- Sampling from Reservoir

If the sample is required for Particle Analysis, **DO NOT** Sample from a reservoir. Sampling from the reservoir for purposes of particle counting **IS NOT RECOMMENDED** by today's standards because fluid in the reservoir does not represent the fluid flowing in the system. There is no excuse for not sampling from a flowing line according to ISO 4021. The reason this part is presented here is to sample fluids from non-hydraulic driven machines such as engine oil sumps and transmission gear boxes. Sampling fluids from these machines is required for material and properties analysis, not for particle analysis.

As shown in Fig. 3.6, a sample is typically taken from between the pump and the filter housing of an offline filter. If no offline filter system is installed, a vacuum type sampling pump is a valid option. In such a case the sample should be drawn 10 cm (4 inches) off the lowest part of the tank.

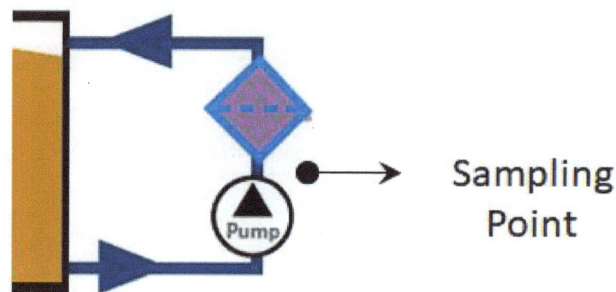

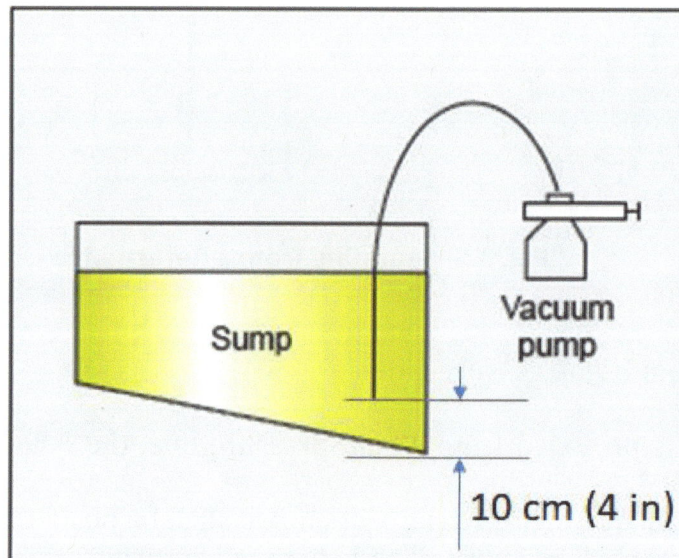

Fig. 3.6- Sampling from a Reservoir

3.2.3- Hydraulic Fluid Sampling Kit

Hydraulic fluid sampling tools should contain the following basic elements:

Vacuum Pump: As shown in Fig. 3.7, a vacuum pump is a necessary tool for extracting an oil sample from the sample port. When used in combination with a sample port adapter, flexible tubing, and a sample bottle the user is able to connect to any sample port for contamination free oil sampling in the most representative locations. Manual and electric pumps are available.

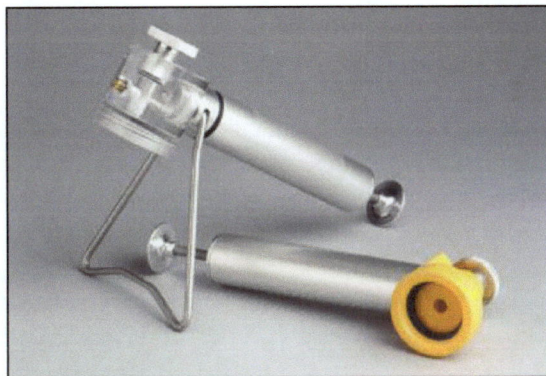

Fig. 3.7- Hydraulic Fluid Sampling Vacuum Pump (www.tricocorp.com)

Sampling Bottle: As shown in Fig. 3.8, a qualified sampling container must satisfy the following conditions:
- **Container:** Glass or plastic bottle with wide mouth.
- **Size:** Approximately 250 mL (4 ounce) size.
- **Cover:** screw-on cap with plastic film between the cap and the bottle.
- **Label:** to record the sampling data.
- **Cleanliness:** Cleaned by filtered air, designated as *"Super Clean"*, and qualified in accordance with **ISO 3722**.

Fig. 3.8- Super Clean Sampling Bottle (www.tricocorp.com)

As shown in Fig. 3.9, *Ultra Clean Vacuum Device* bottles (*UCVD*) are cleaned to an ISO code of 11/9/4 and sealed. Unlike other "super clean bottles", there is no need to open the cap. The bottle may be used in conjunction with a sampling probe, and it also avoids the need for a sampling vacuum pump. The operator simply connects the tubing from the port to the bottle and opens the valve. When finished, bottle valve is just closed.

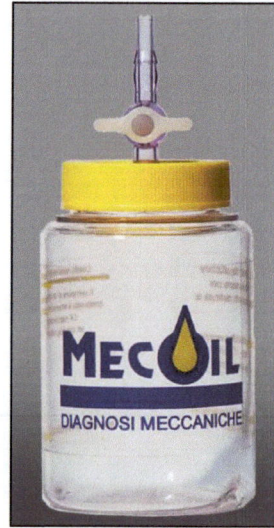

Fig. 3.9- Ultra Clean Sampling Bottle (www. mecoil.net)

Sampling Port: As shown in Fig. 3.10, sample ports are designed to draw samples from the sampling points and to provide superior leak protection. Sample ports are available in several types and sizes to match the varying requirements of manufacturers.

Fig. 3.10- Hydraulic Fluid Sampling Ports (www.tricocorp.com)

Sampling Tubing: As shown in Fig. 3.11, special sampling tubing is used in combination with a sample port, a vacuum pump, and a sample bottle. This tubing must be as clean as the bottle, so it does not add contaminants. A good idea is to pre-flush with system fluid of 10 times tube volume prior to using it to take samples.

Fig. 3.11- Hydraulic Fluid Sampling Tubing (www.tricocorp.com)

There are several portable *Sampling Kits* available in the market. These kits contain the basic elements and some other accessories such as disposable membranes, solvents, etc. Figure 3.12 shows a portable sampling and analysis kit.

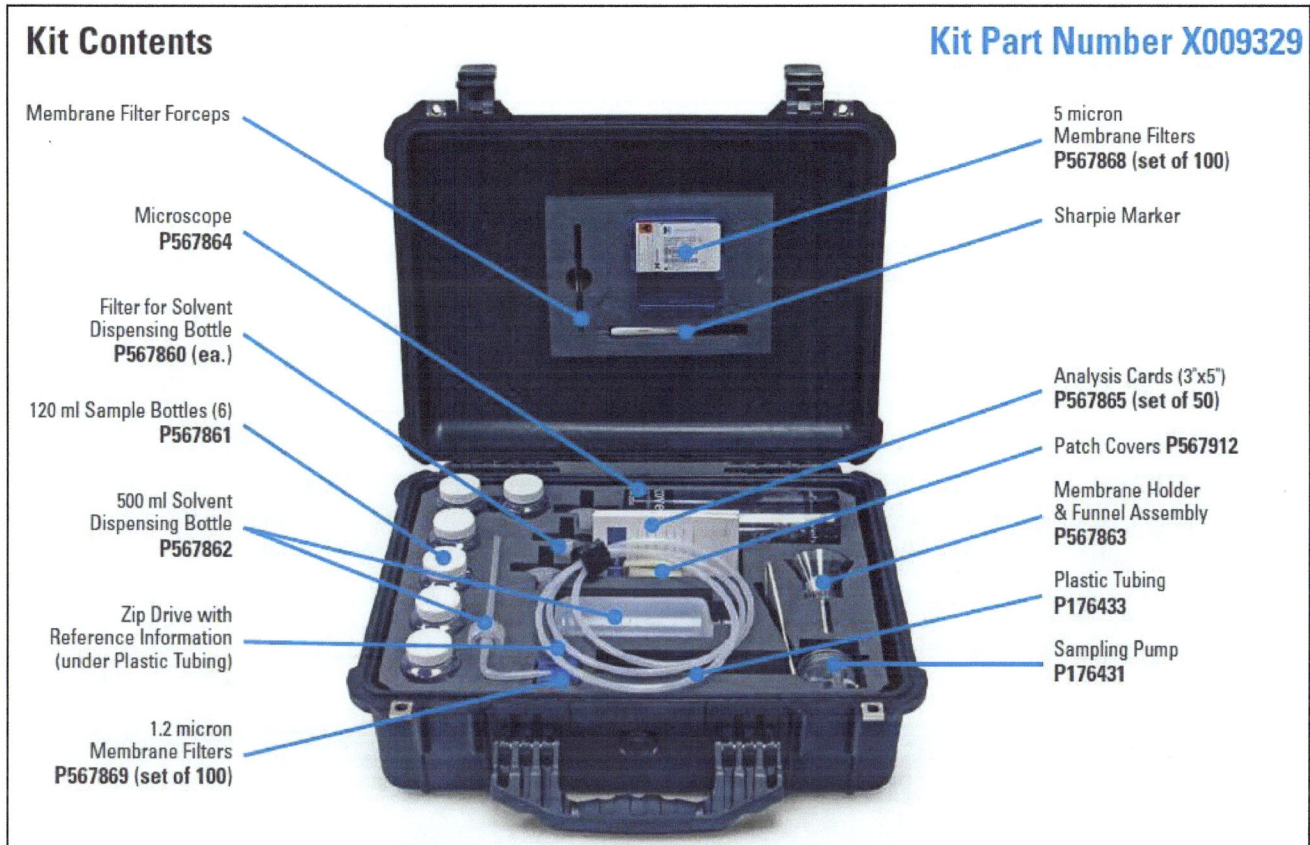

Kit Contents

Kit Part Number X009329

Membrane Filter Forceps

Microscope
P567864

Filter for Solvent
Dispensing Bottle
P567860 (ea.)

120 ml Sample Bottles (6)
P567861

500 ml Solvent
Dispensing Bottle
P567862

Zip Drive with
Reference Information
(under Plastic Tubing)

1.2 micron
Membrane Filters
P567869 (set of 100)

5 micron
Membrane Filters
P567868 (set of 100)

Sharpie Marker

Analysis Cards (3"x5")
P567865 (set of 50)

Patch Covers P567912

Membrane Holder
& Funnel Assembly
P567863

Plastic Tubing
P176433

Sampling Pump
P176431

Fig. 3.12- Hydraulic Fluid Portable Sampling and Analysis Kit (Courtesy of Donaldson)

3.2.4- Hydraulic Fluid Sampling Procedure

Detailed information on obtaining a sample can be found in ISO 4021. However, in brief, to get a representative sample:

- DO NOT sample immediately after oil change or addition of makeup fluids.
- Run the system for at least 30 minutes or until it is warmed-up.
- Shift directional valves several times to ensure that the fluid has been well circulated and is well mixed.

Steps for oil sampling from a kidney loop (Fig. 3.13):
1. Place the oil container beneath the sampling valve.
2. Open and close the valve five times and leave it open.
3. Flush the pipe by draining one liter (one US quart) into the container.
4. Open the sample bottle while keeping the cap in your hand to avoid contaminating it.
5. Place the bottle under the oil flow without touching any other part
6. Fill the bottle to approximately 80% full.
7. Place the cap on the bottle immediately after taking the sample.
8. Close the sampling valve.
9. Fill in label and stick it onto the sample bottle.
10. Pack the sample bottle in plastic bag and cardboard container.

Fig. 3.13- Steps for oil Sampling from a Kidney Loop (Courtesy of C.C. Jensen Inc.)

Steps for oil sampling from a reservoir using a vacuum pump (Fig. 3.14):

1. Assemble the tube with the pump:
   - Cut a suitable piece of tube off the roll.
   - Use new tube every time.
   - Push the tube into the pump head.
   - Always flush tube before taking the sample.

2. Fit the bottle by screwing it onto the pump head.

3. Sample withdrawal:
   - Lower the free end of the plastic tube to 10 cm (4 inches) above the lowest part of the tank, in the center of the tank. Be careful not to let the tube touch the walls or the bottom of the reservoir.
   - Create a vacuum in the bottle by a few pump strokes, and fill the bottle to approximately 80%

4. Close the cap

It is to be reminded that this is NOT RECOMMENDED for purposes of Particle Analysis of hydraulic-driven machines. It is used for sampling gear boxes or transmission fluids.

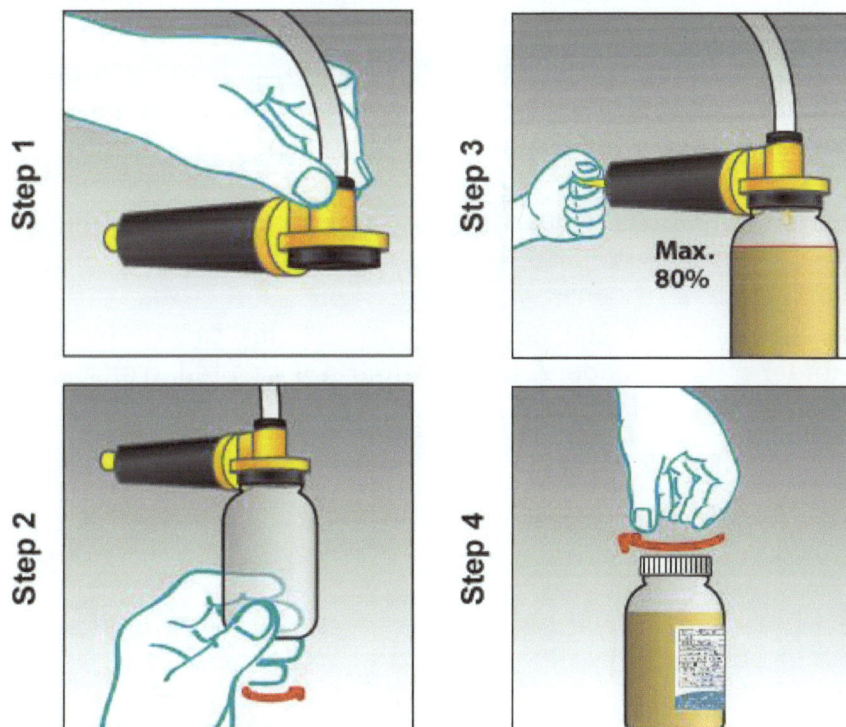

Fig. 3.14- Oil Sampling from a Reservoir using a Vacuum Pump (Courtesy of C.C. Jensen Inc.)

3.3- Hydraulic Fluid Material Analysis

3.3.1- Air Content

Standard methods for measuring air content in the system were explained in Chapter 4.

3.3.2- Water Content

Standard method for measuring water content was explained in Chapter 5.

3.3.3- Solids Content

Knowing the wear metal content of the fluid, helps predict which component may be undergoing irreversible degradation and possible catastrophic failure. This information is used as input for proactive maintenance plans.

Wear Metal Analysis (ASTM D5185): *Atomic Absorption Spectrograph* is performed to determine wear metal content. This test measures the amount of each metallic element, such as; iron, copper, lead, zinc, silicone, aluminum, tin, nickel or chromium, found in a sample. The sample is vaporized over an extremely hot flame. A light of fixed characteristic wavelength, for the metallic element being tested for, is passed through the sample. The amount of light absorbed by the sample indicates the quantity of that metallic element present in the sample. The results are usually recorded as parts per million (ppm) by weight.

Note: This method only looks for particles 5-6 μm or smaller. It is not a substitute for particle counting.

Visual Inspection: Material analysis for solid particles can be performed by observing the fluid sample under a microscope. A trained operator can tell, from what he/she is seeing, the type and the shape of the solid particles.

Table 3.4 and the associated Fig. 3.15 shows some documented observations of particulate contamination.

Sample #	Particle Type	Effect
1	▪ Mainly rust. ▪ White particles. ▪ Additives.	▪ Rapid oil aging. ▪ Pumps and valves breakdown.
2	▪ Oil aging products.	▪ Blocking filters. ▪ Silting-of systems.
3	▪ Metal chips	▪ Pumps and valves breakdown. ▪ Wearing of seals. ▪ Leakage.
4	▪ Particles of bronze, brass, and copper	▪ Pumps and valves breakdown. ▪ Leakage. ▪ Oil aging. ▪ Seal wear.
5	▪ Gel-type residue from filter element	▪ Blocking filters. ▪ Silting of systems.
6	▪ Silicon due to lack of or inadequate, air breather fitter.	▪ Heavy wear in components. ▪ Pumps and valves breakdown. ▪ Wearing of seals. ▪ Leakage.
7	▪ Colored particles (red/brown). ▪ Synthetic particles (blue).	▪ Pumps and valves breakdown. ▪ Wearing of seals.
8	▪ Fibers due to initial contamination, open tank, cleaning clothes, etc.	▪ Blocking of orifices. ▪ Leaking from seat valves.

Table 3.4- Particulate Content Analysis (Courtesy of Hydac)

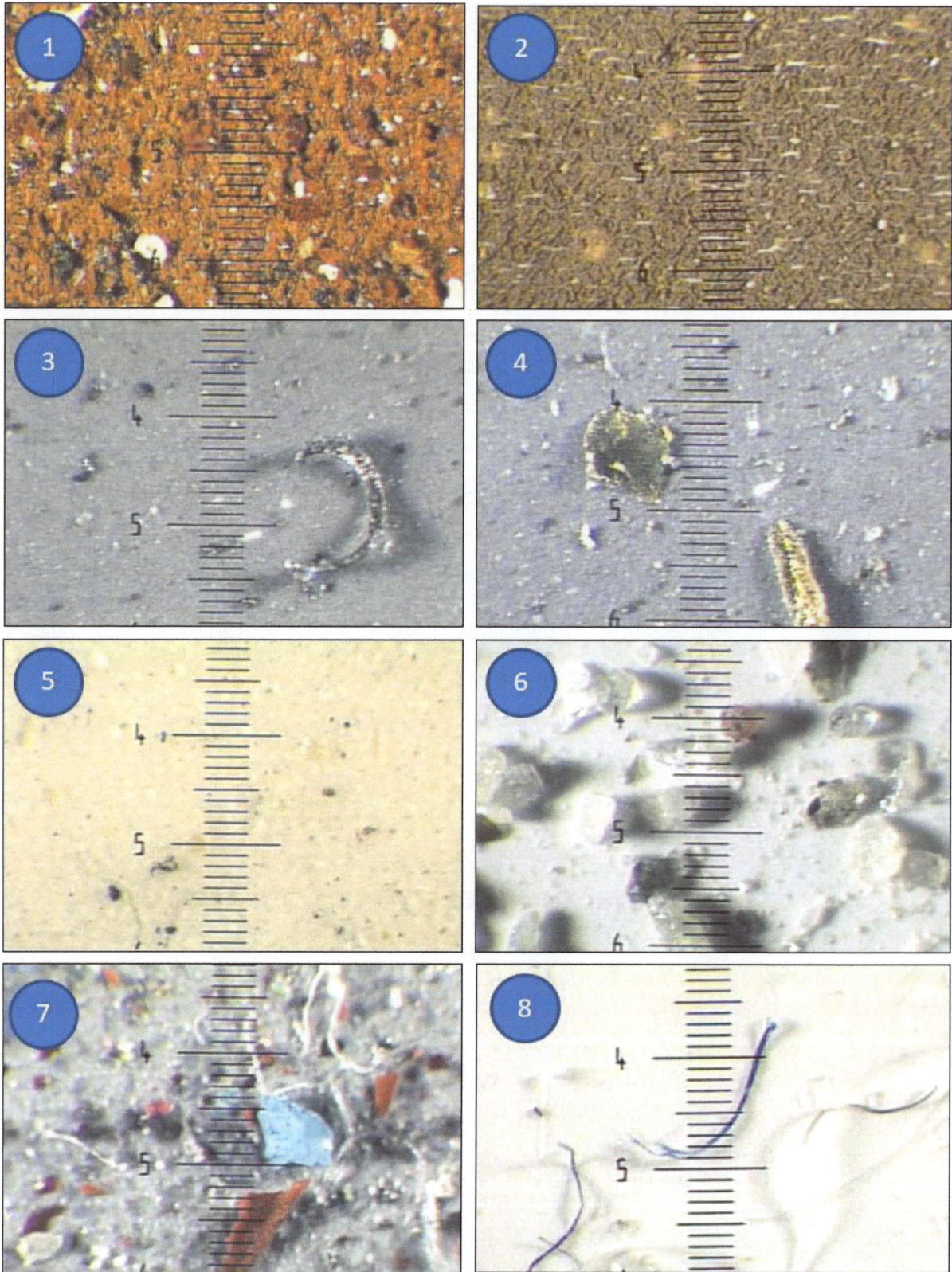

Fig. 3.15- Particulate Content Observation (Courtesy of Hydac)

Table 3.5 and the associated Fig. 3.16 shows some other documented observations of particulate contamination.

Sample #	Particle Type	Source
1	Silica	Most Commonly sand or dust associated with airborne contamination containing hard, translucent particles.
2	Bright Metal	Most commonly products of component wear and fluid breakdown within the system. Visible contaminant will usually appear to contain shiny metallic particles of various colors.
3	Rust	Most commonly seen when water is present in the system. Contaminants contain dull orange or brown particles.
4	Fibers	Most commonly generated from paper and fabric products. Sources of contamination also include cellulose filter media and shop rags.
5	Slit	A very high concentration of silt-size particles and/or additive package ingredients. If the additive package breaks down in this way, it is no longer functioning.
6	Gel	A dense accumulation on the analysis membrane that makes the particle contamination evaluation impossible.

Table 3.5- Particulate Content Analysis (Courtesy of Donaldson)

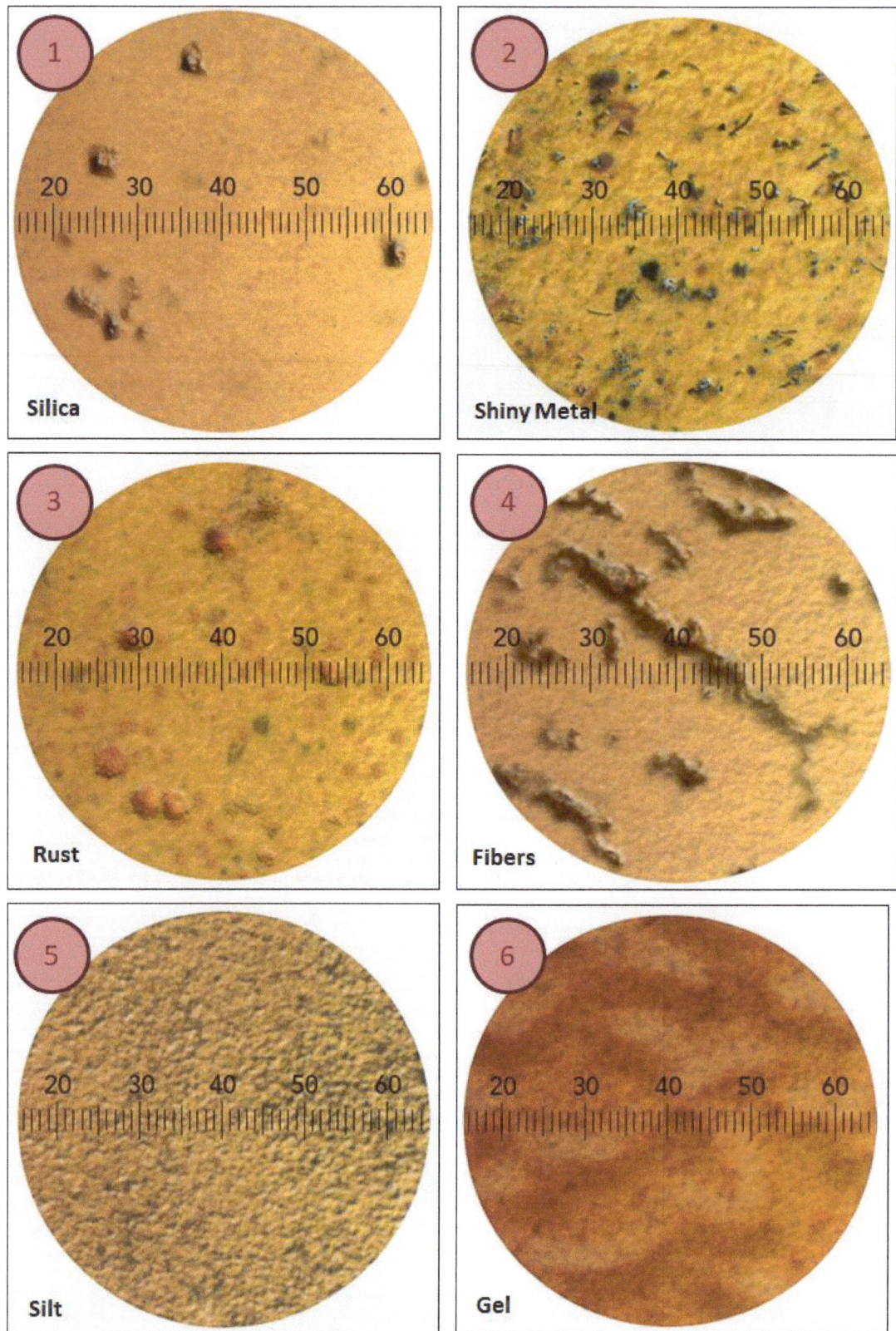

Fig. 3.16- Particulate Content Observation (Courtesy of Donaldson)

Figure 3.17 show some other documented observations of particulate contamination.

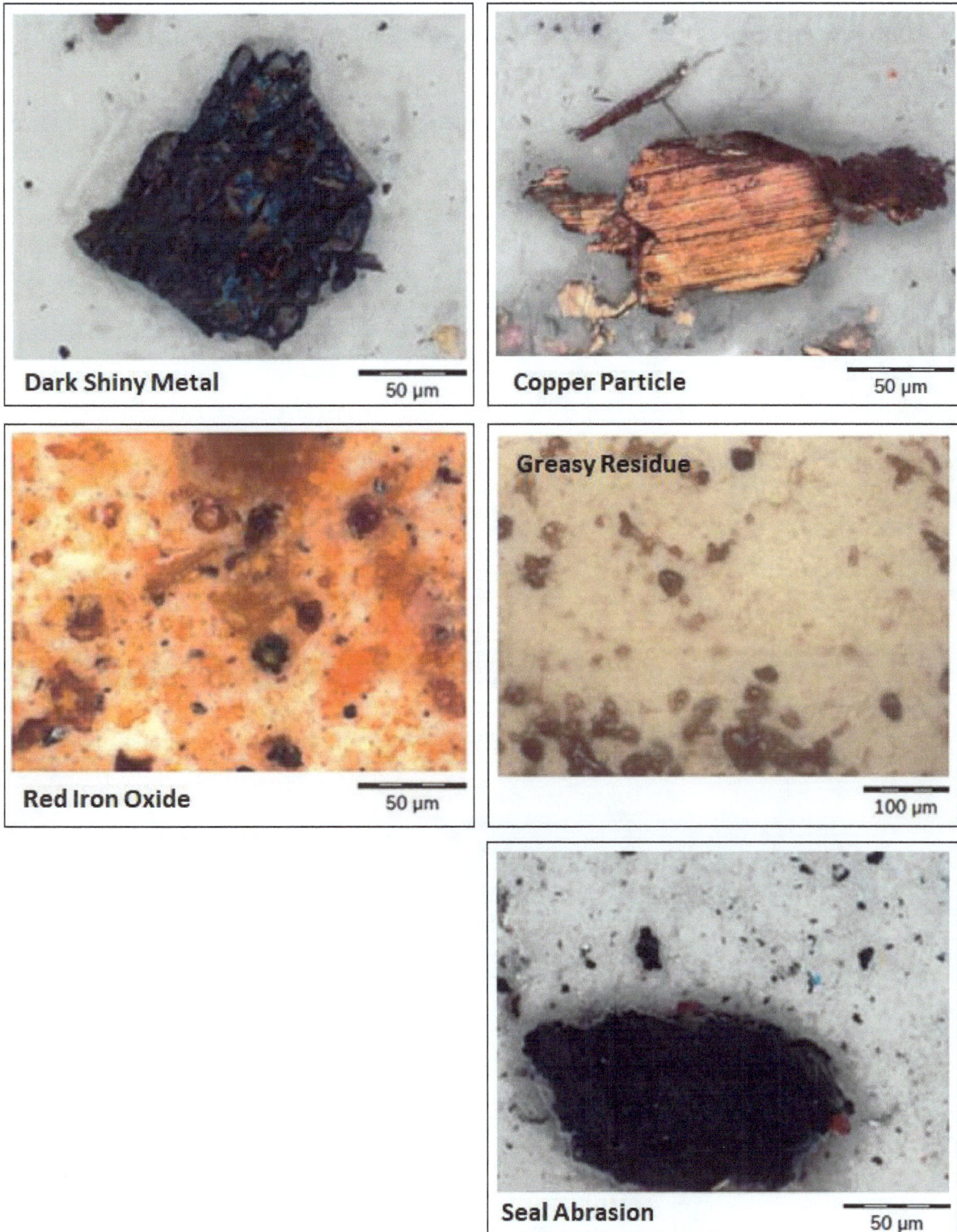

Fig. 3.17- Particulate Content Observation (Courtesy of Bosch Rexroth)

3.4- Hydraulic Fluid Cleanliness Standards

As it has been stated earlier, there are two types of hydraulic fluid analysis, the Material Analysis and the Particle Analysis.

After several statistical and experimental investigations and based on the critical clearances in the hydraulic components, the experts found that there are specific particle sizes that are the most harmful to the hydraulic components.

Therefore, there was a need to standardize the contamination level and use those standards as references to measure the cleanliness level of hydraulic fluids. The following subsections introduce the accepted cleanliness standards.

3.4.1- Two-Code ISO Standard 4406-1987

International Organization for Standardization (ISO) developed the Two-Code **ISO Standard 4406-1987** standard. It is discussed here only as background information as it has been updated in 1999 to be three-code standard. This standard should no longer be used because it was replaced with the ISO 4406-1999 discussed in section 3.4.2.

The code is structured, as shown in Fig. 3.18, from two numbers separated by a slash. These numbers indicate the concentration of particles in each milliliter (1 cc) of a hydraulic fluid sample. The first numerical code is assigned for particle size larger than 5 microns. The second numerical is assigned for particle size larger than 15 microns.

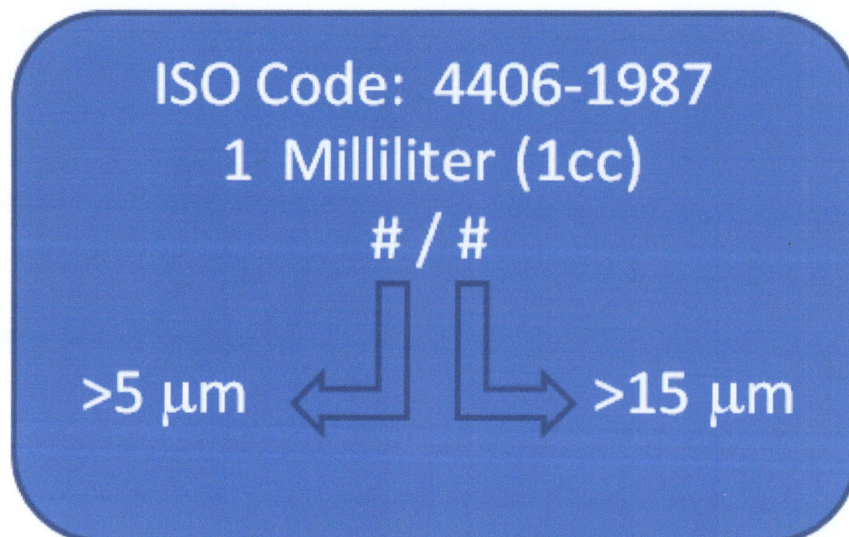

Fig. 3.18- Structure of ISO Code 4406-1987

Table 3.6 shows the particle concentration per the ISO Code 4406-1987. For example, 19/14 cleanliness code indicates 2501-5000 particles at 5μm and 81-160 particles at 15μm per ml of fluid.

Particle Concentration (Particles per milliliter)	Range Number
10,000,000	30
5,000,000	29
2,500,000	28
1,300,000	27
640,000	26
320,000	25
160,000	24
80,000	23
40,000	22
20,000	21
10,000	20
5,000	19
2,500	18
1,300	17
640	16
320	15
160	14
80	13
40	12
20	11
10	10
5	9
2.5	8
1.3	7
0.64	6
0.32	5
0.16	4
0.08	3
0.04	2
0.02	1
0.01	0.9
0.005	0.8
0.0025	0.7

Table 3.6- Particle Concentration per ISO Code 4406-1987

3.4.2- Three-Code ISO Standard 4406-1999

ISO 4406, which was first issued in 1987, was significantly updated in 1999. The Three-Code ISO Standard 4406-1999 is the most common standard in use. The code is structured, as shown in Fig. 3.19, from three numbers separated slashes. Like the old code, these numbers indicate the concentration of particles in each milliliter (1 cc) of a hydraulic fluid sample. Unlike the old code:

- The first numerical code referred to particle size larger than 4 microns.
- The first numerical code referred to particle size larger than 6 microns.
- The first numerical code referred to particle size larger than 14 microns.

In some cases, the code may appear as */18/13. This code means that the particle size less than 4 µm has no considerable effect on the system.

In some cases, also it appears as 12/08/*. This indicates that the particle greater than 14 µm was too few to statically provide an accurate value.

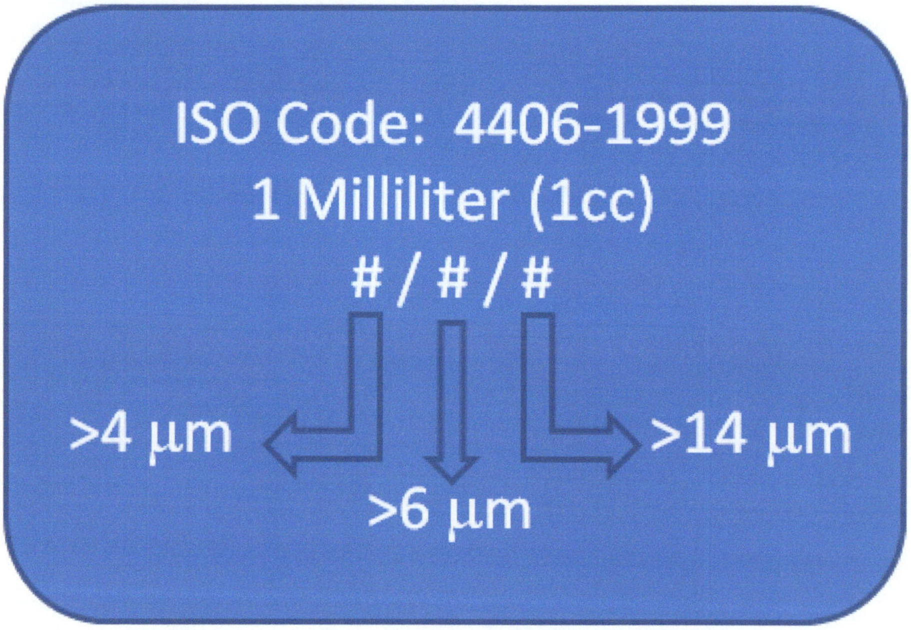

Fig. 3.19- Structure of ISO Code 4406-1999

Table 3.7 shows the particle concentration per the ISO Code 4406-1999. The darkened part of the table is the realistic cleanliness levels in typical hydraulic systems.

For example, cleanliness code 20/18/14 means the particulate concentration in each milliliter of the fluid sample are as follows:

- More than 5,000 and up to and including 10,000 particles of size larger than 4 microns.
- More than 1,300 and up to and including 2,500 particles of size larger than 6 microns.
- More than 80 and up to and including 160 particles of size larger than 14 microns.

ISO 4406-1999 Range Numbers		
	Number of Particles per Millimeter	
Range Number	More Than	Up to and Including
28	1,300,000	2,500,000
27	640,000	1,300,000
26	320,000	640,000
25	160,000	32,000
24	80,000	160,000
23	40,000	80,000
22	20,000	40,000
21	10,000	20,000
20	5,000	10,000.
19	2,500	5,000
18	1,300	2,500
17	640	1,300
16	320	640
15	160	320
14	80	160
13	40	80
12	20	40
11	10	20
10	5	10
9	2.5	5
8	1.3	2.5
7	0.64	1.3
6	0.32	0.64
5	0.16	0.32
4	0.08	0.16
3	0.04	0.08
2	0.02	0.04
1	0.01	0.02
0	0	0.01

Table 3.7- Particle Concentration per ISO Code 4406-1999

Figure 3.20 shows, how to generate the code based on the particles count.

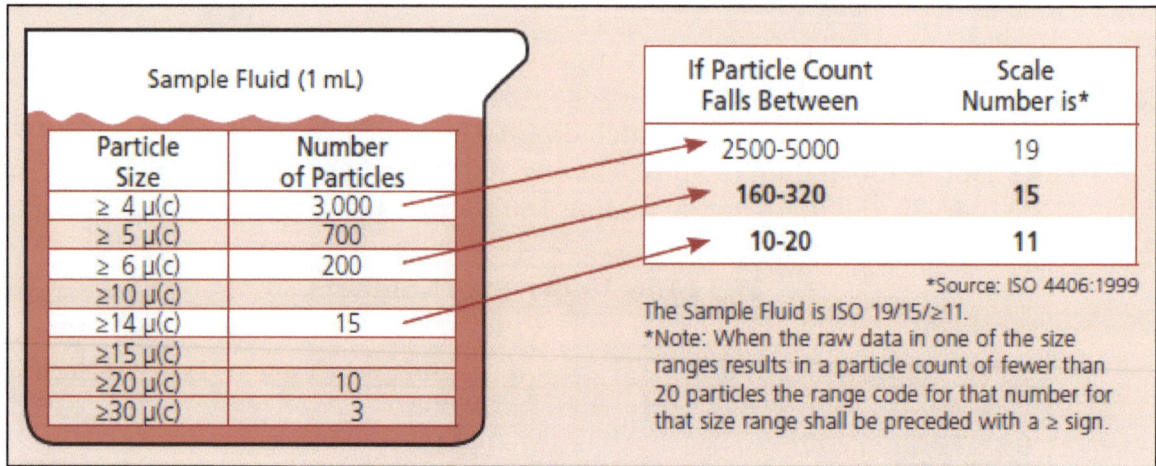

Sample Fluid (1 mL)			If Particle Count Falls Between	Scale Number is*
Particle Size	Number of Particles		2500-5000	19
≥ 4 µ(c)	3,000	→	160-320	15
≥ 5 µ(c)	700		10-20	11
≥ 6 µ(c)	200			
≥10 µ(c)				
≥14 µ(c)	15			
≥15 µ(c)				
≥20 µ(c)	10			
≥30 µ(c)	3			

*Source: ISO 4406:1999
The Sample Fluid is ISO 19/15/≥11.
*Note: When the raw data in one of the size ranges results in a particle count of fewer than 20 particles the range code for that number for that size range shall be preceded with a ≥ sign.

Fig. 3.20- Structure of ISO Code 4406-1999 (Courtesy of Schroeder)

It is mistakenly understood that the new supplied oil is cleaner than what you have in the system! This is usually incorrect because the fluid in the system is continuously filtered. Typically, as shown in Fig. 3.21, new fluid as delivered from the drum, has a cleanliness level of ISO Code 23/21/19. Figure 3.22 also confirms that new fluid is not clean.

ISO Code 23/21/19

Fig. 3.21- Typical Cleanliness ISO Code for New Hydraulic Fluid

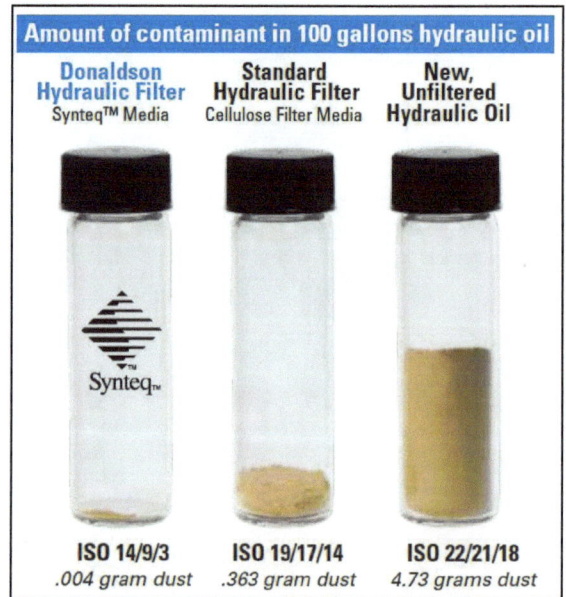

Amount of contaminant in 100 gallons hydraulic oil

Donaldson Hydraulic Filter Synteq™ Media	Standard Hydraulic Filter Cellulose Filter Media	New, Unfiltered Hydraulic Oil
ISO 14/9/3	ISO 19/17/14	ISO 22/21/18
.004 gram dust	.363 gram dust	4.73 grams dust

Fig. 3.22- Amount of Dirt in a Given Volume of Oil (Courtesy of Donaldson)

Figure 3.23 shows typical cleanliness levels during transportation of the fluid between various locations.

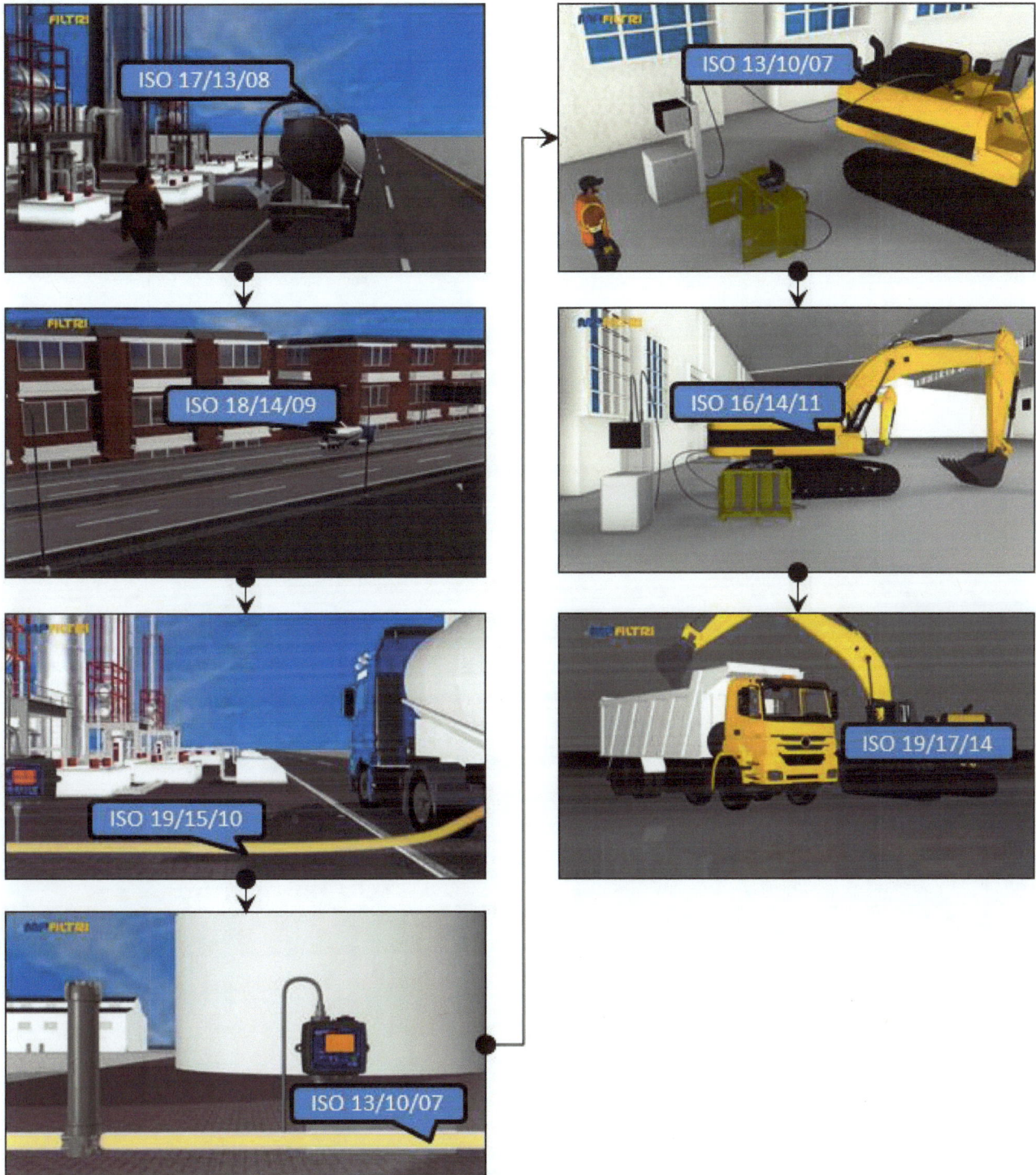

Fig. 3.23- Cleanliness Levels during Transportation of the Fluid between Various Locations (Courtesy of MPFiltri)

The example shown in Fig. 3.24 provides a practical understanding of the importance of keeping the hydraulic fluid clean to the recommended standard. The example also shows the effect of the cleanliness level on component life time and machine productivity. In this example two identical pumps are tested with hydraulic fluids, where one of them is cleaner than the other. The pump operating with a cleaner fluid receives much less dirt during a year and lasts much longer than the other pump.

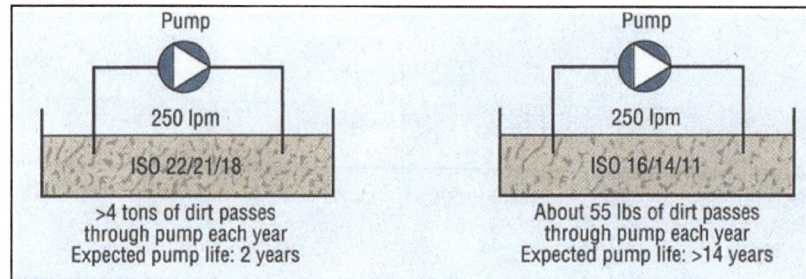

Fig. 3.24- Effect of Cleanliness Level on a Pump Life Time (Hydraulic & Pneumatic Magazine)

Table 3.8 shows the life extension factor due to keeping the oil clean. For example, moving the iso code from 22/20/17 down to 16/14/11 will extend the life time of hydraulic systems and diesel engines 5 times, for rolling bearings 3 times, for journal bearings 4 times, and for gearboxes 2.5 times.

Life Extension Table - Cleanliness Level, ISO Codes

	21/19/16		20/18/15		19/17/14		18/16/13		17/15/12		16/14/11		15/13/10		14/12/9		13/11/8		12/10/7	
24/22/19	2	1.6	3	2	4	2.5	6	3	7	3.5	8	4	>10	5	>10	6	>10	7	>10	>10
	1.8	1.3	2.3	1.7	3	2	3.5	2.5	4.5	3	5.5	3.5	7	4	8	5	10	5.5	>10	8.5
23/21/18	1.5	1.5	2	1.7	3	2	4	2.5	5	3	7	3.5	9	4	>10	5	>10	7	>10	10
	1.5	1.3	1.8	1.4	2.2	1.6	3	2	3.5	2.5	4.5	3	5	3.5	7	4	9	5.5	10	8
22/20/17	1.3	1.2	1.6	1.5	2	1.7	3	2	4	2.5	5	3	7	4	9	5	>10	7	>10	9
	1.2	1.05	1.5	1.3	1.8	1.4	2.3	1.7	3	2	3.5	2.5	5	3	6	4	8	5.5	10	7
21/19/16			1.3	1.2	1.6	1.5	2	1.7	3	2	4	2.5	5	3	7	4	9	6	>10	8
			1.2	1.1	1.5	1.3	1.8	1.5	2.2	1.7	3	2	3.5	2.5	5	3.5	7	4.5	9	6
20/18/15					1.3	1.2	1.6	1.5	2	1.7	3	2	4	2.5	5	3	7	4.6	>10	6
					1.2	1.1	1.5	1.3	1.8	1.5	2.3	1.7	3	2	3.5	2.5	5.5	3.7	8	5
19/17/14							1.3	1.2	1.6	1.5	2	1.7	3	2	4	2.5	6	3	8	5
							1.2	1.1	1.5	1.3	1.8	1.5	2.3	1.7	3	2	4	2.5	6	3.5
18/16/13									1.3	1.2	1.6	1.5	2	1.7	3	2	4	3.5	6	4
									1.2	1.1	1.5	1.3	1.8	1.5	2.3	1.8	3.7	3	4.5	3.5
17/15/12											1.3	1.2	1.6	1.5	2	1.7	3	2	4	2.5
											1.2	1.1	1.5	1.4	1.8	1.5	2.3	1.8	3	2.2
16/14/11													1.3	1.3	1.6	1.6	2	1.8	3	2
													1.3	1.2	1.6	1.4	1.9	1.5	2.3	1.8
15/13/10															1.4	1.2	1.8	1.5	2.5	1.8
															1.2	1.1	1.6	1.3	2	1.6

Legend (within table): Hydraulics and Diesel Engines | Rolling Element Bearings | Journal Bearings and Turbo Machinery | Gearboxes and others

Table 3.8- Effect of Cleanliness Level on Components Lifetime (Courtesy of Noria Corporation)

Hydraulic components and systems manufacturers should define the cleanliness code that must be respected by the end users. If no information is available from manufacturers, Table 3.9 provide some typical guidelines.

Pump/Motors	Target Cleanliness Class
Fixed Gear or Vane	20/18/15
Fixed Piston	19/17/14
Variable Vane	18/16/13
Variable Piston	18/16/13
Drives	
Cylinders	20/18/15
Hydrostatic Drives	16/15/12
Test Rigs	15/13/10
Valves	
Check Valve	20/18/15
Directional Valve	20/18/15
Standard Flow Control Valve	20/18/15
Poppet Valve	19/17/14
Proportional Valve	18/16/13
Servo valve	15/13/10
Bearings	
Anti-Friction Bearing	18/15/12
Transmission	17/15/12
Ball Bearing	15/13/10
Roller Bearing	16/14/11

Table 3.9- Guideline for Cleanliness Levels per ISO 4406-1999

As shown in Table 3.10, for proportional valves, the typical cleanliness code is 18/16/13. This means the maximum allowable particulate concentration in each milliliter of the fluid sample is as follows:

- More than 1,300 and up to and including 2,500 particles larger than 4 microns.
- More than 320 and up to and including 640 particles larger than 6 microns.
- More than 40 and up to and including 80 particles larger than 14 microns.

For a servo valves, the typical cleanliness code is 15/13/10. This means the maximum allowable particulate concentration is as follows:

- More than 160 and up and including to 320 particles larger than 4 microns.
- More than 40 and up to and including 80 particles larger than 6 microns.
- More than 5 and up to and including 10 particles larger than 14 microns.

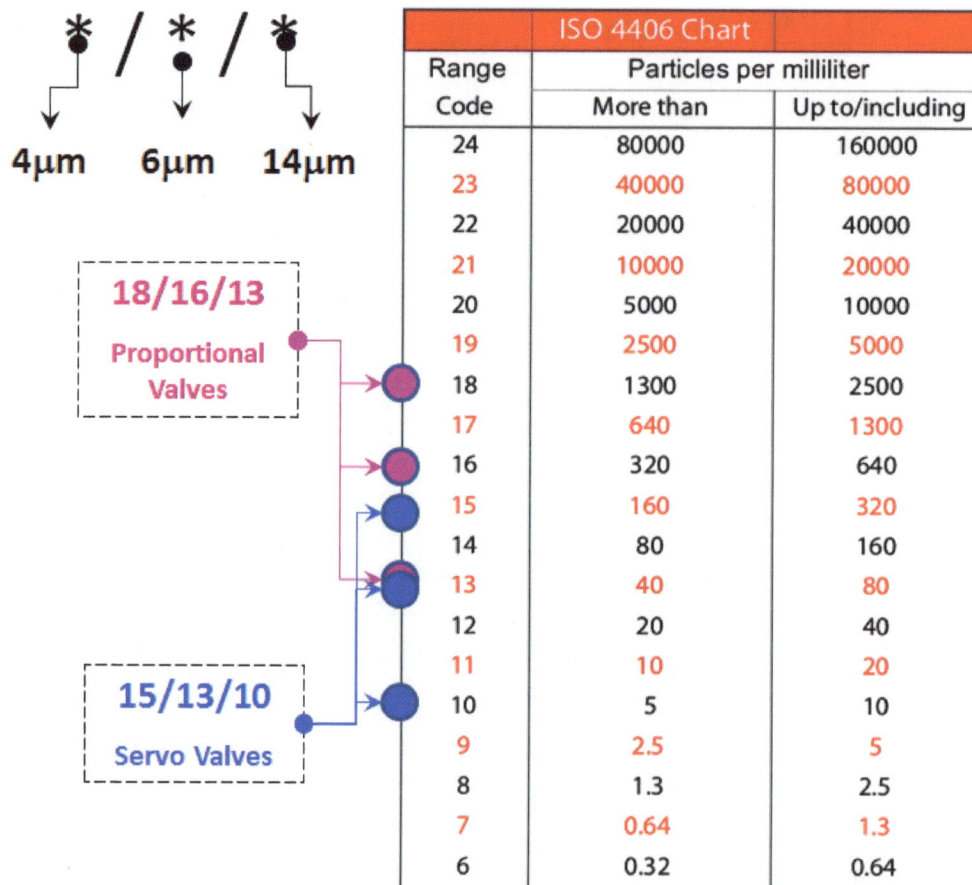

Range Code	ISO 4406 Chart	
	Particles per milliliter	
	More than	Up to/including
24	80000	160000
23	40000	80000
22	20000	40000
21	10000	20000
20	5000	10000
19	2500	5000
18	1300	2500
17	640	1300
16	320	640
15	160	320
14	80	160
13	40	80
12	20	40
11	10	20
10	5	10
9	2.5	5
8	1.3	2.5
7	0.64	1.3
6	0.32	0.64

Table 3.10- Particle Concentration for EH Valves per ISO Code 4406-1999

3.4.3- NAS Standard 1638

National Aerospace Standard (NAS 1638) is a particulate contamination coding system used in the fluid power industry. NAS 1638 became the American National Aerospace Standard in 1964 to control the amount of contamination delivered in aircraft hydraulic components.

Today, use of NAS-1638 is very limited for the sake of ISO 4406 standard. Correlation tables are available to compare NAS cleanliness class versus other standards.

Table 3.11 shows the NAS 1638 Standard. The table shows the size and the concentration (particle counts) in a given volume (100 ml = 100 cc) of a hydraulic fluid sample. It converts the particle counts at various size ranges into convenient broad-base classes. The standard is arranged from the cleanest possible fluid (class 00) to the dirtiest oil (class 12).

NAS class is assigned based on **highest** particle number among the individual range of particle sizes. For example, if in a 100 mi sample of fluid:
- For particle size 5-15 µm, 1000 particles were found.
- For particle size 15-25 µm, 356 particles were found.
- For particle size 25-50 µm, 126 particles were found.
- For particle size 50-100 µm, 45 particles were found.
- For particle size > 100 µm, 16 particles were found.
- Contamination class is "Class 6"

Contamination Class	Particle Size in µm (in 100 ml)				
	5-15	15-25	25-50	50-100	>100
00	125	22	4	1	0
0	250	44	8	2	0
1	500	89	16	3	1
2	1,000	178	32	6	1
3	2,000	356	63	11	2
4	4,000	712	126	22	4
5	8,000	1,425	253	45	8
6	16,000	2,850	506	90	16
7	32,000	5,700	1,012	180	32
8	64,000	11,400	2,025	360	64
9	128,000	22,800	4,045	720	128
10	256,000	45,600	8,100	1,440	256
11	512,000	91,200	16,200	2,880	512
12	1,024,000	182,400	32,400	5,760	1,024

Table 3.11- Particle Concentration per NAS 1638

3.4.4- SAE Standard AS 4059(E)

Society of Automotive Engineering developed the standard *SAE* **AS 4059(E)**. Like NAS 1638, The SAE 4059 (E) cleanliness classes are based on particle size and concentration. Unlike NAS 1638, particle sizes are labeled with letters (A - F), and method of evaluating the particle size is part of evaluating the contamination level. Table 3.12 and the following examples show how to use this standard:

Example 1: Cleanliness Class is (AS 4059:6). This means that maximum count for all sizes of particles should not absolutely exceed the number indicated for class 6.

Example 2: Cleanliness Class is (AS 4059 :6 B). This means that maximum count for particles size B particles should not exceed the number indicated for class 6, i.e. maximum of 19,500 particles of a size of 5 μm or 6 μm$_{(c)}$ depends on the method of measuring.

Example 3: Cleanliness Class is (AS 4059 :7 B / 6 C). This means that:
- Maximum number of particles Size B (5 μm or 6 μm$_{(c)}$) = 38,900 / 100 ml.
- Maximum number of particles Size C (15 μm or 14 μm$_{(c)}$) = 3,460 / 100 ml.

Example 4: Cleanliness Class is (AS 4059:6 B-F). This means that maximum count for particles size range B-F should not exceed the number indicated for class 6.

	Maximum Particle Concentration* (particles/100ml)					
ISO 4402 *	> 1 μm	> 5 μm	> 15 μm	> 25 μm	> 50 μm	> 100 μm
ISO 11171**	> 4 μm$_{(C)}$	> 6 μm$_{(C)}$	> 14 μm$_{(C)}$	> 21 μm$_{(C)}$	> 38 μm$_{(C)}$	> 70 μm$_{(C)}$
Size Coding	A	B	C	D	E	F
000	195	76	14	3	1	0
00	390	152	27	5	1	0
0	780	304	54	10	2	0
1	1,560	609	109	20	4	1
2	3,120	1,220	217	39	7	1
3	6,250	2,430	432	76	13	2
4	12,500	4,860	864	152	26	4
5	25,000	9,730	1,730	306	53	8
6	50,000	19,500	3,460	612	106	16
7	100,000	38,900	6,920	1,220	212	32
8	200,000	77,900	13,900	2,450	424	64
9	400,000	156,000	27,700	4,900	848	128
10	800,000	311,000	55,400	9,800	1,700	256
11	1,600,000	623,000	111,000	19,600	3,390	1,020
12	3,200,000	1,250,000	222,000	39,200	6,780	

* ISO 4402 or Optical Microscope. Particle size is based on longest dimension
** ISO 11171 or Electron Microscope. Particle size is based on projected area equivalent diameter
Table 3.12- Particle Concentration per SAE AS4059(E)

3.4.5- Contamination Standards Cross-Reference

If a cleanliness level of certain standard is not available, it still can be approximately defined through a cross-reference among the other standards. Table 3.13 and the associated Fig. 3.25 provide an example of cross reference between various standard.

Sample #	NAS 1638	ISO 4406: 1999	SAE AS 4059
1	Class 3	14/12/9	Class 4
2	Class 4	15/13/10	Class 5
3	Class 5	16/14/11	Class 6
4	Class 6	17/15/12	Class 7
5	Class 7	18/16/13	Class 8
6	Class 8	19/17/14	Class 9
7	Class 9	20/18/15	Class 10
8	Class 10	21/19/16	Class 11
9	Class 11	22/20/17	Class 12
10	Class 12	23/21/18	Class 13

Table 3.13- Approximate Cross-Reference for Contamination Classes (Courtesy of Hydac)

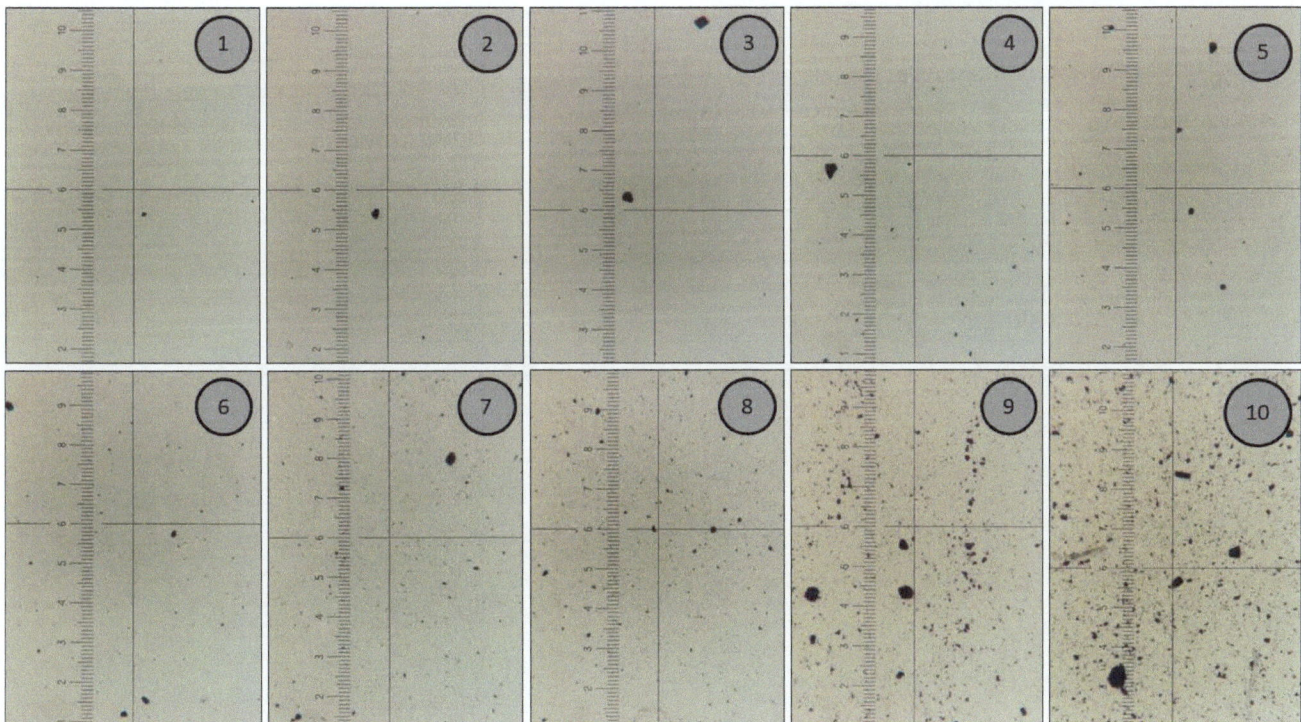

Magnification: x100 (1 Scale Mark = 10 microns)

Fig. 3.25- Samples of Fluid Contamination Levels (Courtesy of Hydac)

Table 3.14 shows another cross-reference between the ISO 4406-1987, NAS 1638, and SAE 749 standard which was not presented in this textbook. It is to be mentioned that the 2-codes ISO standard and the SAE 749 standard are no longer in use and they are stated here for historical purposes.

Table 3.14- Approximate Cross-Reference for Contamination Classes (Courtesy of Donaldson)

ISO 4406 CODE	NAS 1638 CLASS	SAE 749 CLASS
11/8	2	—
12/9	3	0
13/10	4	1
14/9		
14/11	5	2
15/9		
15/10		
15/12	6	3
16/10		
16/11		
16/13	7	4
17/11		
17/14	8	5
18/12		
18/13		
18/15	9	6
19/13		
19/16	10	
20/13		
20/17	11	
21/14		
21/18	12	
22/15		
22/17		

Table 3.15 shows another cross-reference where additional standards are included such as MIL-STD and Gravimetric standards.

ISO 4406:1999	SAE AS 4059	NAS 1638-01/196	MIL-STD 1246A 1967	ACFTD Gravimetric Level-mg/L
24				
23/20/18		12		
22/19/17	12	11		
21/18/16	11	10		
20/17/15	10	9	300	
19/16/14	9	8		
18/15/13	8	7	200	1
17/14/12	7	6		
16/13/11	6	5		
15/12/10	5	4		0.1
14/11/9	4	3	100	
13/10/8	3	2		
12/9/7	2	1		0.01
11/8/6	1	0		
10/7/5	0	00		
8/7/4	00		50	
5/3/01			25	
2/0/0			5	

Table 3.15- Approximate Cross-Reference for Contamination Classes (Courtesy of Schroeder)

3.5- Hydraulic Fluid Particle Analysis

3.5.1-Visual Inspection

After obtaining a sample a visual inspection can be made. Visual inspection is the first and easiest qualitative contamination test. No equipment is required and can be done in field. By visual comparison between new and used fluid samples, an initial assessment can be made and to some extent fluid cleanliness can be qualitatively judged. Holding the sample up to the light will reveal any particles larger than 40 microns. If particles can be seen, the fluid is very dirty. As a minimum, it must be filtered and may be even changed. If, by naked eyes, you can't see solid contaminants this means further analysis is required.

Oil color and odor are good signs for a quick field test.

Figure 3.26 provides guidelines for visual inspection of hydraulic fluids:

1. Signs of oxidation and sludge formation in the reservoirs and around filters.
2. Signs of thermal degradation or darkened oil due to varnish formation.
3. Signs of foaming where oil color tends to be milky or cloudy.
4. Signs of Stable oil-water emulsion and fluid could also be milky or cloudy.
5. Signs of oil change is overdue where oil is very dark.

Fig. 3.26- Hydraulic Fluid Visual Inspection

3.5.2- Silt Index Test

As shown in Fig. 3.27, Silt is very fine particles generated from continuous erosion of metal parts inside hydraulic components. *Silt Index Test* is not a very famous method today and was phased out for the sake of other advanced and more accurate methods. The way it works is a sample of fluid is forced through a porous filter. The silting index is calculated based on difference in the pressure during passing the first and the second half of the sample.

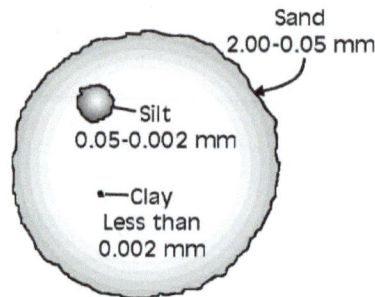

Fig. 3.27- Relative Size of Silt

3.5.3-Patch Test

Patch Test is another qualitative contamination test. Figure 3.28 shows the device used to perform the patch test.

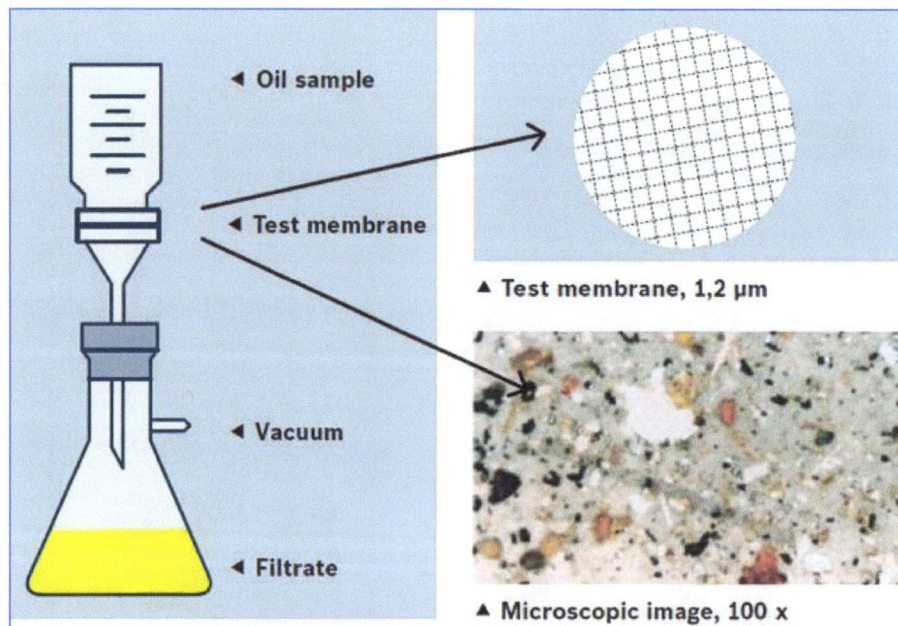

Fig. 3.28- Patch Test Device (Courtesy of Bosch Rexroth)

As shown in Fig. 3.29, Usually the patch test device and the required accessories including the index membrane paper and the *Fluid Cleanliness Comparison Guide* are available as a kit.

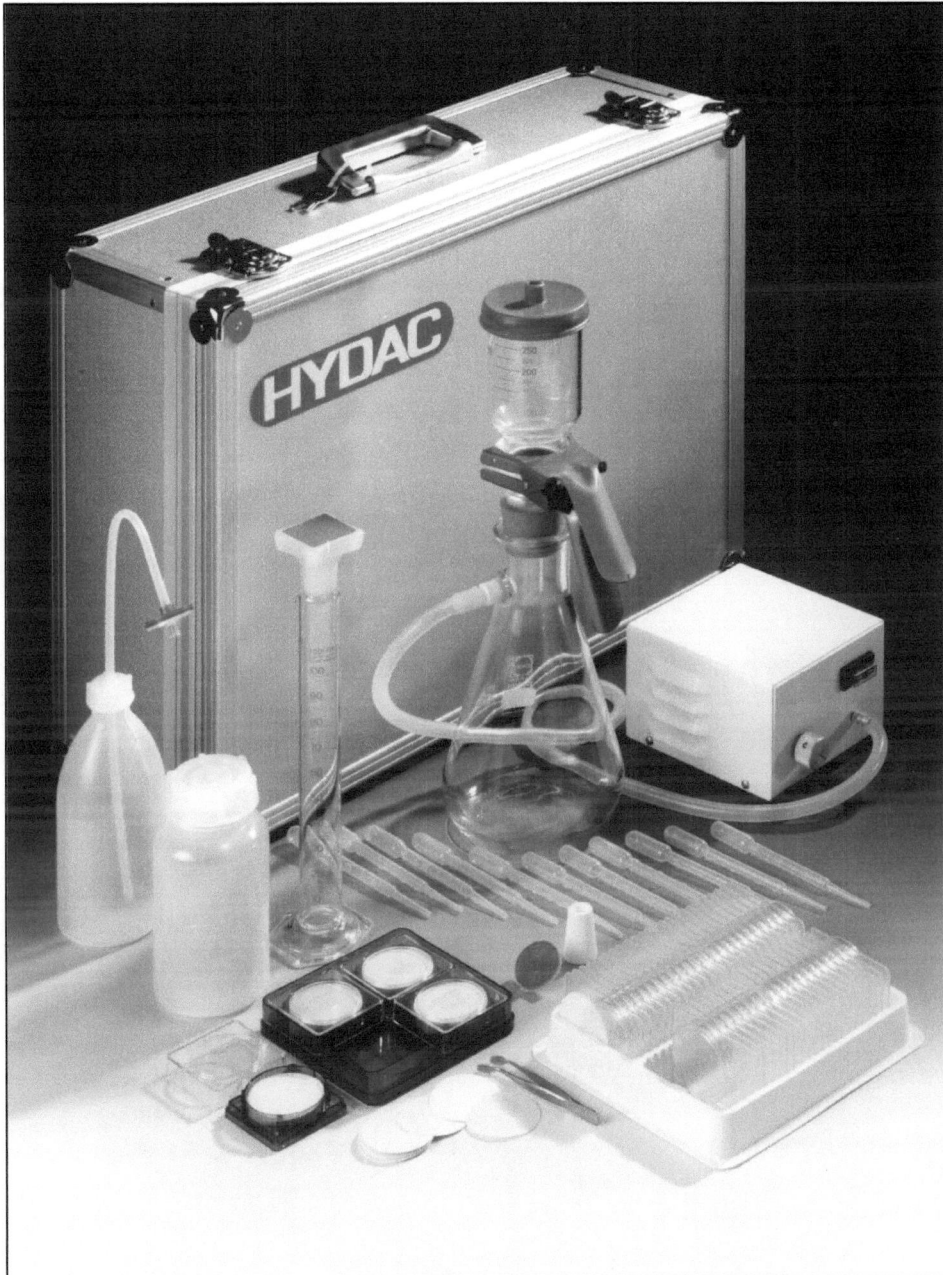

Fig. 3.29- Hydraulic Fluid Portable Patch Test Kit (Courtesy of Hydac)

Basic Steps for Patch Test:

- Assemble the pump, the funnel, and clamp on empty lower flask.
- Flush the fitter assembly with pre-filtered solvent.
- Place a patch membrane on filter holder.
- Dilute the oil sample with filtered solvent and mix vigorously.
- Turn on the vacuum pump.
- As shown in Fig. 3.30, pour the fluid sample into the funnel and fill to the 25 ml level.

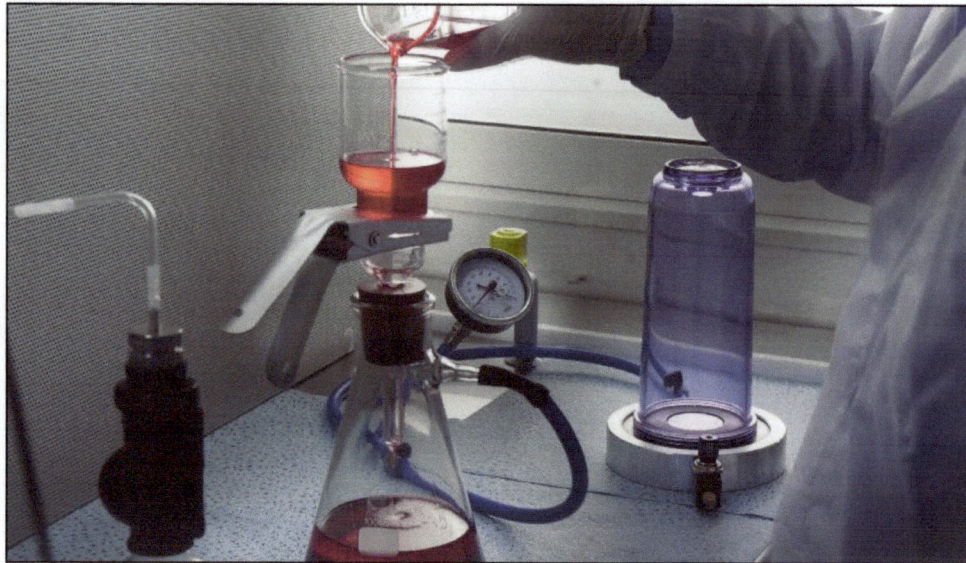

Fig. 3.30- Performing Patch Test (Courtesy of MPFiltri)

- When the sample passes completely through the patch membrane, remove membrane with forceps, and air dry it.
- Place on clean index card and immediately cover with adhesive analysis laminated cover.
- Inspect for debris.
- As shown in Fig. 3.31, color and shade of the membrane patch indicates the category of contamination such as Normal (1), Abnormal (2), and Critical (3).

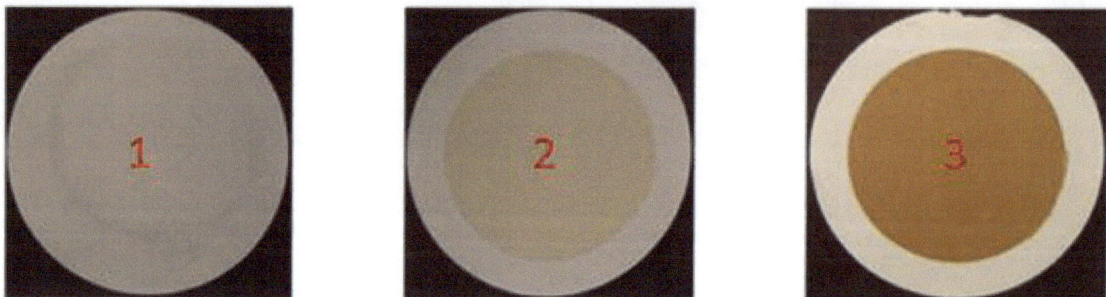

Fig. 3.31- Observations from Patch Test (Courtesy of MPFiltri)

3.5.4- Gravimetric Analysis (ISO 4405)

Gravimetric Analysis is a standard test method referred to as **(ISO 4405).** ISO 4405 describes the cleaning of the equipment being used and the procedure of performing the test. It also describes the preparatory procedures for the analysis membranes. In addition to the same apparatus used for the patch test, gravimetric analysis uses a sensitive weight scale and filter membrane whose weight has been previously defined. The way it works is passing a known volume (100 ml) of oil sample through the filter membrane using a vacuum pump. The cleanliness level is based on the difference between the weight of membrane before and after passing the sample of fluid through it. Results are given as mg/l.

3.5.5- Microscopic Particle Counting (ISO 4407)

ISO 4407 contains a description of *Microscopic Particle Counting*. It is also referred to as *Optical* or *Visual* particle counting. The way it works is similar to the patch and the gravimetric tests. The only difference is that, as shown in Fig. 3.32, a special membrane filter is used that has an average pore size less than 1 μm and grid markings. After the oil sample is passed through the membrane, it is dried and then taken to be viewed under microscope as shown in Fig. 3.33.

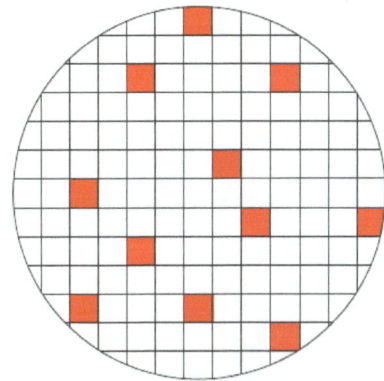

Fig. 3.32- Scaled Paper Membrane for Microscopic Particle Counting (Courtesy of Hydac)

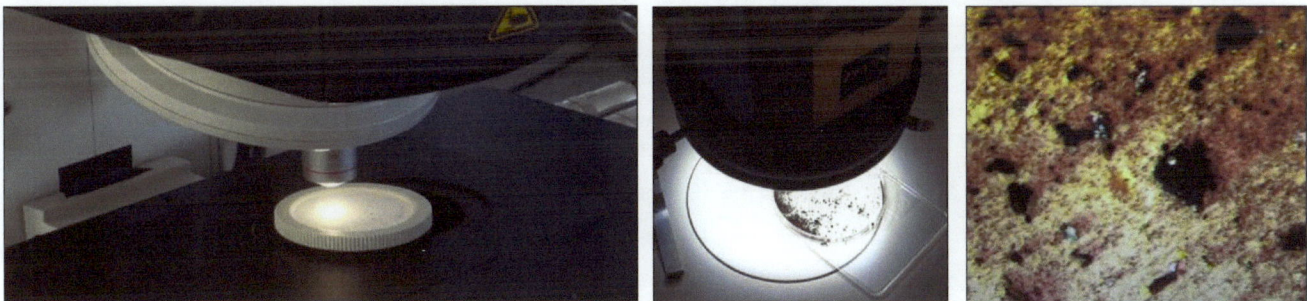

Fig. 3.33- Microscopic Particle Counting (Courtesy of MPFiltri)

The viewed membrane is compared with reference library of photos that represents various levels of contamination. The experience of the operator is important in obtaining accurate results. Figures 3.34 through 3.42 form a reference library for obtaining the cleanliness level based on ISO 4406 standard.

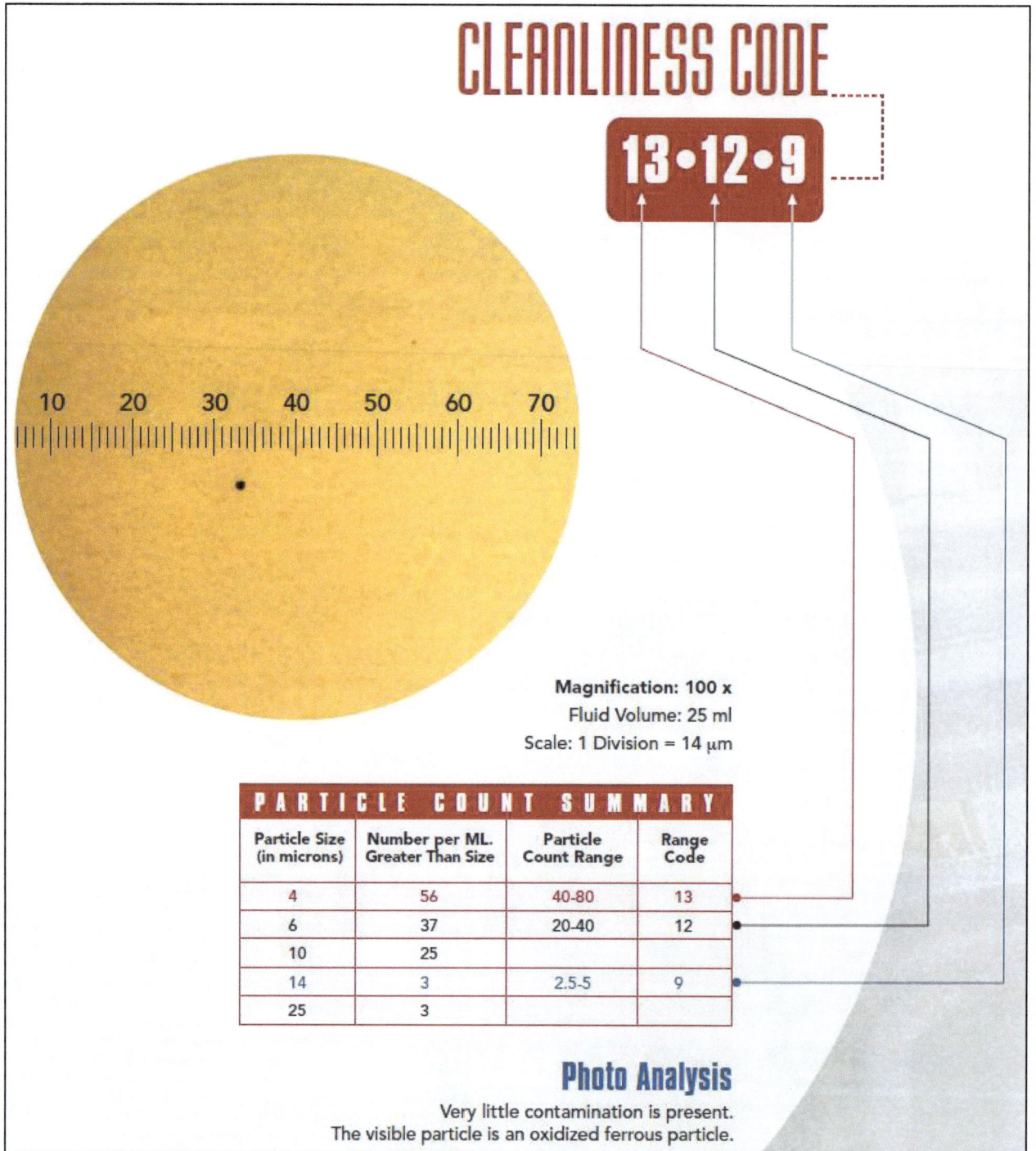

CLEANLINESS CODE

13•12•9

10   20   30   40   50   60   70

Magnification: 100 x
Fluid Volume: 25 ml
Scale: 1 Division = 14 µm

PARTICLE COUNT SUMMARY

Particle Size (in microns)	Number per ML. Greater Than Size	Particle Count Range	Range Code
4	56	40-80	13
6	37	20-40	12
10	25		
14	3	2.5-5	9
25	3		

Photo Analysis

Very little contamination is present.
The visible particle is an oxidized ferrous particle.

Fig. 3.34- Reference Photo (1) for Microscopic Particle Counting (Courtesy of Donaldson)

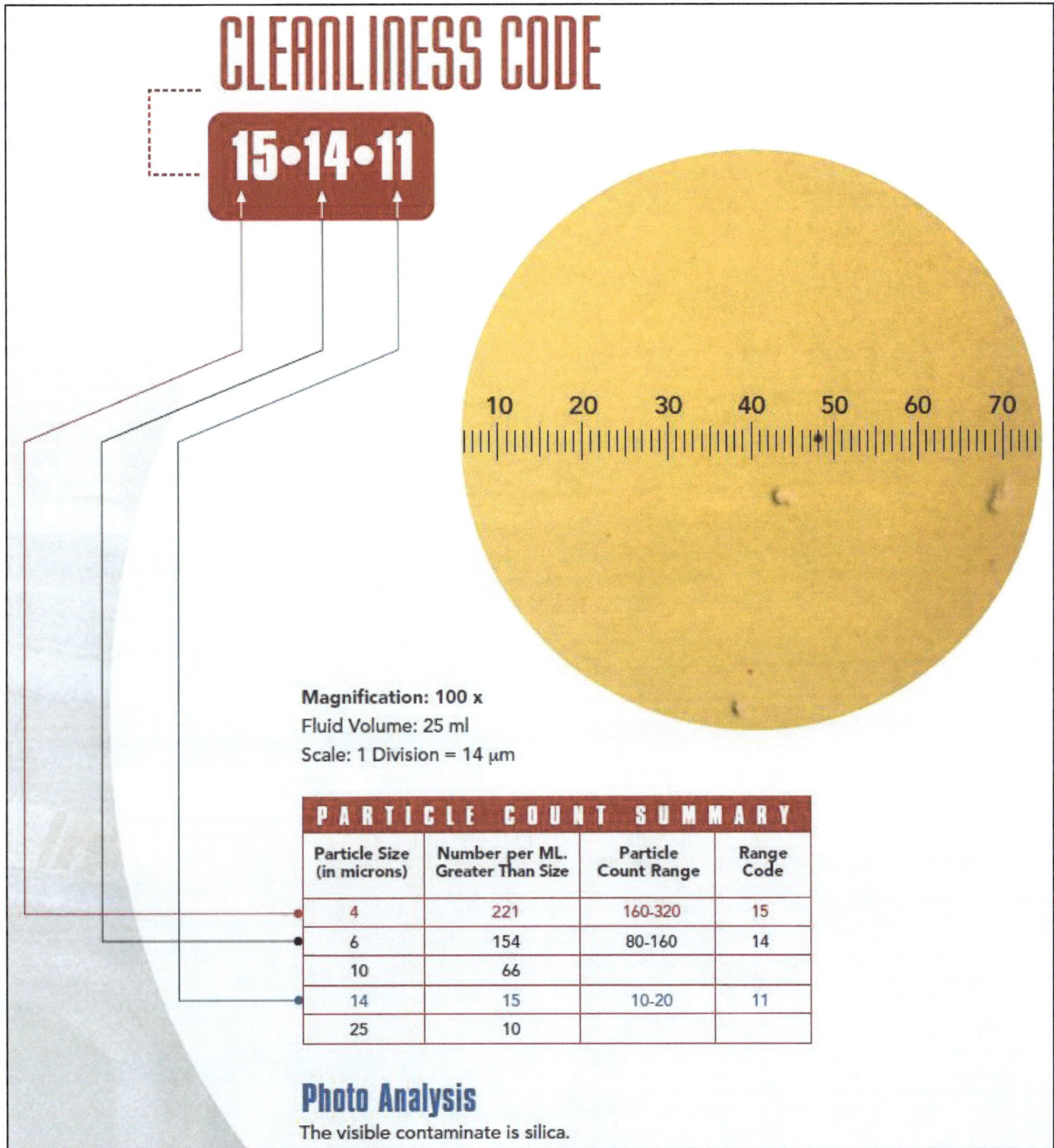

Fig. 3.35- Reference Photo (2) for Microscopic Particle Counting (Courtesy of Donaldson)

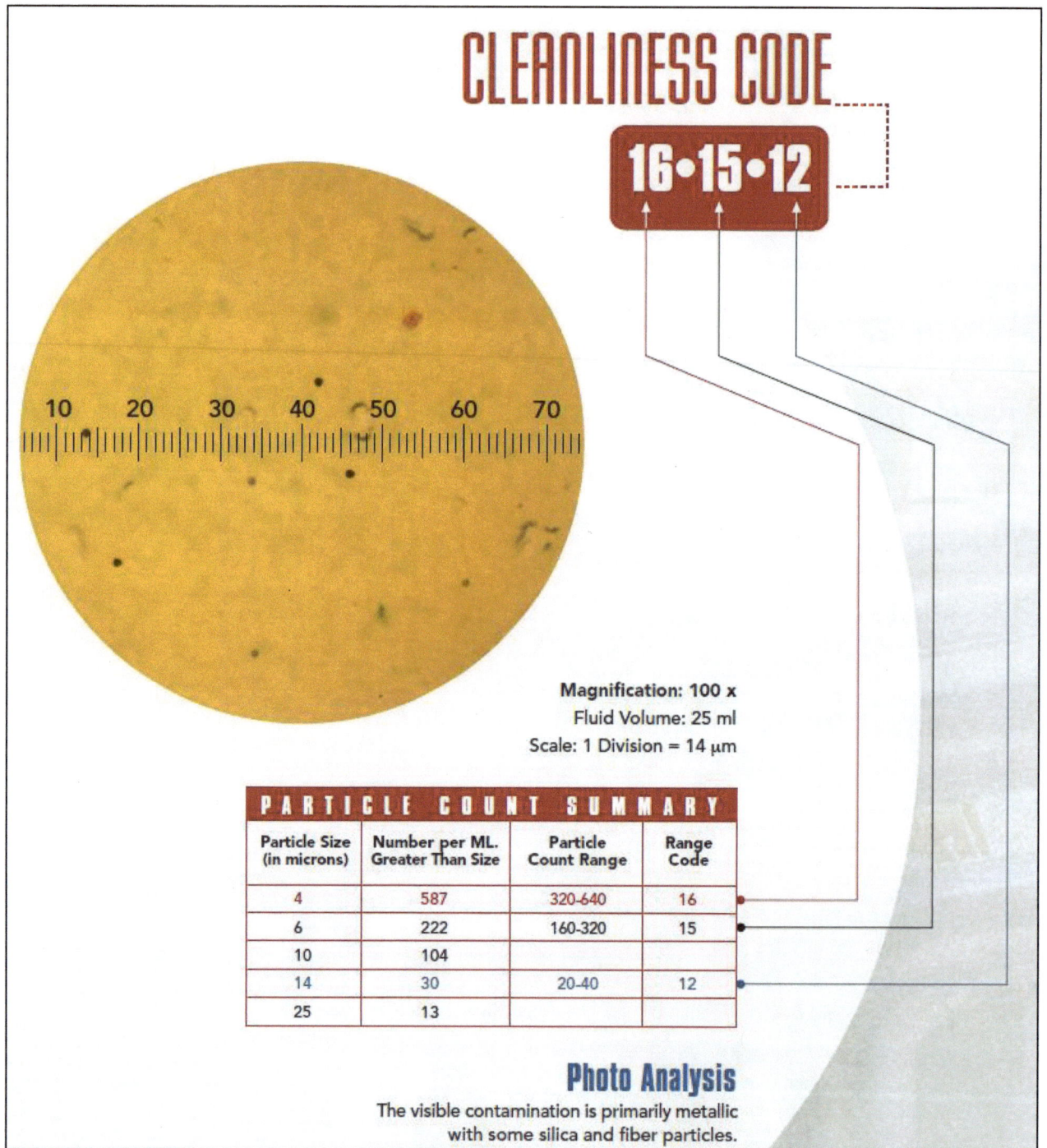

CLEANLINESS CODE

16•15•12

Magnification: 100 x
Fluid Volume: 25 ml
Scale: 1 Division = 14 μm

PARTICLE COUNT SUMMARY

Particle Size (in microns)	Number per ML. Greater Than Size	Particle Count Range	Range Code
4	587	320-640	16
6	222	160-320	15
10	104		
14	30	20-40	12
25	13		

Photo Analysis

The visible contamination is primarily metallic with some silica and fiber particles.

Fig. 3.36- Reference Photo (3) for Microscopic Particle Counting (Courtesy of Donaldson)

CLEANLINESS CODE

18•16•13

Magnification: 100 x

Fluid Volume: 25 ml

Scale: 1 Division = 14 μm

PARTICLE COUNT SUMMARY

Particle Size (in microns)	Number per ML. Greater Than Size	Particle Count Range	Range Code
4	1,978	1,300-2,500	18
6	396	320-640	16
10	230		
14	60	40-80	13
25	24		

Photo Analysis

The visible contamination is primarily silica with some metallic, oxidized ferrous and rust particles.

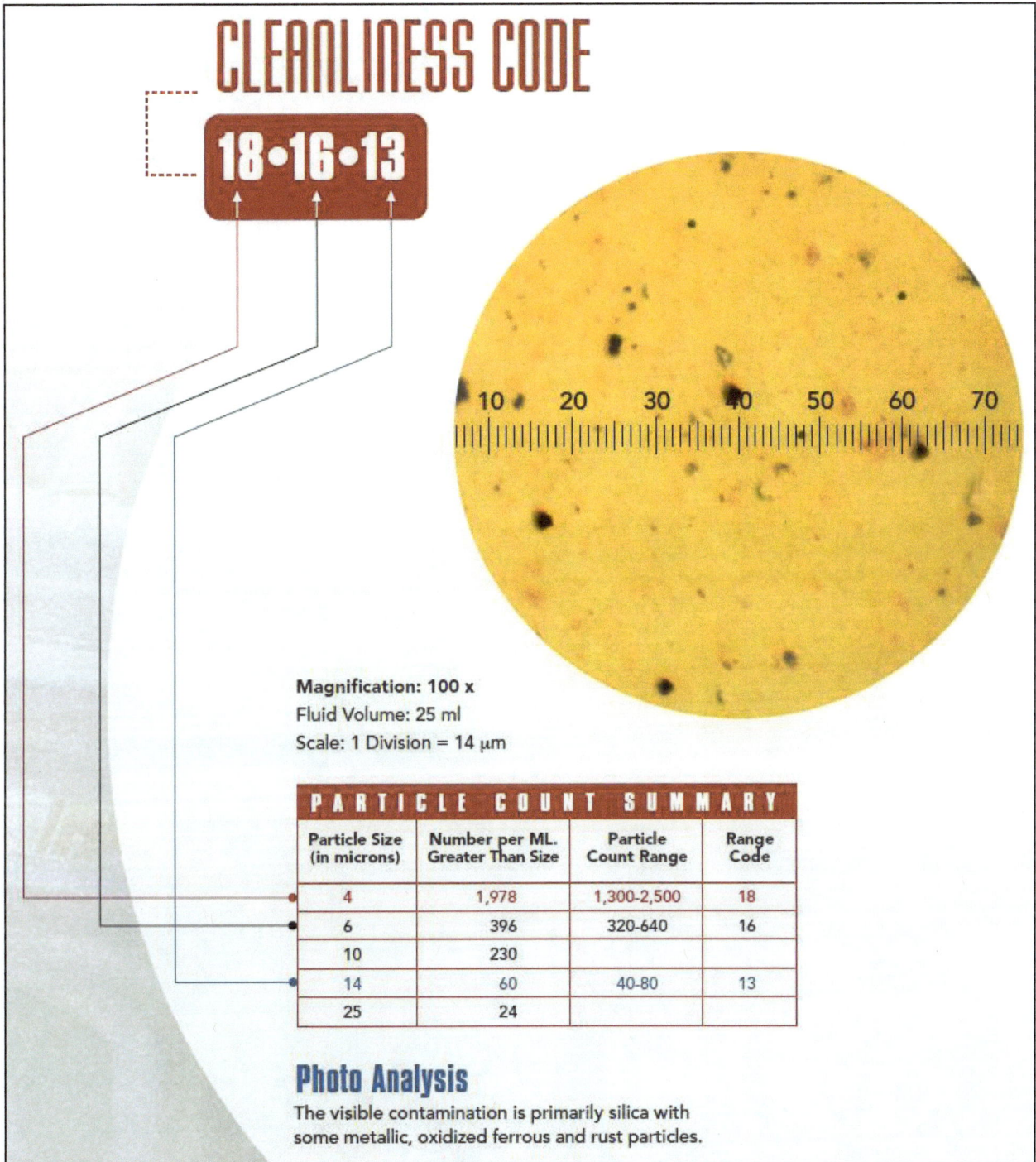

Fig. 3.37- Reference Photo (4) for Microscopic Particle Counting (Courtesy of Donaldson)

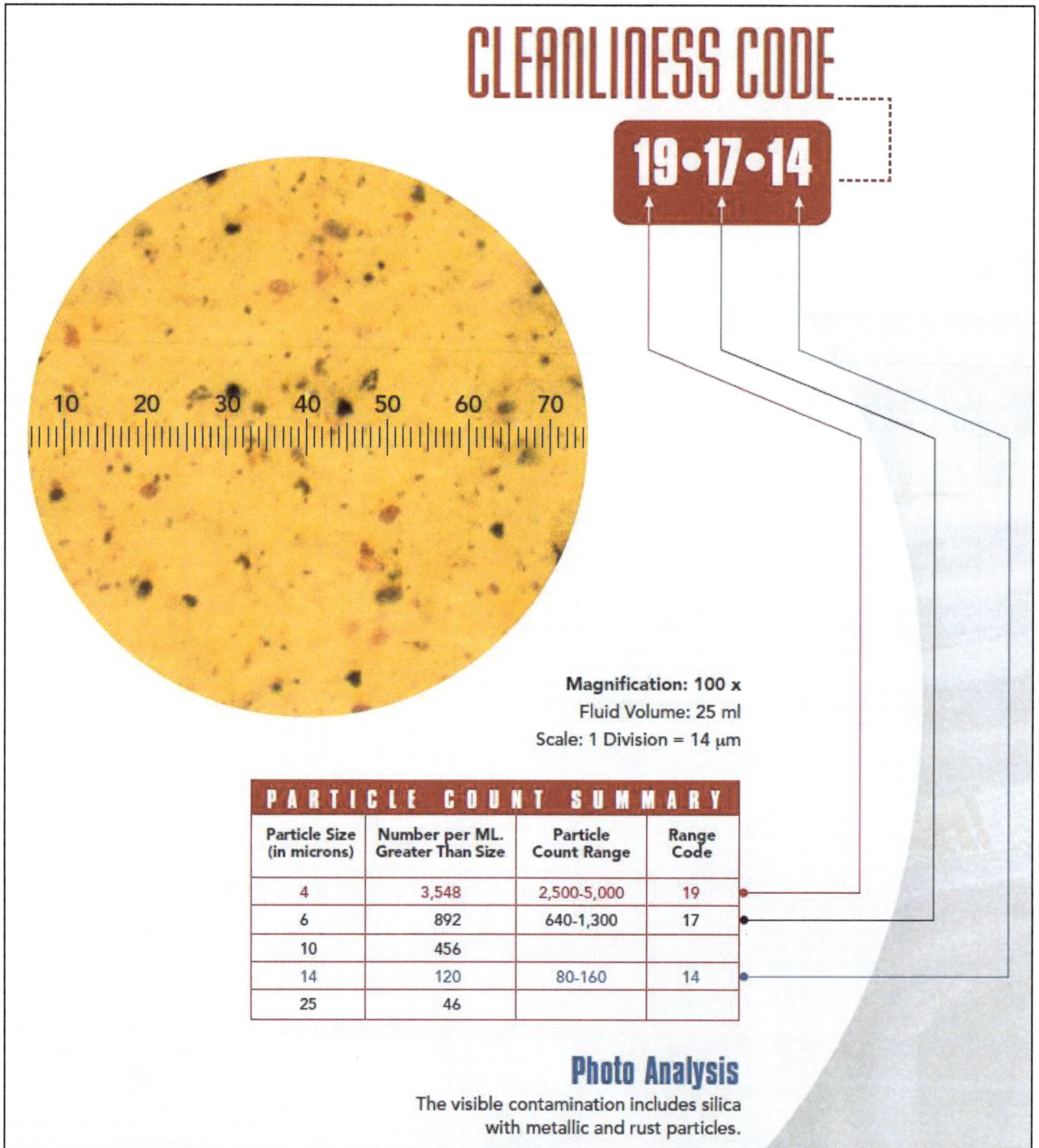

CLEANLINESS CODE

19•17•14

Magnification: 100 x
Fluid Volume: 25 ml
Scale: 1 Division = 14 μm

PARTICLE COUNT SUMMARY

Particle Size (in microns)	Number per ML. Greater Than Size	Particle Count Range	Range Code
4	3,548	2,500-5,000	19
6	892	640-1,300	17
10	456		
14	120	80-160	14
25	46		

Photo Analysis
The visible contamination includes silica with metallic and rust particles.

Fig. 3.38- Reference Photo (5) for Microscopic Particle Counting (Courtesy of Donaldson)

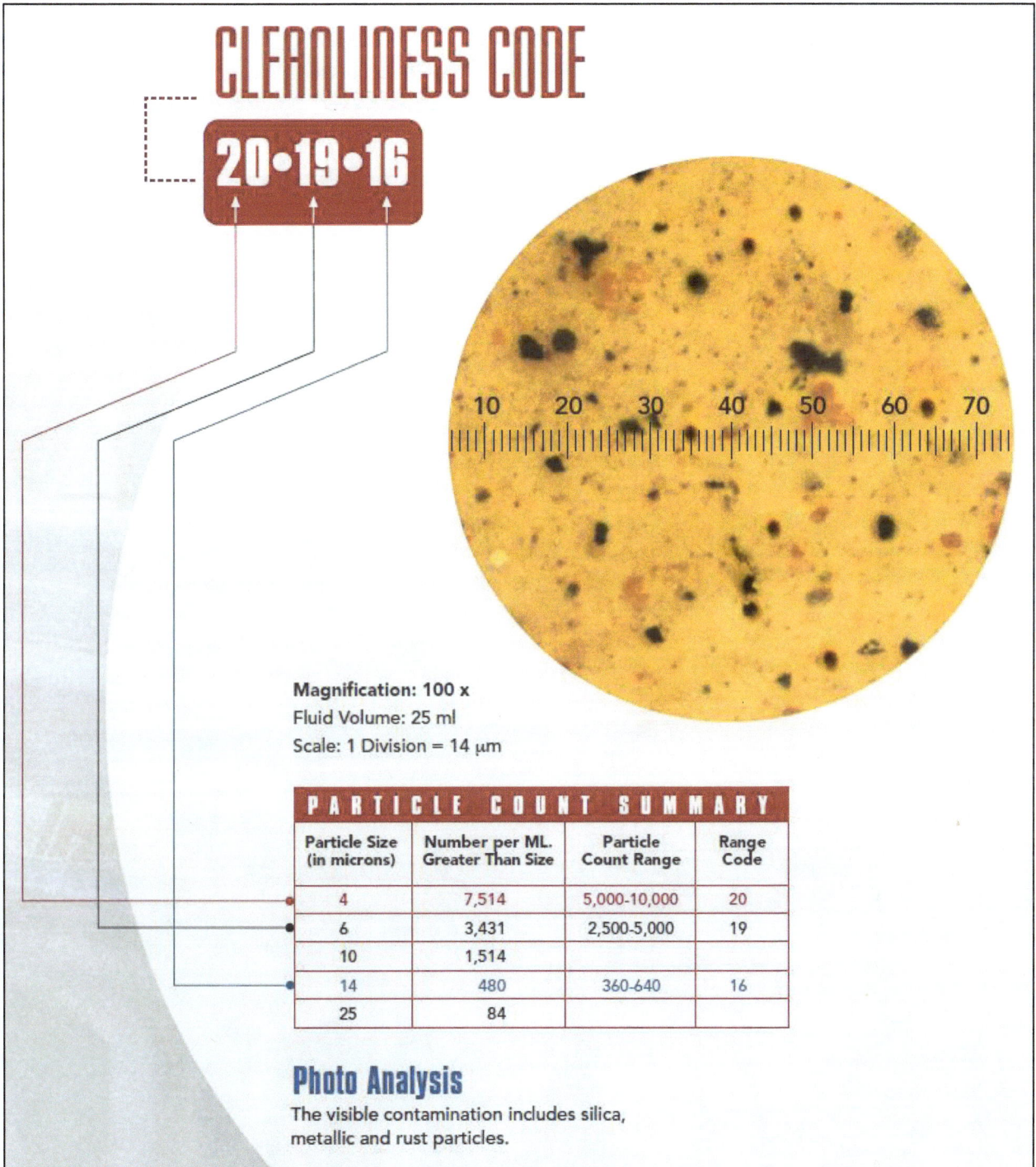

CLEANLINESS CODE

20•19•16

Magnification: 100 x
Fluid Volume: 25 ml
Scale: 1 Division = 14 μm

PARTICLE COUNT SUMMARY

Particle Size (in microns)	Number per ML. Greater Than Size	Particle Count Range	Range Code
4	7,514	5,000-10,000	20
6	3,431	2,500-5,000	19
10	1,514		
14	480	360-640	16
25	84		

Photo Analysis

The visible contamination includes silica, metallic and rust particles.

Fig. 3.39- Reference Photo (6) for Microscopic Particle Counting (Courtesy of Donaldson)

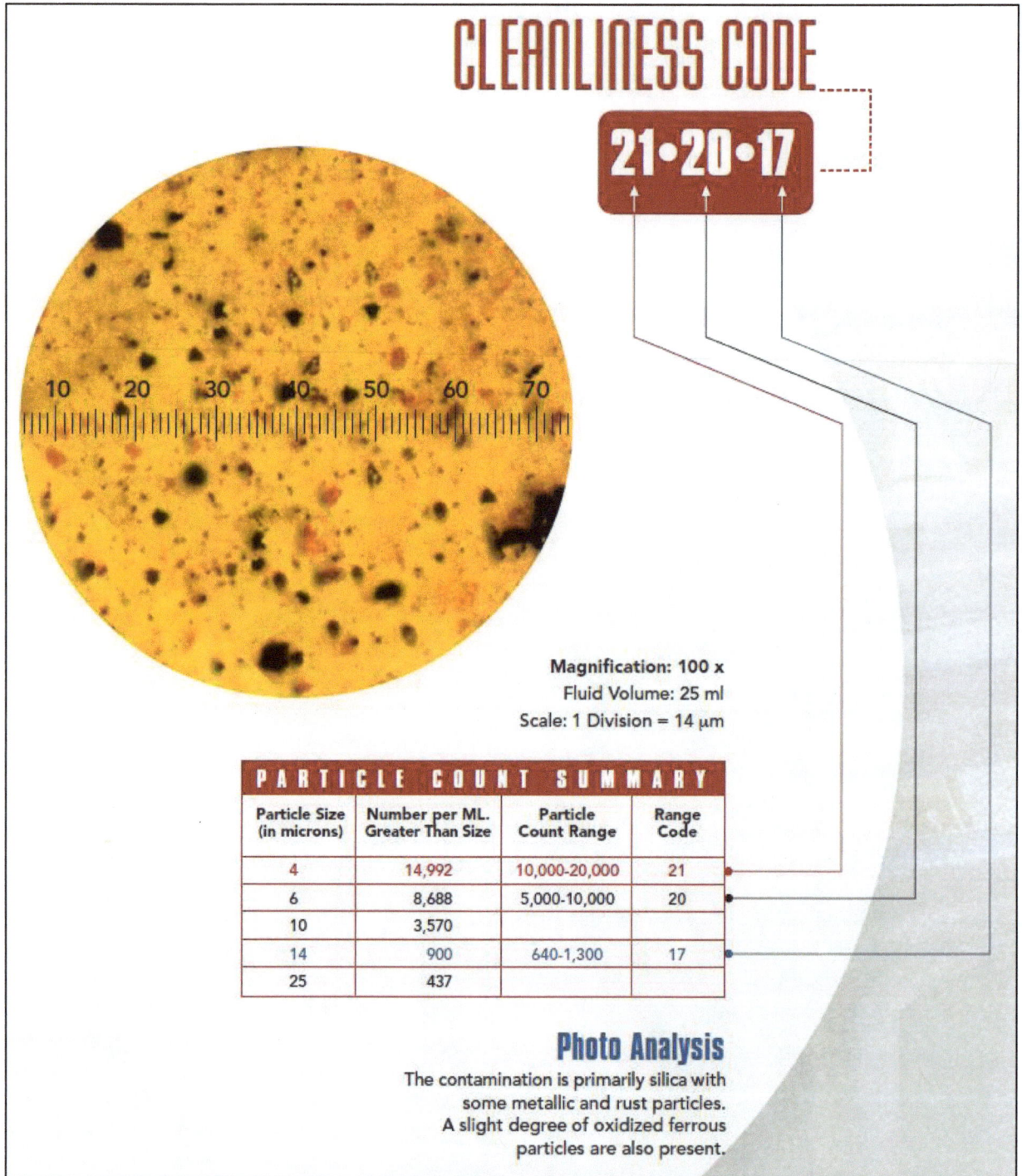

CLEANLINESS CODE

21•20•17

Magnification: 100 x
Fluid Volume: 25 ml
Scale: 1 Division = 14 μm

PARTICLE COUNT SUMMARY

Particle Size (in microns)	Number per ML. Greater Than Size	Particle Count Range	Range Code
4	14,992	10,000-20,000	21
6	8,688	5,000-10,000	20
10	3,570		
14	900	640-1,300	17
25	437		

Photo Analysis

The contamination is primarily silica with
some metallic and rust particles.
A slight degree of oxidized ferrous
particles are also present.

Fig. 3.40- Reference Photo (7) for Microscopic Particle Counting (Courtesy of Donaldson)

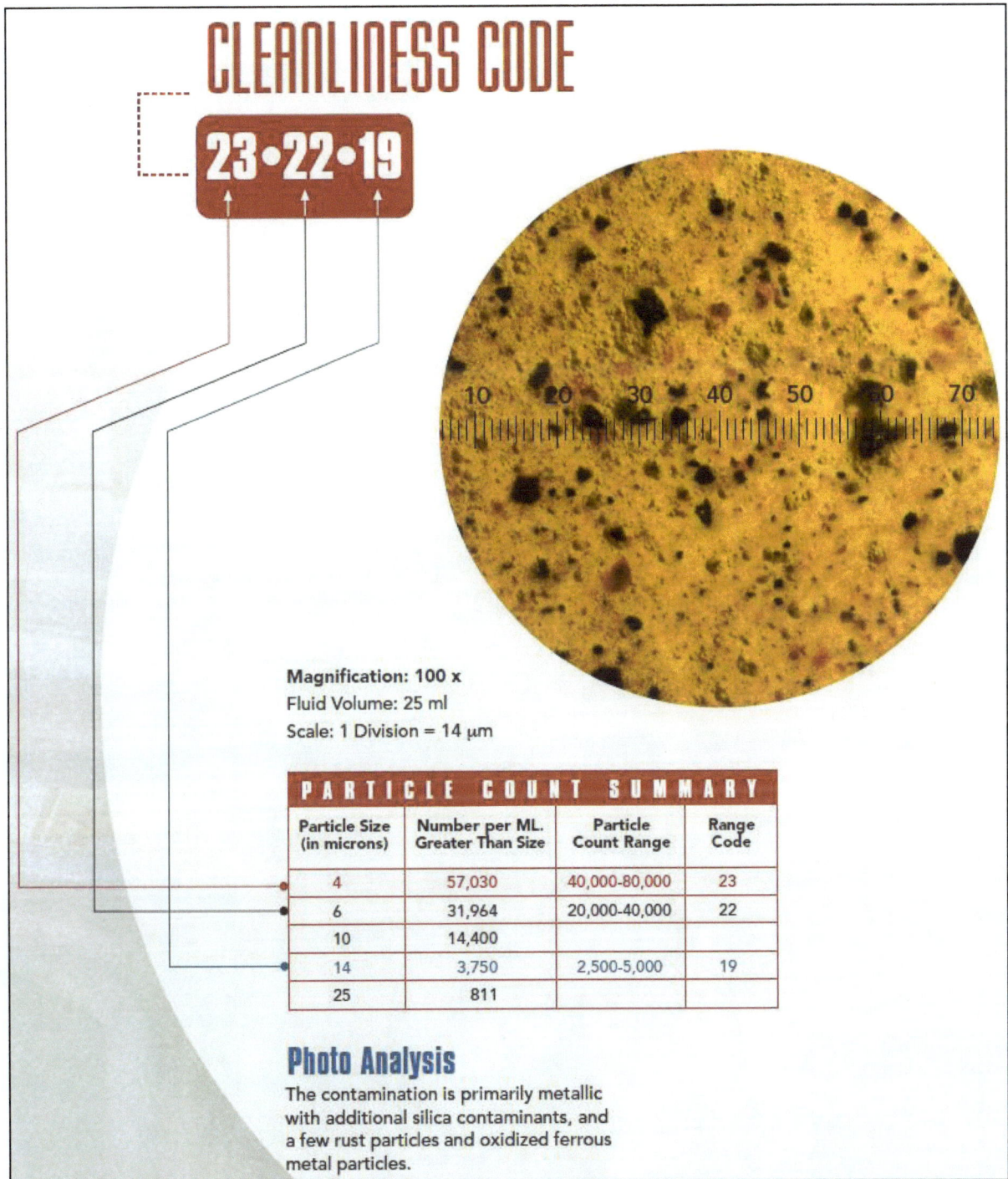

CLEANLINESS CODE

23•22•19

Magnification: 100 x
Fluid Volume: 25 ml
Scale: 1 Division = 14 μm

PARTICLE COUNT SUMMARY

Particle Size (in microns)	Number per ML. Greater Than Size	Particle Count Range	Range Code
4	57,030	40,000-80,000	23
6	31,964	20,000-40,000	22
10	14,400		
14	3,750	2,500-5,000	19
25	811		

Photo Analysis

The contamination is primarily metallic with additional silica contaminants, and a few rust particles and oxidized ferrous metal particles.

Fig. 3.41- Reference Photo (8) for Microscopic Particle Counting (Courtesy of Donaldson)

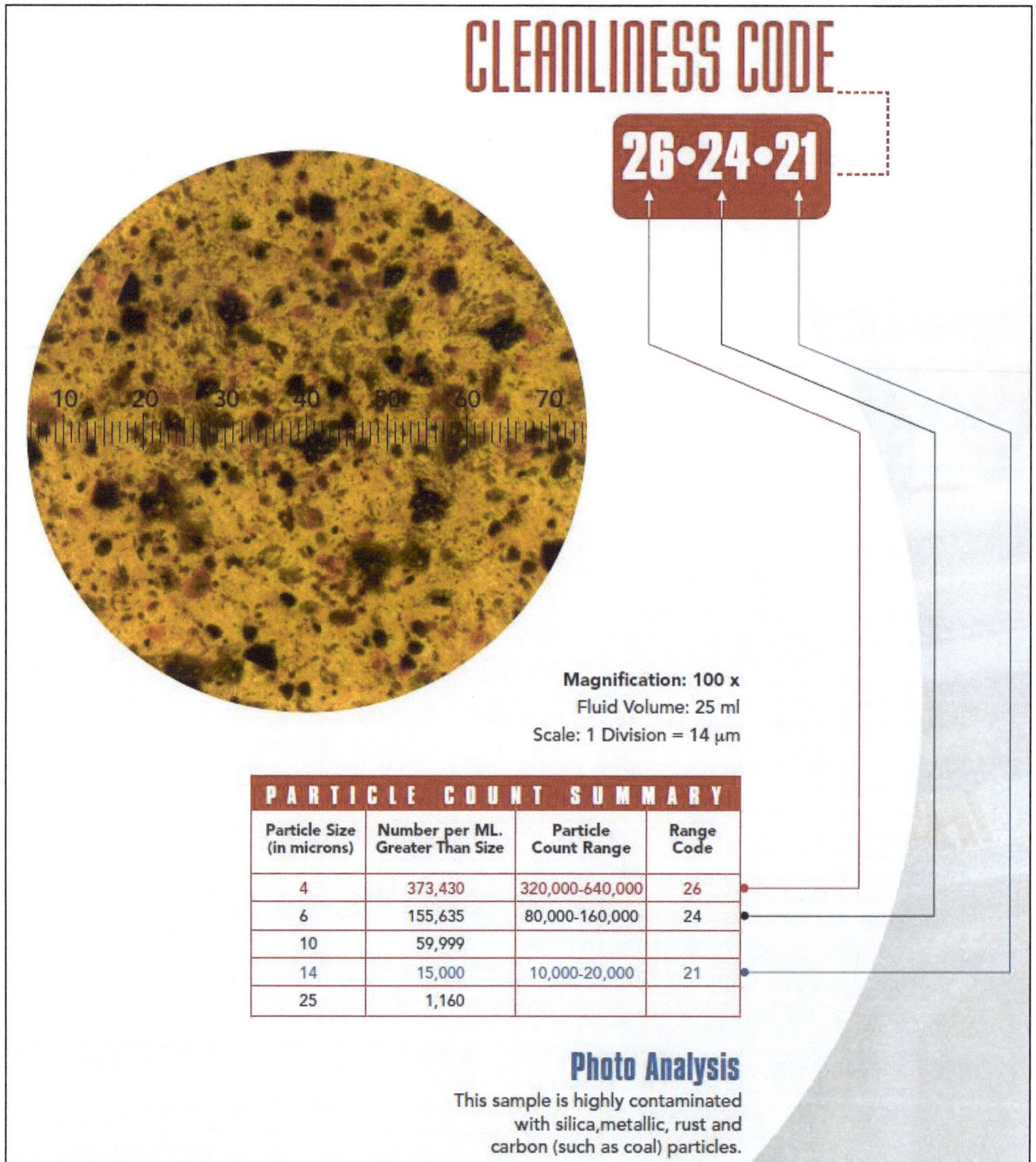

CLEANLINESS CODE

26•24•21

Magnification: 100 x
Fluid Volume: 25 ml
Scale: 1 Division = 14 μm

PARTICLE COUNT SUMMARY

Particle Size (in microns)	Number per ML. Greater Than Size	Particle Count Range	Range Code
4	373,430	320,000-640,000	26
6	155,635	80,000-160,000	24
10	59,999		
14	15,000	10,000-20,000	21
25	1,160		

Photo Analysis
This sample is highly contaminated with silica, metallic, rust and carbon (such as coal) particles.

Fig. 3.42- Reference Photo (9) for Microscopic Particle Counting (Courtesy of Donaldson)

3.5.6- Automatic Particle Counting (ISO 11500:2008)

Automatic particle counters, based on its ability and design sophistication, are classified as follows:

- Particle Counters.
- Particle Monitors.
- Particle Classifiers.

8.5.6.1- Automatic Particle Counters (APC)

ISO 11500:2008 specifies an automatic particle-counting procedure for determining the number and sizes of particles present in hydraulic-fluid bottle samples of clear, homogeneous, single-phase liquids using an *Automatic Particle Counter* (*APC*). This standard is applicable to the monitoring of the cleanliness level of fluids circulating in hydraulic systems, the progress of a flushing operation, the cleanliness level of support equipment and test rigs and the cleanliness level of packaged stock.

The method defined in this standard is the most common and accurate fluid analysis in use today. As shown in Fig. 3.43, it works based on the light-extinction principle. By passing a known volume of fluid sample between a light transmitter and a detector of an optical sensor, the sensor counts the particles and capture the shadow of each individual particle.

Images of a particle are used for purposes of identifying the particle size. As sown in Fig. 3.44, there are two standards for identifying a particle size as follows:

- **ISO 4402:1991.** This standard defines the particle size based on the longest cord within the body of the particle. It is **no longer in use** and replaced by the ISO 11171.
- **ISO 11171:1999.** This standard defines the particle size based on diameter of the equivalent circular area.

Results are reported based on the standards that were loaded into the microprocessor of the electronic particle counter. In some old APCs, air bubbles or water drops are counted as particles that affects the accuracy of the results.

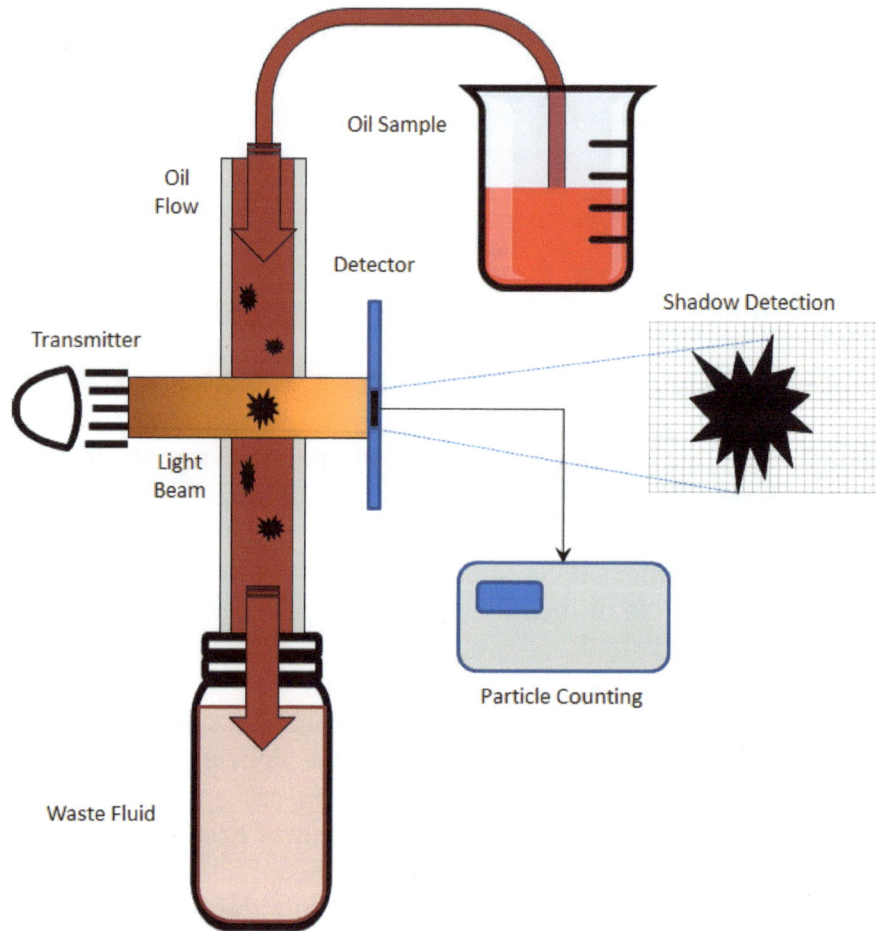

Fig. 3.43- Electronic Particle Counting Concept of Operation

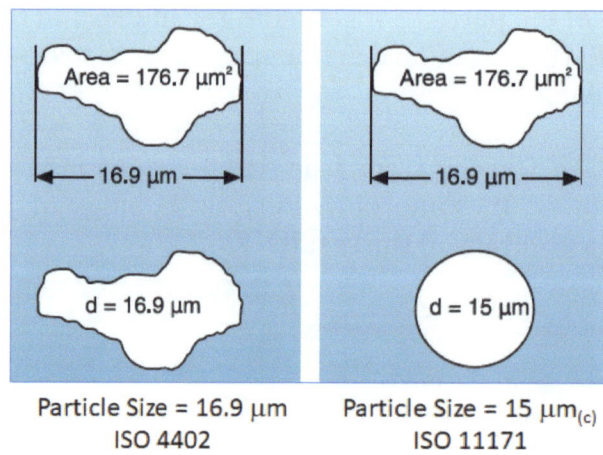

Fig. 3.44- Methods of Identifying the Particle Size (Courtesy of Hydac)

Particle counters can work offline, referred to as *Desktop APC*, on a given sample or it can be portable for use in field. Figure 3.45 shows the Flow Control Unit FCU 1000 from Hydac as an example of a portable electronic particle counter for use online in the field.

Fig. 3.45- Portable Electronic Particle Counter FCU 1000 (Courtesy of Hydac)

Figure 3.46 shows connecting the FCU 1000 directly to the sampling point on the machine (left) or to fluid sampling bottle (right). For further details, the specific manufacturers manual must be reviewed for proper operating conditions. The unit is programmable to report the results in various contamination standard including ISO, NAS, and SAE.

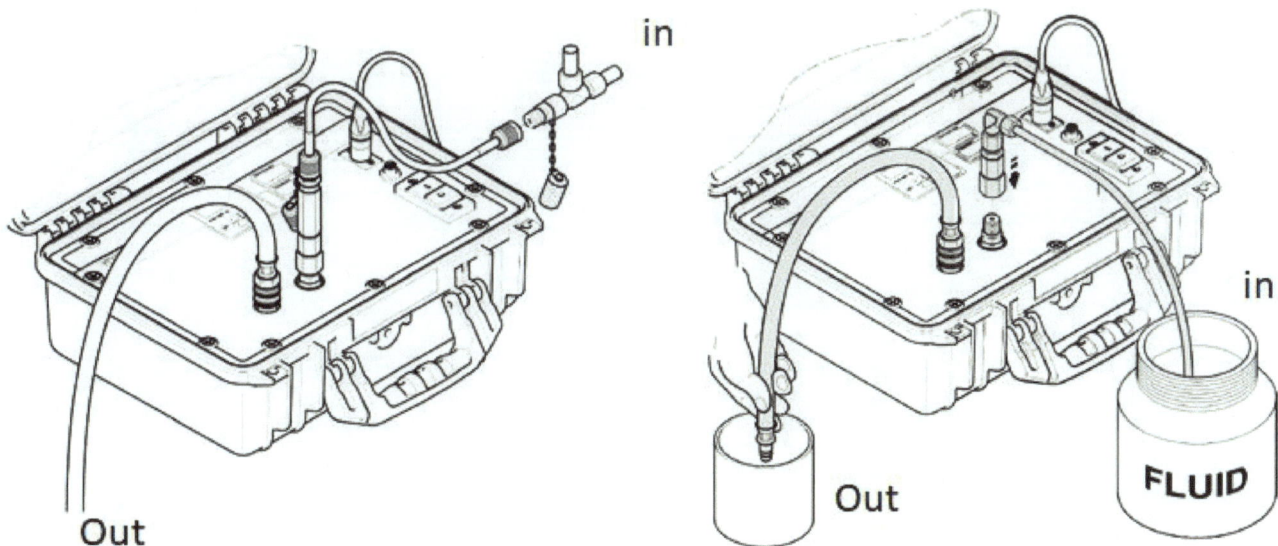

Fig. 3.46- Connecting the FCU 1000 to a Fluid Source (Courtesy of Hydac)

As shown in Fig. 3.47, The FCU 1000 unit offers sharing the measured values, via data interface, with a PC or the customer system. Data can also be uploaded to a USB or transferred wirelessly through Bluetooth connection. As shown in Fig. 3.48, results can be readout on a digital screen of a hand-held unit.

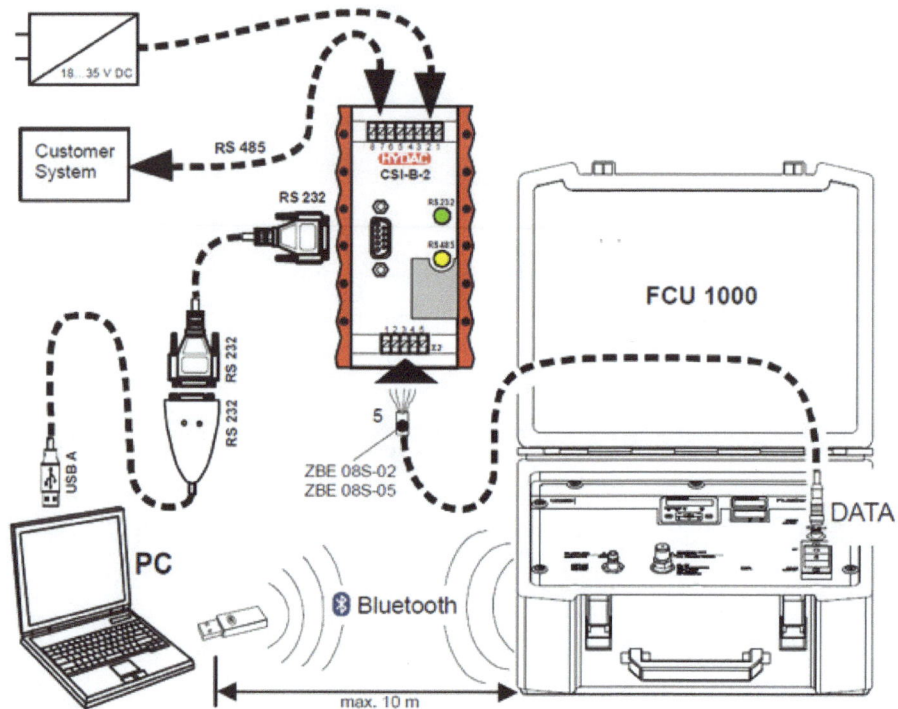

Fig. 3.47- FCU 1000 Shares the Data with a PC and Customer System (Courtesy of Hydac)

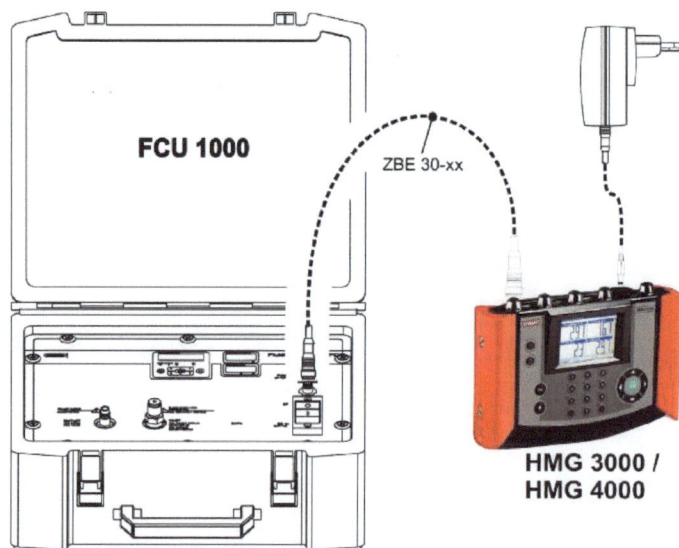

Fig. 3.48- FCU 1000 Shares the Data with a Hand-Held Unit (Courtesy of Hydac)

3.5.6.2- Particle Monitors

Particle Monitors are simpler than the APCs and provide only the cleanliness level in ISO Code. Figure 3.49 shows an example of *Inline Contamination Monitor* (*ICM*) from MPFilrti. The ICM automatically measures and displays particulate contamination, moisture and temperature levels in various hydraulic fluids. It can be used as a standalone device or controlled by external PC.

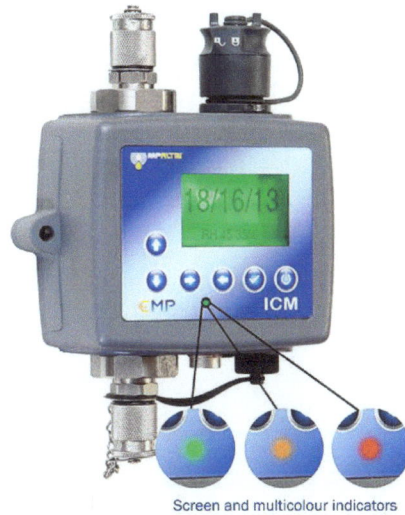

**Fig. 3.49- Inline Contamination Monitor (ICM)
(Courtesy of MPFiltri)**

The ICM is designed specifically to be mounted directly to systems, where ongoing measurement or analysis is required, and where space and costs are limited. As shown in Fig. 3.50, the ICM can be assembled on either the pressure line or the return line of a hydraulic system.

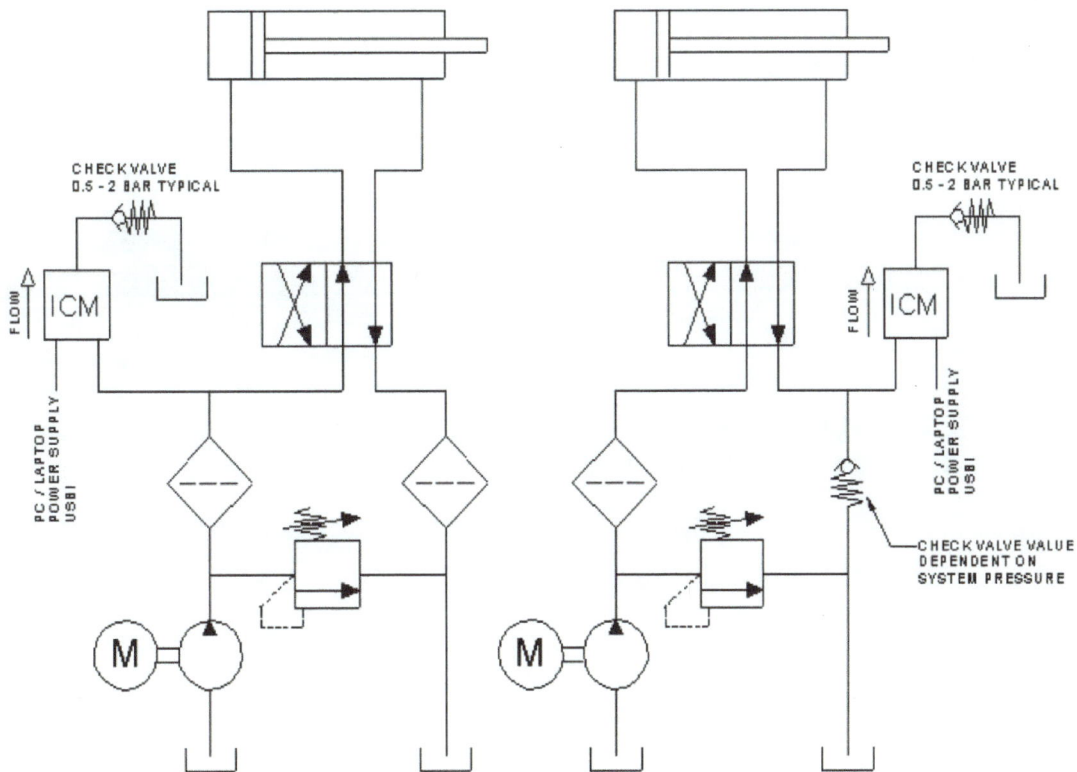

Fig. 3.50- ICM Connected to either Pressure Line or Return Line (Courtesy of MPFiltri)

3.5.6.3- Particle Classifiers

Particle Classifiers are the high end of particle analysis devices. In addition to obtaining the cleanliness level, they can capture images of the particles for wear analysis. Figure 3.51 shows various particle shadows and wear classification based on analytical data.

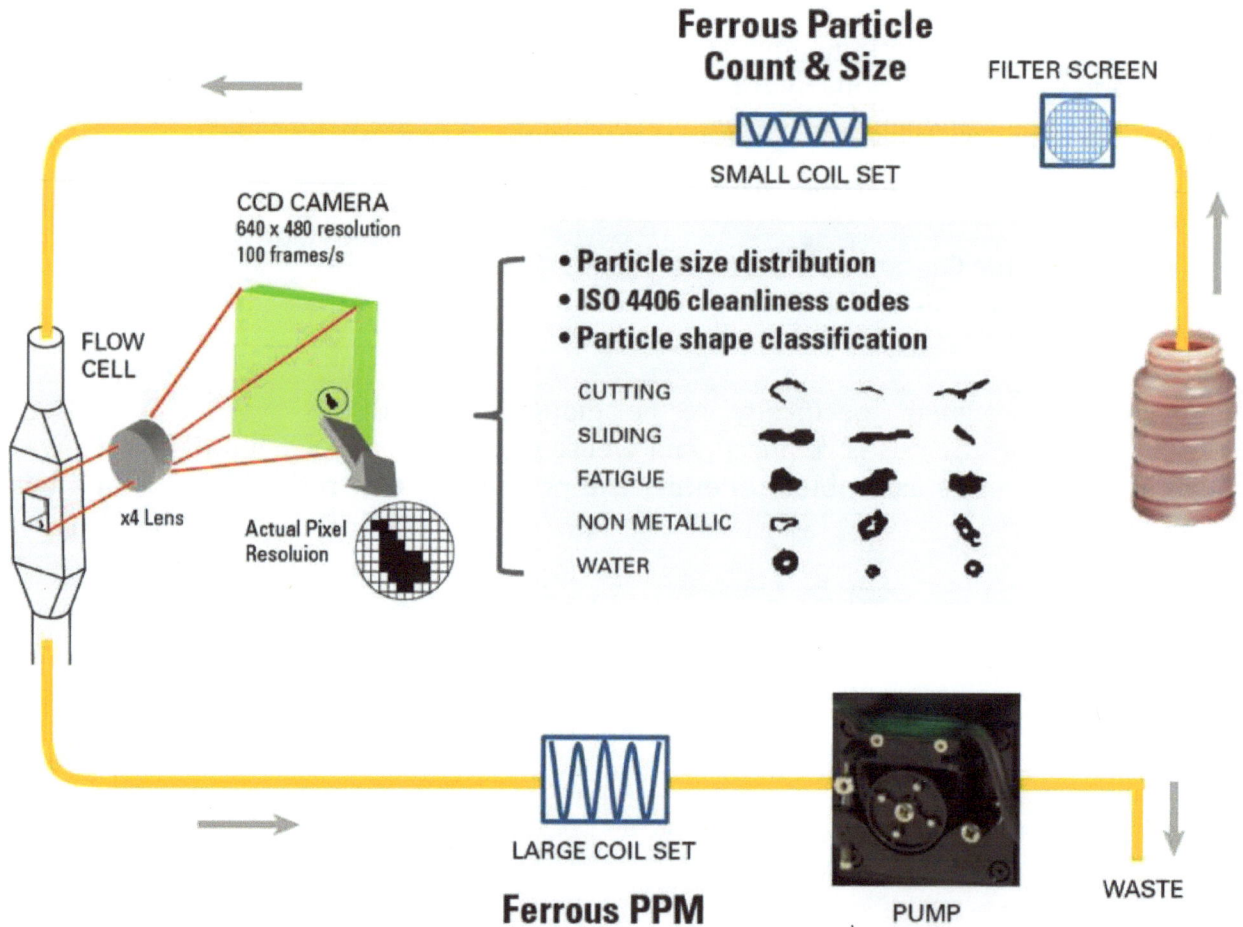

Fig. 3.51- Methods of Analyzing Wear Particles (Courtesy of Spectro Scientific)

As shown in Fig. 3.52, actual photos of various particles under a microscope confirm the wear analysis based on the shadow of wear particles.

Figure 3.53 shows an example of a particle classifier. This instrument is capable of obtaining the contamination class and analyzing the wear particles from ferro metals as well.

Fig. 3.52- Particulate Wear Analysis (Courtesy of Bosch Rexroth)

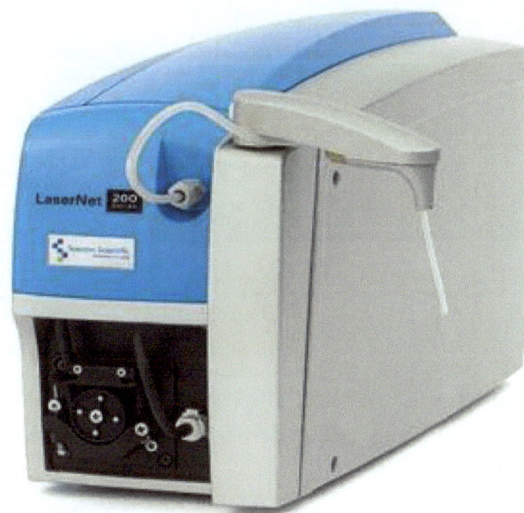

Fig. 3.53- Particulate Wear Analyzer (Courtesy of Spectro Scientific)

3.5.6.4- Calibration of Automatic Particle Counters

As shown in Fig. 3.54, to ensure the accuracy of measurements, a particle counter should be frequently calibrated according to ISO 11171. The calibration method is to pass through the counter a hydraulic fluid sample of known volume and contamination class. The results from the counters under calibration are compared to the readings from a reference counter.

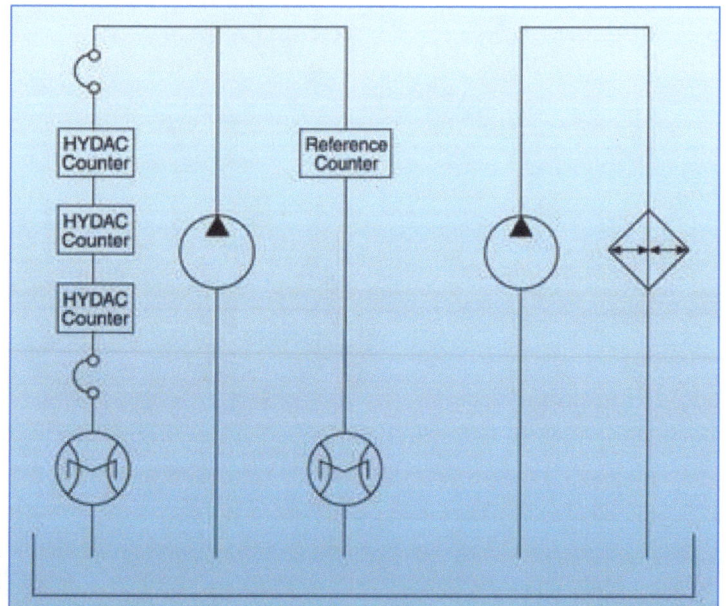

**Fig. 3.54- Particle Counter Calibration
(Courtesy of Hydac)**

A special *Test Dust* is used to form the sample for calibration as follows:

ACFTD (Air Cleaner Fine Test Dust): *ACFTD* was the first test dust from Arizona (USA) and it was made of ground silica granules ranging from 0 to 100 µm. It was marketed in the 1960s and is used until 1990 under ISO 4402. As in the early 1990s the US company which had the patent stopped producing it. Therefore, it is **no longer in use and replaced by MTD.**

MTD (ISO Medium Test Dust): ISO 12103 led to the following four different categories of test dust

- ISO 12103-A1   UFT (ISO Ultra Fine Test Dust).
- ISO 12103-A2   FTD (ISO Fine Test Dust).
- ISO 12103-A3   MTD (ISO Medium Test Dust).
- ISO 12103-A4   CTD (ISO Course Test Dust).

As shown in Fig. 3.55, starting 1997, *MTD* test dust was considered for use in ISO 11171:1999

The **ISO 11943:1999** calibration standard covers the calibration of automatic online particle counters for fluids using MTD 12103 Dust.

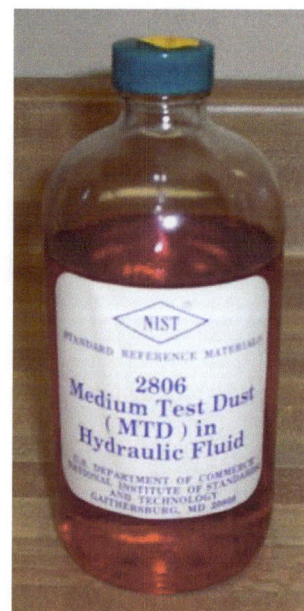

Fig. 3.55- Fluid Sample for Particle Counter Calibration

Figure 3.56 shows the amount of dirt in oil to create 10 gallons of ISO 20/18/13 oil.

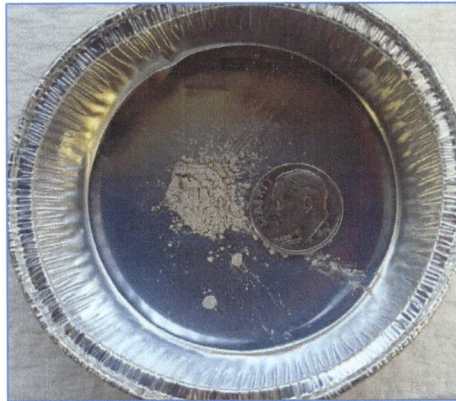

Fig. 3.56- Amount of Dirt to Create 10 Gallons of ISO 20/18/13 (Courtesy of MSOE)

3.6- Interpretation of Fluid Analysis Report

Replacing oil based on time or operation hours is expensive. Basing oil changes on condition is best. How much 'life' remains in a fluid can be seen by looking at the base oil and additive package during an oil analysis. As a rule of thumb, the additive level in used oil has to be at least 70% of the additive level of new oil (Ref. Noria Corporation). It is therefore vital to sample every incoming fluid drum/batch to establish the base line. This will also help to prevent a faulty oil batch being used. (Ref. CJC).

A good oil analysis report will answer the following key questions:
- Is the fluid suitable for further use?
- What level of contaminants are evident?
- Are the base fluid properties and additives still intact?
- Has a critical wear situation developed?
- Are seals, breathers and filters operating effectively?
- Is fluid degradation speeding up?
- Could a severe varnish problem occur soon?

At a minimum, an oil analysis should include:
- Viscosity.
- Particle counts and ISO Code 4406.
- Moisture/water content in ppm.
- Acidity level.
- Element analysis (wear and additives level).

Other results may also be important, depending on the application.

Reading a fluid analysis report can be an overwhelming and sometimes seemingly impossible task without an understanding of the fluid properties and the various types of contamination. This section introduces interpretation for several fluid analysis reports.

The first report, shown in Table 3.16, presents the spectrometric analysis for different wear metals and additives in addition to fluid viscosity, water content, and Tan #.

For the wear metals, the report indicates one of three levels for each element analyzed:
- Low (L): Low when compared with acceptable limits.
- Normal (N): Within acceptable limits.
- High (H): Above the normal level, indicating significant component wear, but is not at the critical stage.

SPECTRUM ANALYSIS		
Wear Metals And Additives	ppm By Weight	Status
Iron	120.0	H
Copper	510.0	H
Chromium	< 1.0	N
Lead	< 1.0	N
Aluminum	1.0	N
Tin	< 1.0	N
Silicon	< 1.0	N
Zinc	423.0	N
Magnesium	< 1.0	N
Calcium	540.0	H
Phosphorus	10.0	L
Barium	1.0	N
Boron	< 1.0	N
Sodium	< 1.0	N
Molybdenum	< 1.0	N
Silver	< 1.0	N
Nickel	< 1.0	N
Titanium	< 1.0	N
Maganese	< 1.0	N
Antimony	< 1.0	N
L = LOW N= NORMAL H = HIGH		

Viscosity Analysis ASTM D445	
SSU @ 100° F: 100.0	cst 40° C: 21.6

Water Analysis ASTM D1744
Water Content (ppm): 101.0

Neutralization Analysis - ASTM D974
TAN: 0.1

Remarks
1. Please check spectro-metric analysis abnormal conditions

Table 3.16- Fluid Analysis Report, Example 1 (Excerpted from Lightening Reference Handbook)

The second report, shown in Table 3.17, presents an idea of what is acceptable, cautiously acceptable, and the critical results.

Oil analysis log book			
Parameter	Baseline	Caution	Critical
Particle count ISO 4406	15/13/10 (pre-filtered)	17/15/12	19/17/15
Viscosity (cSt)	32	low 29 high 35	low 25 high 38
Acid number (AN, mg KOH/g)	0.5	1.0 - 1.5	above 1.5
Moisture (KF in ppm)	100	200 - 300	above 300
Elements (in ppm) Fe	7	10 - 15	above 15
Al	2	20 - 30	above 30
Si	5	10 - 15	above 15
Cu	5	30 - 40	above 40
P	300	220	150 and less
Zn	200	150	100 and less
Oxidation (FTIR)	1	5	above 10
Ferrous Density (PQ, WPC, DR)	-	15	above 20

Table 3.17- Example of Analysis Log Book (Courtesy of C.C. Jensen Inc.)

As shown in Fig. 3.57, some of the fluid analysis reports use color-coded sliding scale to indicate the status of the measured values. It tells at a glance whether the analysis results are in the normal range or the overall degree to which problems have been detected. This sale is defined as follows:

1. One means that at least one or more items have exceeded initial flagging points but are still considered as minor.
2. Two means a trend is developing.
3. Three indicates that simple maintenance and/or diagnostics are recommended.
4. Four denotes that failure is likely going to occur if maintenance is not performed.

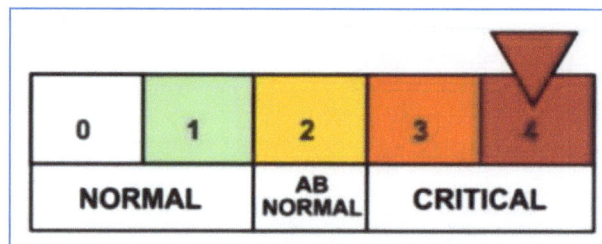

Fig. 3.57- Color-Coded Sliding Scale (Courtesy of Donaldson)

Note: every system will have its norm based o application and operating conditions. Therefoe, it is important to do trend analysis comparing results to current one and look for major changes in levels which can be used to predict impeding problems or failures so that appropriate corrective action can be taken before it occure.

Eventually it can be concluded that, particle Analysis is most often done by optical or automatic particle counting. Table 3.18 shows the features of the common contamination tests.

Method	Units	Advantages	Limitations
Patch Test	visual comparison	Fast and qualitative	Not quantitative
Gravimetric Analysis	mg/Liter	Identifies the total amount of contamination	Can't identify the particle size
Microscopic Particle Counting	number/ml	Provides accurate size and number	Sample preparation and time
Electronic Particle Counting	number/ml	Fast and repeatable results	Counts water as particles

Table 3.18- Features of Contamination Tests

Chapter 4

Fluidic Contamination

Objectives

This chapter covers the sources of hydraulic fluids fluidic contamination. For each source, the chapter explains how the system performance will be affected and possible recommendations to minimize such consequences.

Brief Contents

4.1- Sources of Fluidic Contamination in Hydraulic Fluids
4.2- Forms of Water Contamination in Hydraulic Fluids
4.3- Standard Test Methods for Measuring Water Content in Hydraulic Fluids
4.4- Effects of Fluidic Contaminants
4.5- Best Practices to Minimize Fluidic Contamination

Chapter 4 – Fluidic Contamination

4.1- Sources of Fluidic Contamination in Hydraulic Fluids

As shown in Fig. 4.1, for example, water can enter the system as free water because of rain (1), defective cylinder rod wipers (2), external cleaning by water jet (3), making up the reservoir with a contaminated fluid (4), condensation of water vapor from the atmosphere through a vented reservoir when a hot system cool down at night (5), defective oil-water heat exchangers (6).

Other *Fluidic Contaminations* may result from:
- Mixing of incompatible hydraulic fluids.
- Residual fluids from flushing or pickling process.
- Other fluids used in the vicinity of the hydraulic system such as paints, cleaning solvents, metal working fluids and coolants.

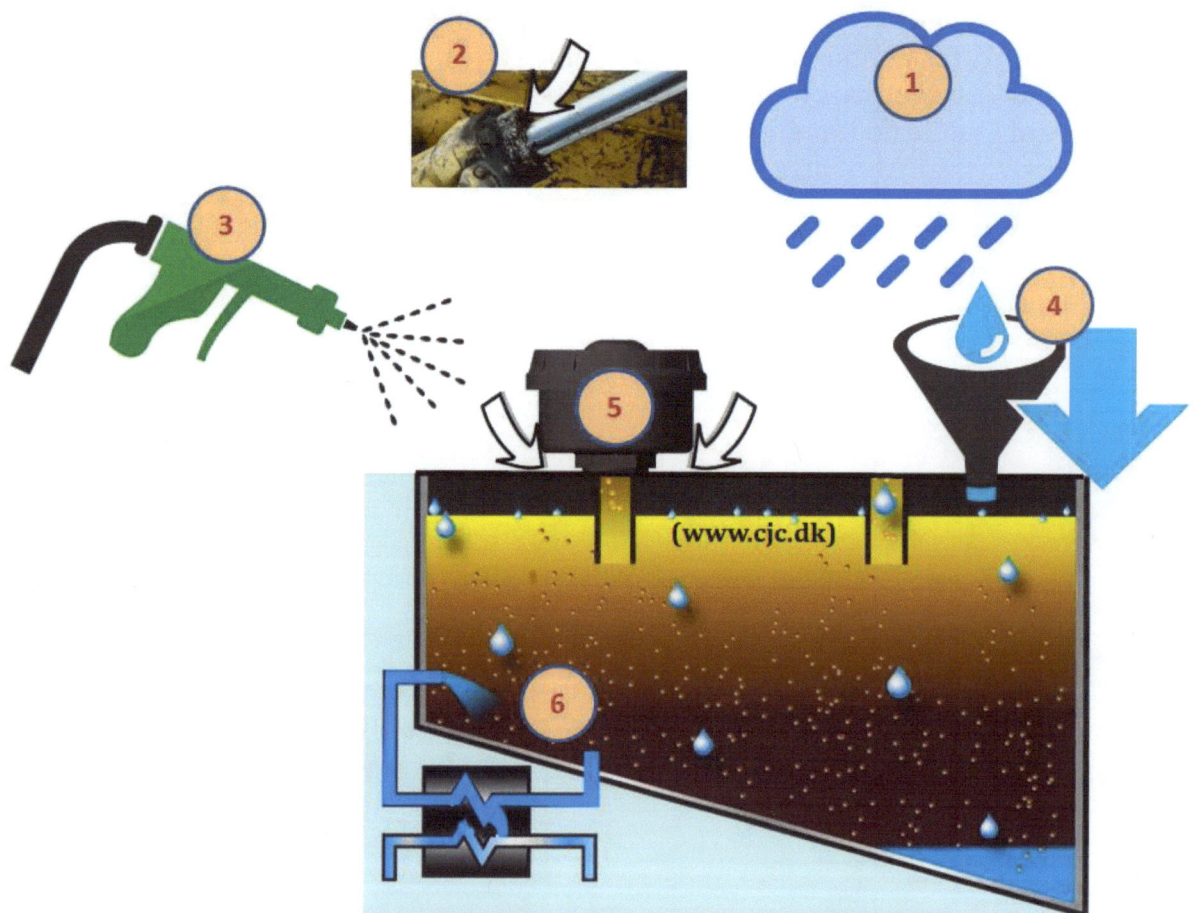

(www.cjc.dk)

Fig. 4.1- Sources of Contamination by Free Water

4.2- Forms of Water Contamination in Hydraulic Fluids

Different hydraulic fluid types have different water absorptive capacity, some fluid types can dissolve more water by integrating it in the molecular structure, others less. The absorptive capacity of a hydraulic fluid depends on the fluid temperature, its molecular structure, and the additive packages of the fluid. Water content in a hydraulic fluid is measured as percentage or *part per million* (ppm). Moving the decimal point four spaces to the right converts percent to ppm. For example, 1% water = 10,000 ppm. As shown in Fig. 4.2, the saturation level of a hydraulic fluid is the *Maximum Water Content* that can dissolve within the molecular structure of the hydraulic fluid at an identified *Critical Temperature*. Table 4.1 shows the saturation level for different hydraulic fluids at 20 °C (68 °F). All numbers in this table are only rough guides, which strongly differ in dependency with the used base oil, additive packages and the application of the hydraulic system.

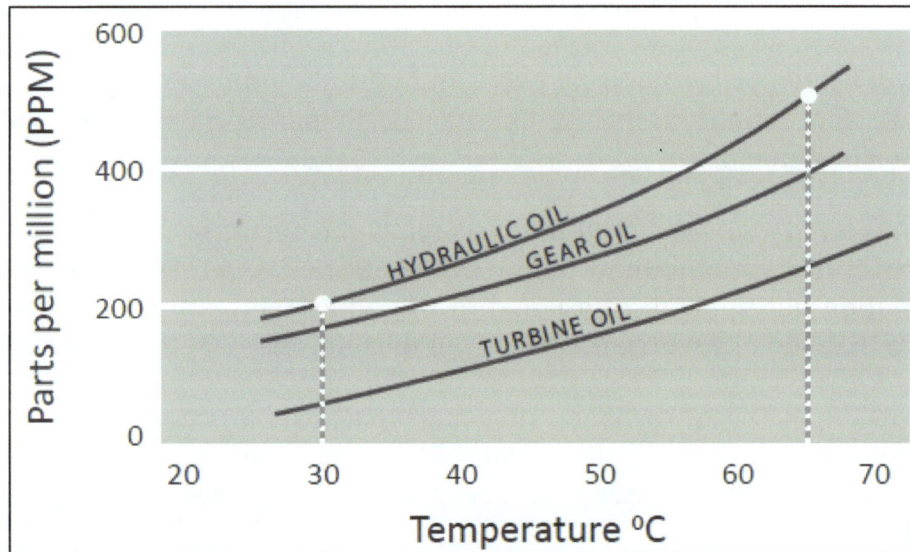

Fig. 4.2- Saturation Level of Different Hydraulic Fluids (Courtesy of C.C. Jensen Inc.)

Fluid Type	Critical Water Content (ppm)
Mineral oil (HLP)	200 - 500
Biodegradable oil (HEES)	700
Fire resistant fluid (HFC=Water in Glycol Emulsion)	> 4000

Table 4.1- Saturation Level of Different Hydraulic Fluids at 20 °C (68 °F)

Figure 4.3 shows the appearance of a hydraulic fluid with various ppm of water contamination.

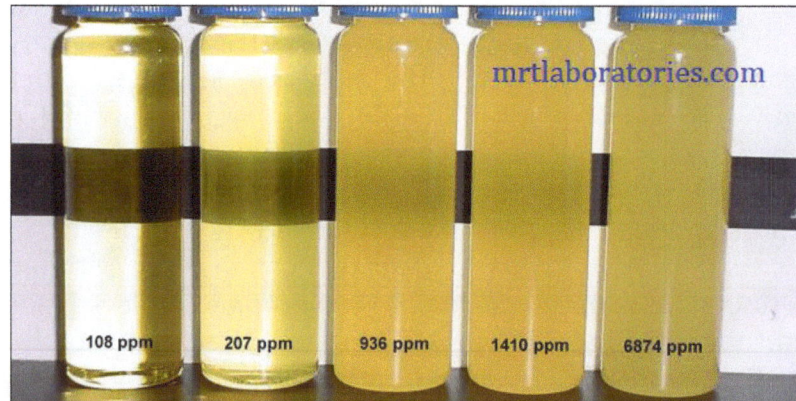

Fig. 4.3- Various Levels of Contamination by Water in Oil

As shown in Fig. 4.4, water typically exists in hydraulic fluids as dissolved or free water. The following are the definitions based on the ISO Standard 5598 "fluid Power Systems and Components Vocabulary".

Dissolved Water: *Dissolved* (*Emulsified*) water is the result of water droplets dispersed at a molecular level in hydraulic fluid below the saturation level. Dissolved water is not visible when in solution but appears as a cloud in the oil as temperature is lowered to the critical temperature that begins to force the water out of solution.

Free Water: When the absorbance of water reaches the saturation point, residual water separates from the fluid forming *Free Water*. This water will usually settle to the bottom of the reservoir and should be removed by periodic draining. Free water is more harmful than dissolved water.

New Oil   Free Water   Emulsified
                 In Oil   Water In Oil

Fig. 4.4- Forms of Water in a Hydraulic Fluid

4.3- Standard Test Methods for Measuring Water Content in Hydraulic Fluids

There are several methods to determine water content. These can be differentiated based on whether the content of water is dissolved or in the free form.

4.3.1- Karl-Fischer Method (**ISO760 - ASTM D6304 – DIN 51777**)

The *Karl-Fischer* method is a well-established technique, used to determine the total water content of oils. Measuring the total water content means, there is no possibility to distinguish between the dissolved and free water. Because dissolved water is less harmful than free water, without knowledge of the saturation point (respective limit of solubility) of the fluid in use, it can be difficult to interpret the results from the Karl Fischer method when the concentration is less 500 ppm. As shown in Fig. 4.5, Karl Fischer is based on titration using electrochemical device consisting of two components, the Karl Fischer titrator and an integrated oven.

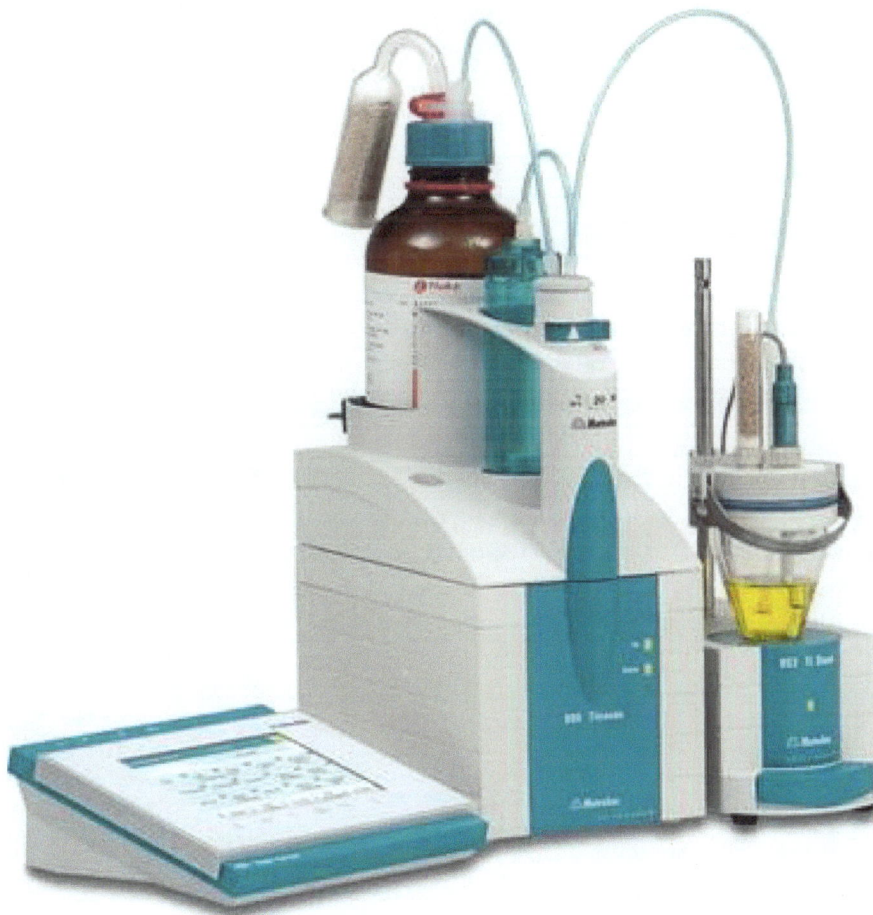

Fig. 4.5- KF Titrator (www.metrohm.com)

4.3.2- Fourier Transform Infrared **(FTIR) (ASTM E2412)**

IR analysis is based upon the same principle as a microwave oven. Microwave ovens transmit radiation through food. Water molecules in the food absorb the particular "micro" wavelengths transmitted by the oven. When water absorbs these specific wavelengths of energy it causes the food to heat up. Carbohydrates, fat, protein, plastic, paper and glass do not absorb microwave radiation.

The *Fourier Transform Infrared* (*FTIR*) is a chemical-free measurement method. As shown in Fig. 4.6, a typical FTIR *spectrometer* device consists of a radiative source of infrared (IR) and a detector. In this technique, the sample that needs to be analyzed is positioned between the detector and the radiative source. The IR light beam is allowed to travel via a sample. The detector is used to collect the transmitted light.

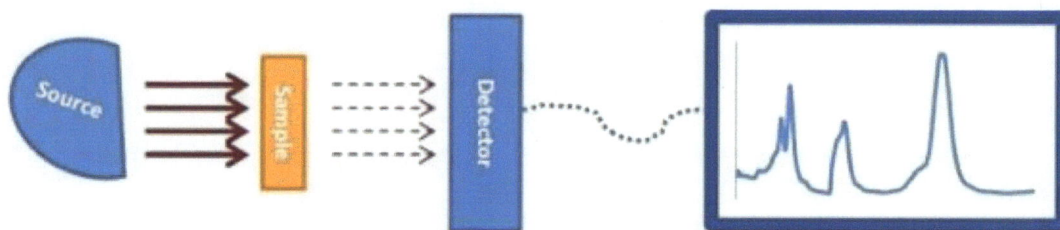

Fig. 4.6- Schematic of Typical Spectrometer (Courtesy of Spectro Scientific)

The FTIR device compares the spectrum of the contaminated oil sample versus fresh oil sample. By calculating the area between the two spectrums along the wave number range, it is possible to determine the water content.

Water contamination, additive depletion and oxidation debris are absent from the oil if the spectra of the new and used fluids are identical. Interpretation of changes in the spectra is done with an understanding of the specific chemistry involved with fluid degradation and oxidation. As shown in Fig. 4.7, the IR light absorbed by pure water can be identified by a peak in the IR spectrum at about wavelength 3400cm^{-1}. Figure 4.8 shows FTIR results for various types of fluidic contamination. **ASTM D7214** provide instructions of the test.

4.3.3- Centrifuge

This method is applicable for water contents greater than 0.1% (1000 ppm). The spinning of the sample in the centrifuge causes the higher density water to collect at the bottom of the centrifuge tube. The volume of the water is compared to the total volume of sample placed in the centrifuge tube.

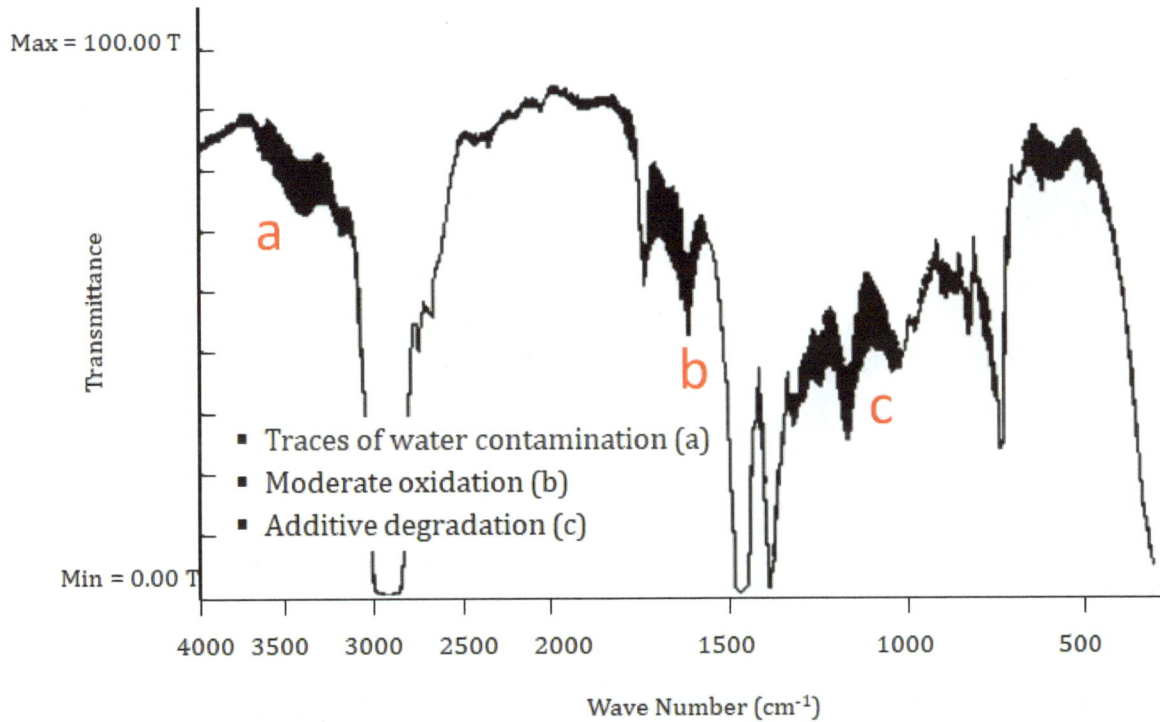

Max = 100.00 T

Transmittance

- Traces of water contamination (a)
- Moderate oxidation (b)
- Additive degradation (c)

Min = 0.00 T

Wave Number (cm⁻¹)

Fig. 4.7- Measurement of Water using FTIR method (Courtesy of MSOE)

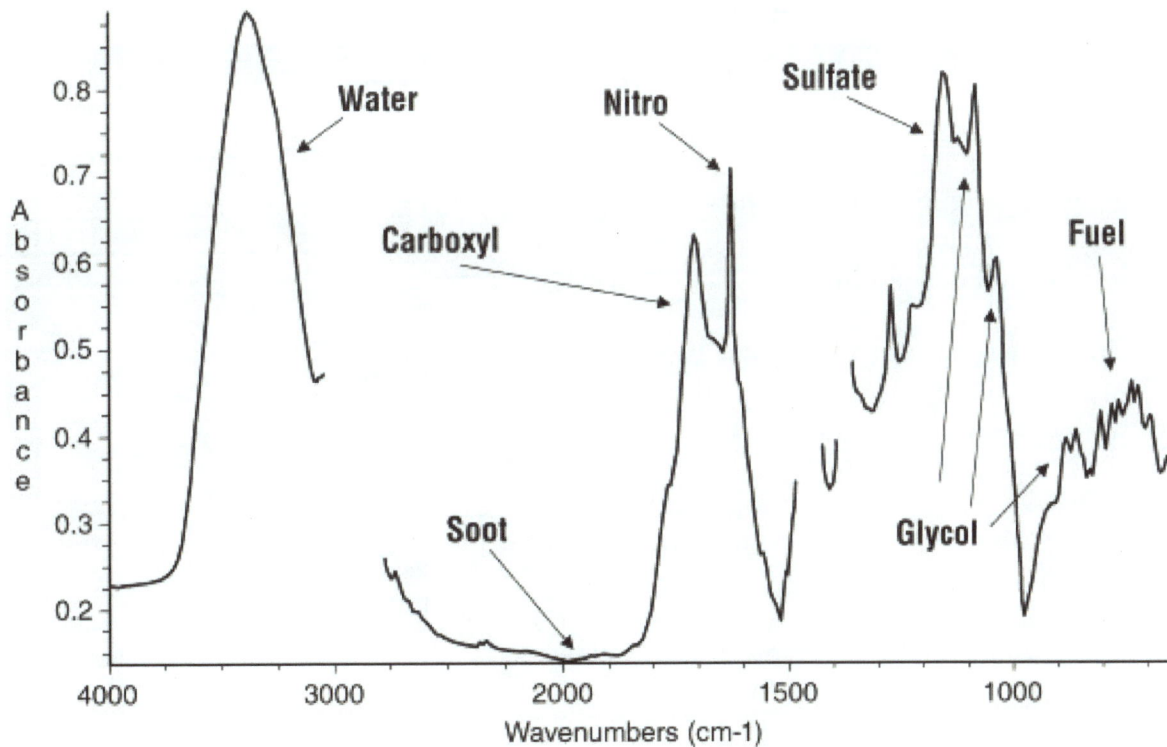

Fig. 4.8- FTIR Analysis Results for Various Fluidic Contamination (Courtesy of Spectro Scientific)

4.3.4- Crackle Test

The "Crackle" test can be done in routine analysis programs and onsite to determine if an oil sample is contaminated with water. This method depends on the fact that water boils at a lower temperature than oil and when a contaminated oil sample is heated, water violently changes phase from liquid to vapor creating a popping noise. This test is just a qualitative test to provide a yes or no answer to the question of water contamination. It does not provide an accurate percentage of water content.

In a simple procedure, as it has been reported by Noria Corporation, maintain surface temperature of a hot plate at 135°C (300°F). Using a clean dropper, place a drop of oil on the hot plate.

As shown in Fig. 4.9, interpretation of the test results are as follows:

- If no crackling or vapor bubbles are produced after a few seconds, no free or emulsified water is present.
- If very small bubbles (0.5 mm) are produced but disappear quickly, approximately 0.05 percent to 0.1 percent water is present.
- If bubbles approximately 2 mm are produced, gather to center of oil spot, enlarge to about 4 mm, then disappear, approximately 0.1 percent to 0.2 percent water is present.
- For moisture levels above 0.2 percent, bubbles may start out about 2 to 3 mm then grow to 4 mm, with the process repeating once or twice. For even higher moisture levels, violent bubbling and audible crackling may result.
- Hot plate temperatures above 135°C (300°F), induce rapid scintillation that may be undetectable.
- Different base stocks, viscosities and additives will exhibit varying results. Certain synthetics, such as esters, may not produce scintillation.

For safety, wear protective eyewear, long sleeves, and the test should be performed in a well-ventilated area.

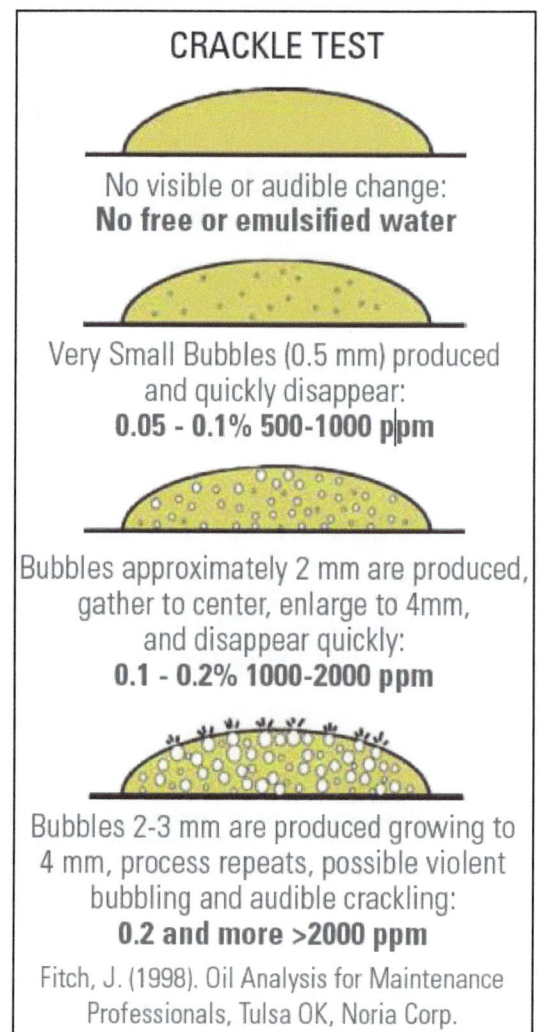

CRACKLE TEST

No visible or audible change:
No free or emulsified water

Very Small Bubbles (0.5 mm) produced and quickly disappear:
0.05 - 0.1% 500-1000 ppm

Bubbles approximately 2 mm are produced, gather to center, enlarge to 4mm, and disappear quickly:
0.1 - 0.2% 1000-2000 ppm

Bubbles 2-3 mm are produced growing to 4 mm, process repeats, possible violent bubbling and audible crackling:
0.2 and more >2000 ppm

Fitch, J. (1998). Oil Analysis for Maintenance Professionals, Tulsa OK, Noria Corp.

Fig. 4.9- Crackle Test (Courtesy of Spectro Scientific)

4.4- Effects of Fluidic Contaminants

The following matter of facts explores the effect of fluidic contamination:

- Oil is considered contaminated when the water content exceeds the saturation level.
- However, water more than 0.5% by volume in a hydrocarbon-based fluid accelerates degradation.
- As shown in Fig. 4.10, The degree of damage depends on form of water, amount of water content, and for how long.

- Form of the water.
- % of water content.
- For how long.

Fig. 4.10- Factors Affect the Degree of Damage due to Fluidic Contamination

- As shown in Fig. 4.11, contamination by water has the same set of effects like gaseous contamination. Despite that, unlike gaseous contamination, system damage and loss of performance due to fluidic contaminants occurs over an extended period.

The following sections show damages due to fluidic contamination.

Fluid Appearance (1): Figure 4.12 shows (on the left) oil with small amount of free water in the bottom and (on the right) after shacking by hand for 30 seconds. As shown in the figure, oil that is contaminated by water in an emulsified state has cloudy/milky appearance and smell of bacteria.

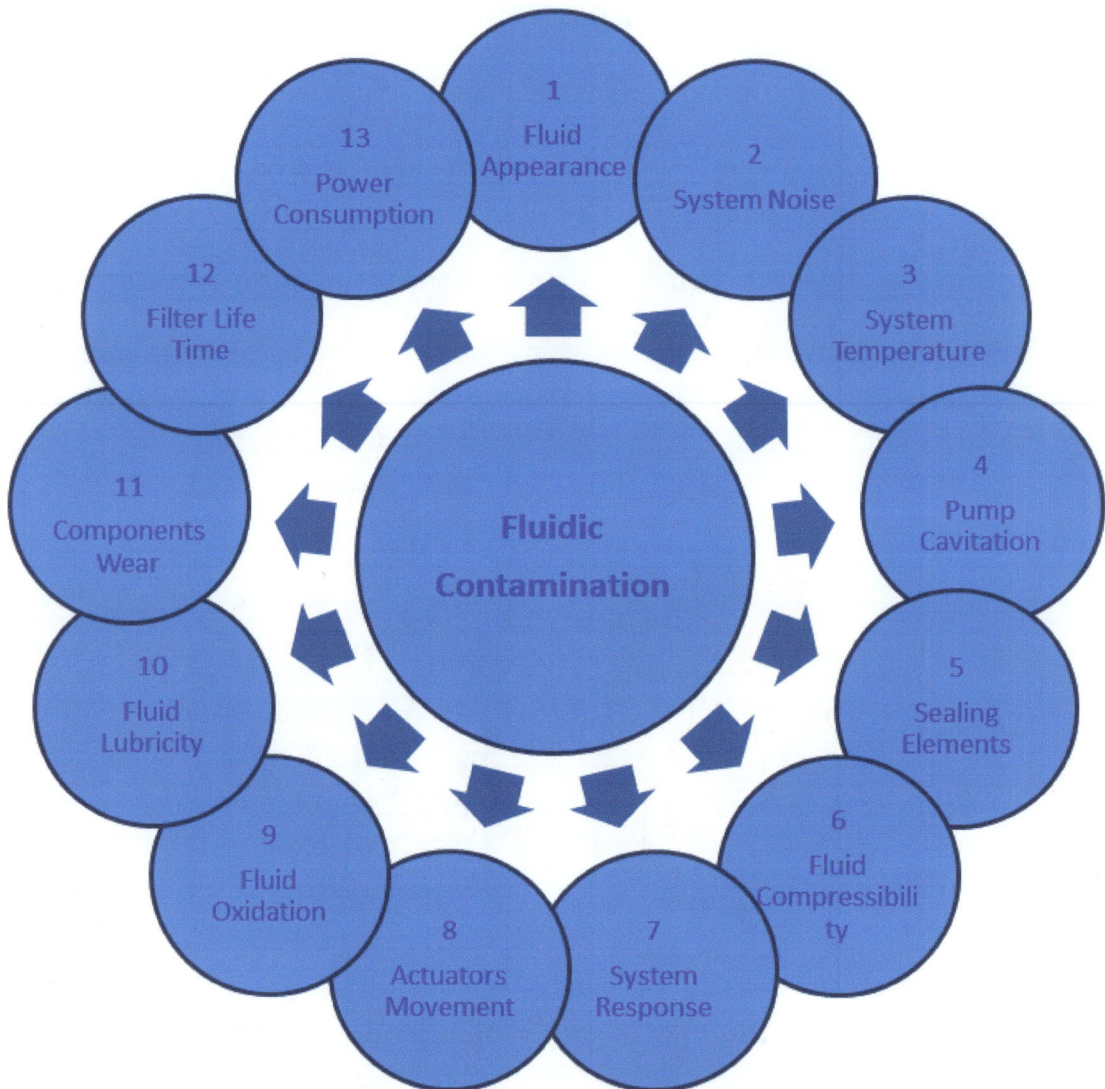

Fig. 4.11- Effects of Fluidic Contamination

Fig. 4.12- Milky/Cloudy Appearance of Hydraulic Fluid Contaminated by Water

System Noise (2): Increased system noise and vibration due to lack of lubrication.

System Temperature (3): water boils at a lower temperature than oil.

Pump Cavitation (4): Water evaporation increases possibility of pump cavitation.

Sealing Elements (5): Water content affects sealing performance of sealing elements.

Fluid Compressibility (6): Contaminated oil has higher equivalent bulk modulus because water has higher bulk modulus than oil.

System Response (7): System response may be affected due to oil degradation.

Actuators Movement (8): Stick-Slip movements of actuators due to oil degradation.

Fluid Oxidation (9): Increased rate of oil oxidation due to water contamination.

Fluid Lubricity (10): Oil loses its ability to lubricate because of water content.

Components Wear (11): Increased wear rate and failure due to lack of lubrication, rust, and corrosion. Reduced components service life. As shown in Fig. 4.13, service life of bearing surfaces reduced below 50% for 0.05% water content 500 parts per million (ppm) contamination by water.

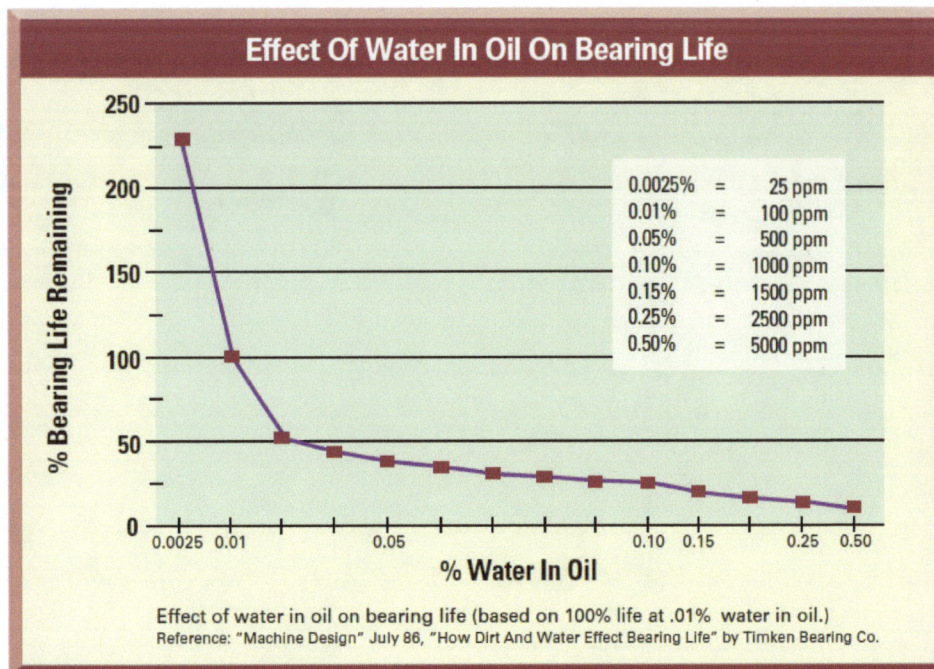

Effect Of Water In Oil On Bearing Life

0.0025%	=	25 ppm
0.01%	=	100 ppm
0.05%	=	500 ppm
0.10%	=	1000 ppm
0.15%	=	1500 ppm
0.25%	=	2500 ppm
0.50%	=	5000 ppm

Effect of water in oil on bearing life (based on 100% life at .01% water in oil.)
Reference: "Machine Design" July 86, "How Dirt And Water Effect Bearing Life" by Timken Bearing Co.

Fig. 4.13- Effect of Water Content on Bearing Life (Courtesy of Parker)

Filter Service Life (12): Reduced filter service life because of sludge formation.

Power Consumption (13): Higher power consumption due to loss of system performance.

Contamination by water has the following additional effects:

Water Icing: Water icing in cold weather results in forming hard crystals.

Bacterial Contamination: In high water-based and biodegradable fluids, the water can tend to support biological growth and generate organic contamination and microbes. It will be seen as accumulation of green microbes sticking on the inside surfaces of the reservoir.

Additives: Decrease of additive performance and increase of additive depletion.

Oil Degradation: Mixing of incompatible hydraulic fluids can result in fluid contamination. For example:
- Mixing Phosphate Ester or brake fluid with Mineral oil creates acids and sludge. As a result, seals will swell, filters will become clogged, critical orifices plugged, and spool valves become sluggish.
- Mixing volatiles (such as diesel fuel, gasoline or solvents) with hydraulic oil reduce oil's viscosity. Consequently, damage may occur to the system due to lack of lubrication.
- Mixing water with certain automatic transmission fluids can cause sludge and small hard crystalline particles to form.

4.5- Best Practices to Minimize Fluidic Contamination

4.5.1- Preventive Practices to Minimize Fluidic Contamination

As previously stated, preventing practices are much more cost effective than removing the contamination. The following practices should be seriously considered to prevent water ingression to the oil.

New Hydraulic Fluid:
- DO NOT mix oils without previously investigating compatibility.
- DO NOT use oils additives that are not necessary for the application.
- Use fluid with high hydrolytic stability to minimize fluid chemical degradation when contaminated by water.
- Compare the oil in operation to fresh oil regularly in order to discover any sudden appearance of water, air or other contaminants.

In Service Hydraulic Fluid: Continuous removal of water out of a hydraulic fluid can improve hydraulic system reliability and is considered as a *Life Extension Method* (*LEM*). Table 4.2 has been developed to give an estimate of how a life time of a machine can be extended by controlling the amount of water content in a hydraulic fluid. For example, if water content in a hydraulic fluid is reduced from 2,500 ppm to 156 ppm, machine life is extended by a factor of 5.

Current moisture level, ppm	\multicolumn{9}{c}{LEM - Moisture Level — Life Extension Factor}
	2	3	4	5	6	7	8	9	10
50,000	12,500	6,500	4,500	3,125	2,500	2,000	1,500	1,000	782
25,000	6,250	3,250	2,250	1,563	1,250	1,000	750	500	391
10,000	2,500	1,300	900	625	500	400	300	200	156
5,000	1,250	650	450	313	250	200	150	100	78
2,500	625	325	225	156	125	100	75	50	39
1,000	250	130	90	63	50	40	30	20	16
500	125	65	45	31	25	20	15	10	8
260	63	33	23	16	13	10	8	5	4
100	25	13	9	6	5	4	3	2	2

1% water = 10,000 ppm. | Estimated life extension for mechanical systems utilizing mineral-based fluids

Example: By reducing average fluid moisture levels from 2,500 ppm to 156 ppm, machine life (MTBF) is extended by a factor of 5

Table 4.2- Life Extension of a Machine (Courtesy of C.C. Jensen Inc.)

After Flushing: After flushing and pickling process, system must be dried by blowing clean hot dry air into the transmission line.

Water Content Sensors: *Relative Humidity* (RH) sensors are used to provide early warning about the water content in the hydraulic fluids before the situation becomes critical. Relative humidity sensors can read water content and transmits the result continuously to the user's control system as a key component in the predictive maintenance of plant and machinery. Typically, when the RH exceeds 80%, the fluid condition is compromised a Karl Fischer test should be performed.

Water content sensors are available in different styles. Figure 4.14 shows a typical low-cost, in-line (left) monitoring solution for measuring dissolved water content in hydraulic, lubricating and insulating fluids. The other style is Offline sensor (right) for checking water contents level during routine maintenance.

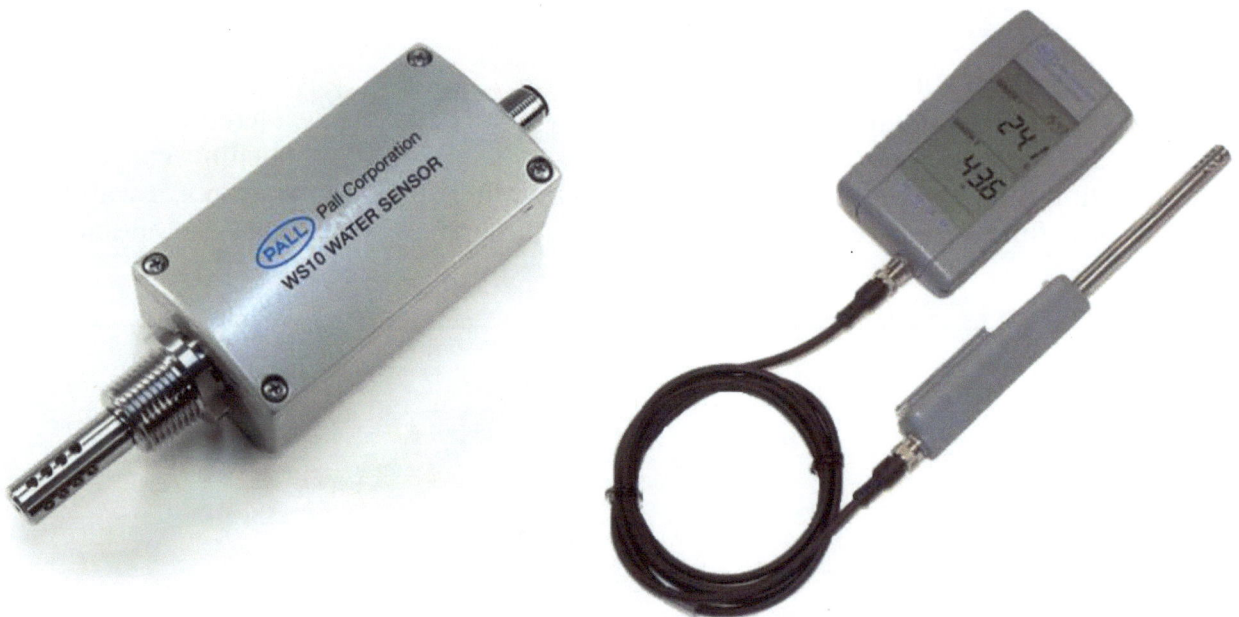

Fig. 4.14- Water Content Sensors (Courtesy of Pall Corporation)

Operational Actions:
- Avoid high-pressure sprays around seals, shafts, fill ports and breathers when washing machines.
- Maintain seals in steam and heating/cooling water systems.
- Chanel water to divert water flow away from reservoir breathers and top covers.
- Use and maintain high-quality rod wiper seals for hydraulic cylinders.
- Prevent water from entering new oil by storing drums indoors.
- Periodically drain water from low points in system.

Closed and Pre-Pressurized Reservoir: Closed reservoirs may be a solution in highly humid environments such as offshore and marine applications.

Desiccant Filter Breather: Install Desiccant Filter Breather on the reservoir to absorb the moisture from the air entering the reservoir. Such filters are available in different styles but all work almost same way. Figure 4.15 shows a typical example from industry.

1. Secondary Filter Element.
2. Visual Indicator
3. Water Vapor Adsorbent
4. Rugged Housing
5. Integrated Stand pipe
6. Foam Pad
7. Quad Check-Valves.
8. Filter Element removes airborne contamination to 0.3-micron absolute and stops free water.

Fig. 4.15- Desiccant Filter Breather (www.descase.com)

4.5.2- Curative Practices to Remove Fluidic Contamination

Normal filtration will not remove water. If hydraulic fluid is contaminated by water, serious and immediate action for water removal is needed. Such action may vary from a simple less expensive to high cost advanced techniques. The best choice of a technique for water removal depends on the volume of the contaminated oil, the form of water content whether dissolved or free water, and the level of contamination by water. The following sections provide most common methods of water removal.

For example, small amount of free water content can be removed by using absorptive breathers or active venting systems.

For large quantities of water, vacuum dehydration, coalescence, and centrifuges are appropriate techniques for its removal. However, as each of these techniques operates on different principles, they have various levels of water removal effectiveness.

Table 4.3 below provides comparative information on these techniques and their relative effectiveness. Care should be taken to apply the best technique to a given situation and its demands for water removal.

	Usage	Prevents Humidity Ingression	Removes Dissolved Water	Removes Free Water	Removes Large Quantities of Free Water	Limit of Water Removal
Adsorptive Passive Breather	prevention	Y				n/a
Active Venting System	prevention and removal	Y	Y	Y		down to <10% saturation
Water Absorbing Cartridge Filter	removal			Y		only to 100% saturation
Centrifuge	removal			Y	Y	only to 100% saturation
Coalescer	removal			Y	Y	only to 100% saturation
Vacuum Dehydrator	removal		Y	Y	Y	down to ~20% saturation

Table 4.3 - Water Prevention and Removal Techniques (Courtesy of Donaldson)

4.5.2.1- Water Removal Techniques for Small Water Contents

Fluid Replacement: As a matter of fact, water can't be removed 100% out of the oil. Therefore, if the quantity of the contaminated oil is small (< 500 gallon =2000 liter), it is recommended to replace it and flush the system.

Periodic Disposal of Free Water by Gravity: If the quantity of the contaminated oil is large, keep the machine at rest for minimum of 2 hours then drain the settled water at the bottom of the tank. It might take a lot longer depending on fluid and amount of water. Oil is heated in an open tank to help evaporate the residual water.

Active Venting System: The method of *Active Venting System* is also known as *Head Space Dehumidification*. This method involves circulating dehumidifying air from the reservoir head space. Water in the oil migrates to the dry air in the head space and is eventually removed by the dehumidifier. Small air flow [approximately 4 standard cubic feet per minute (SCFM)] of desiccant hot dry air (with low dew point) is required. The side effect is air in oil increases.

Adsorption: Free water *adsorption* means accumulating the free water on the surface of some adsorbent material such as silica. That can happen by circulating the contaminated oil through desiccant filters that adsorb water. Filter can be replaced quickly and easily, but this method is expensive and not effective for large quantity of oil. Alternatively, a special water adsorbent is placed in the reservoir. Figure 4.16 shows a typical 12-inch-long water adsorbent is contained in a stainless-steel housing tube that is stored in the tank retrievable via a stainless-steel tether. This method is applicable for small-sized reservoirs. Side effect of that is some of the absorbing material may migrate to the fluid and fine filtration is required to remove it.

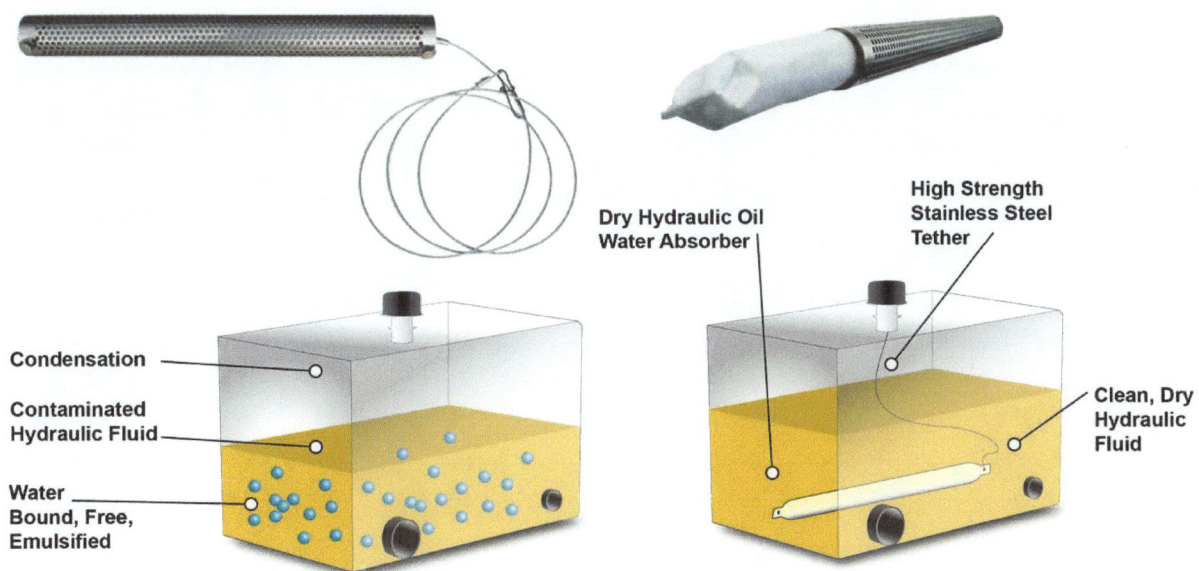

Fig. 4.16- Desiccant Filter Element (www.centerlinedistribution.com)

Absorption (Coalescence): Free water *absorption* means trapping and accumulating water particles by passing the contaminated oil through a special water filter separator.

As shown in Fig. 4.17, A *CJC Filter Separator* is installed offline. The oil is pumped from the lowest point of the oil reservoir and enters the CJC filter separator. The water aggregates in droplets sinking down in the bottom of the filter separator then automatically removed through a water discharge system.

The CJC filter separator removes particles, oxidation and water in one and the same operation. The clean and dry oil is returned to the system and the contamination is removed continuously.

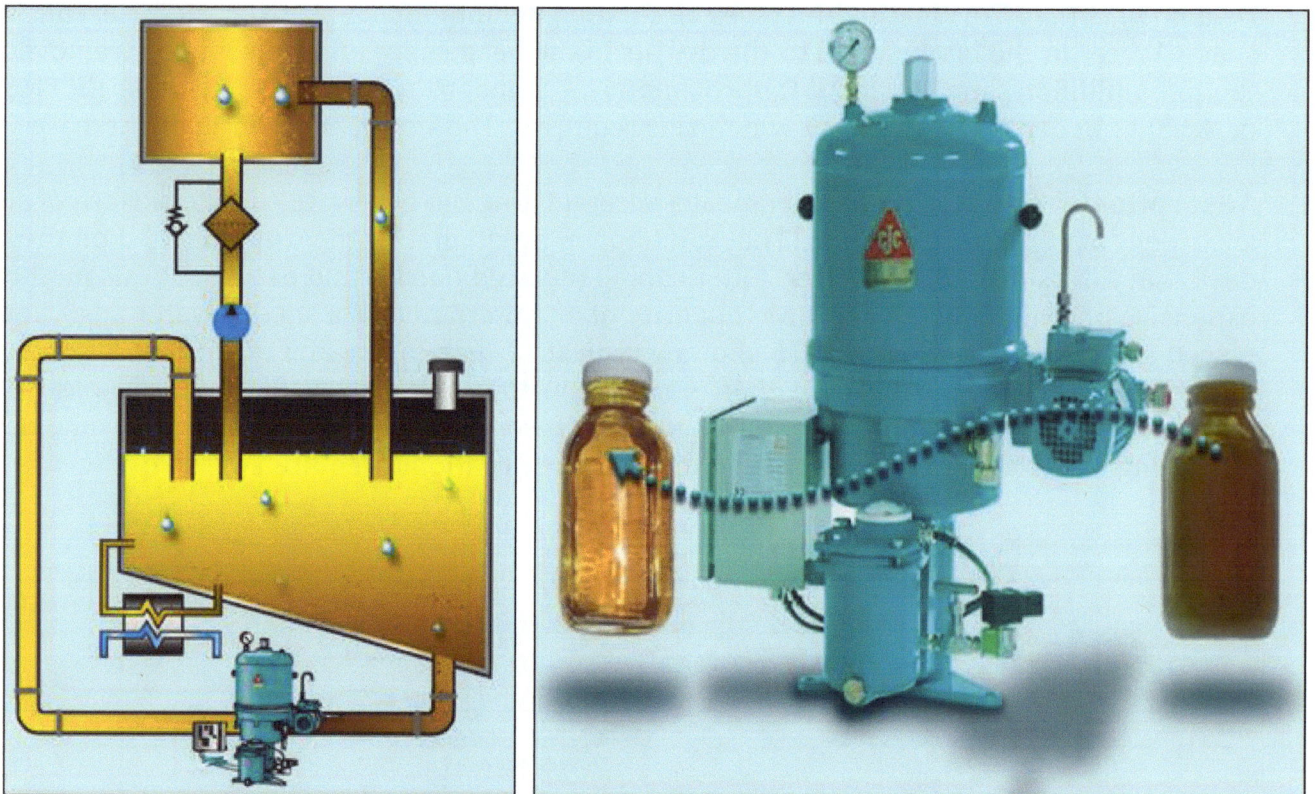

Fig. 4.17- CJC Coalescence Filter Separator (Courtesy of C.C. Jensen Inc.)

Figure 4.18 shows a cutaway of the CJC water separator with the filter element inside. Such a filter element must be replaced based on manufacturer recommendations or at least once a year or when pressure drop exceeds 2 bar.

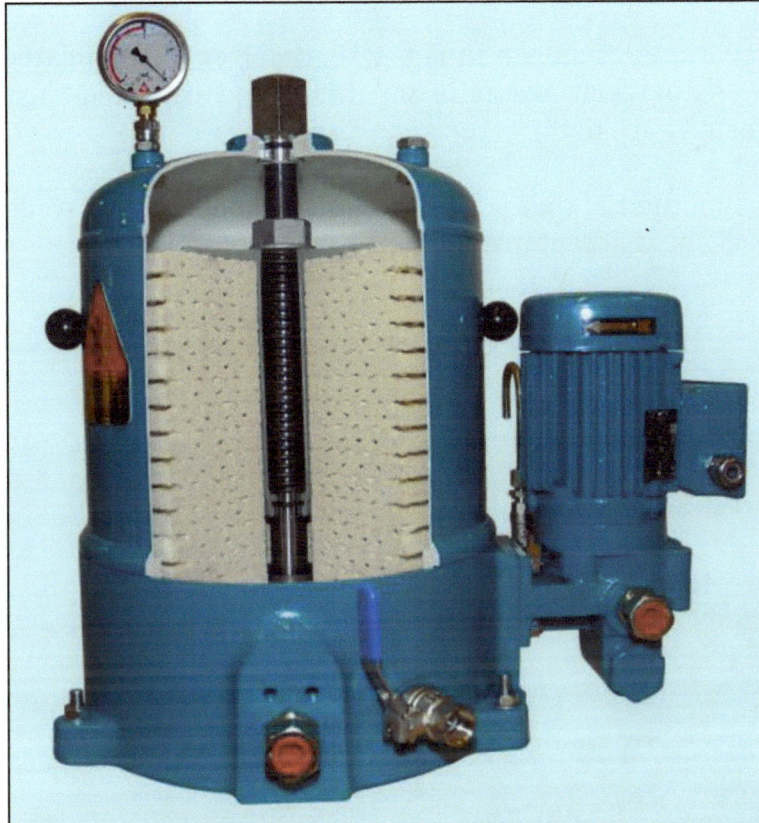

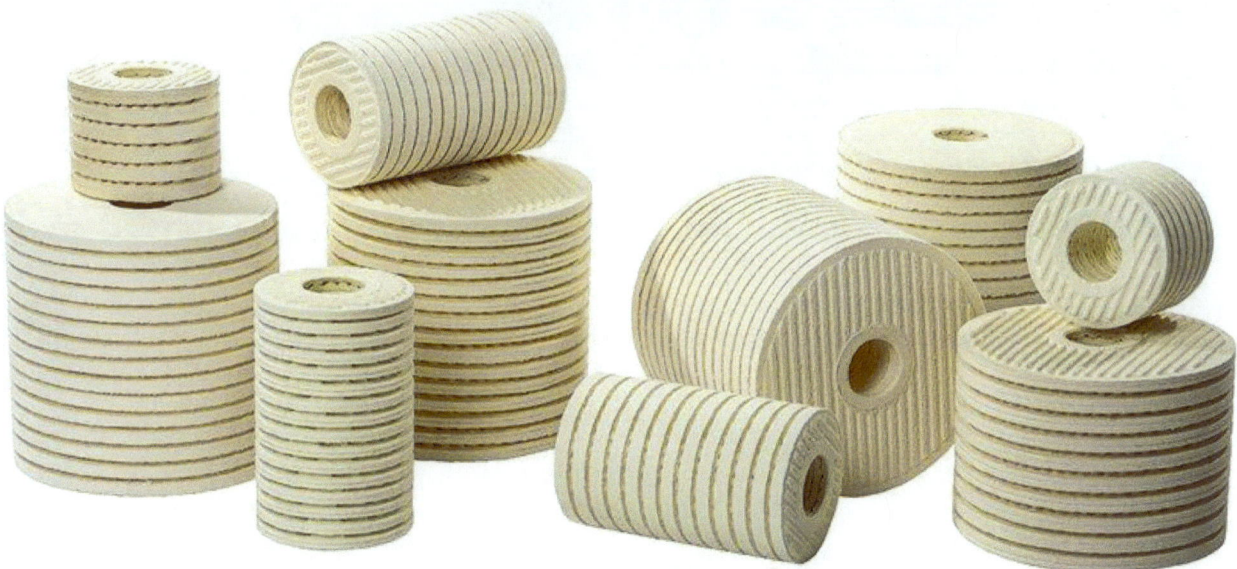

Fig. 4.18- CJC Filter Separator and Filter Elements (Courtesy of C.C. Jensen Inc.)

4.5.2.2- Water Removal Techniques for Large Water Contents

Centrifugal Water Separators: This technique is used to separate free water from contaminated oil. The operating principle of *Centrifugal Water Separators* is based on the fact water has higher mass density than hydraulic fluids.

The *centrifuge* separator, as shown in Fig. 4.19, receives contaminated fluid and subjects it to centrifugal force. As a result, water is separated and collected through the wall while the cleaned oil is directed back to the reservoir.

Side effect of this method is that some oil additive may be removed. Oil should be tested to verify additive package is acceptable for continued use.

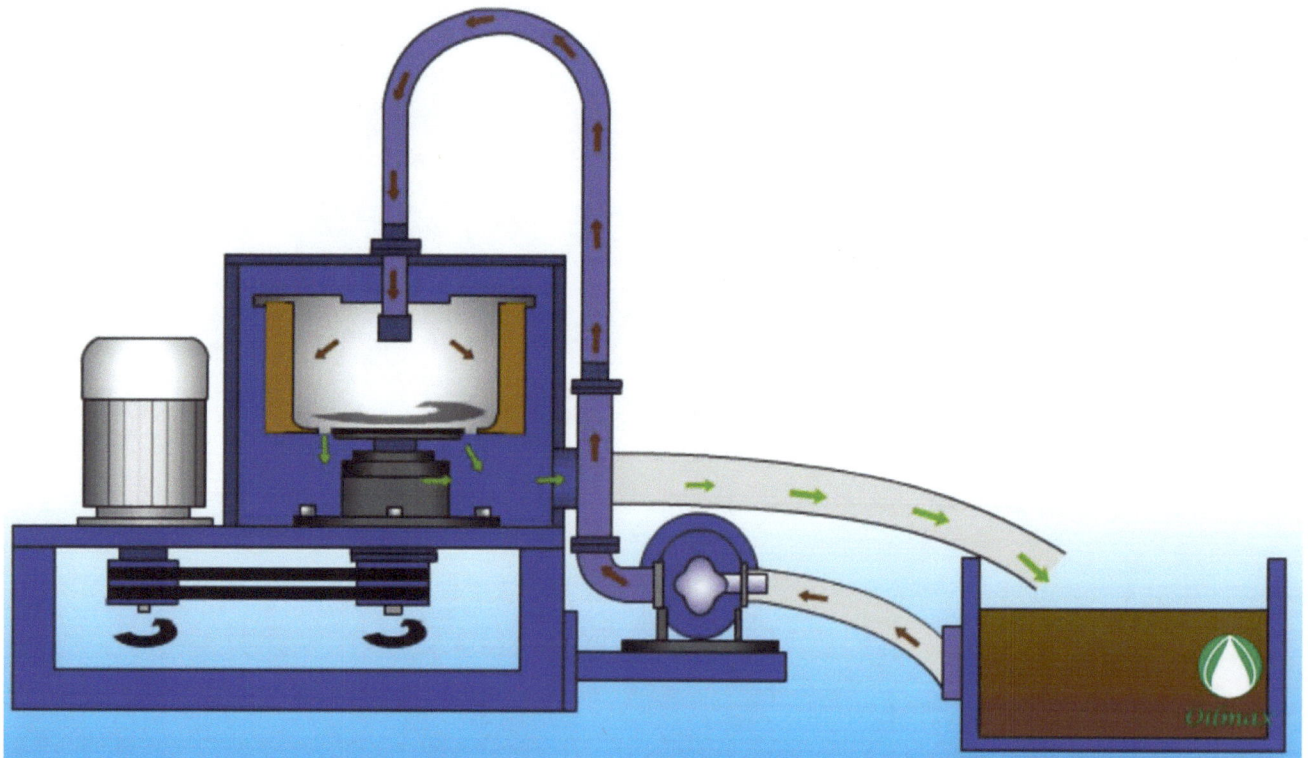

Fig. 4.19- Concept of Operation of Centrifugal Water Separator (www.oilmax.com)

Mass Transfer Vacuum Dehydrator: Removing free water is never enough. An alternative technique to separate free and emulsified water is to use *Mass Transfer Vacuum Dehydration*. Figure 4.20 shows a Pall-branded portable oil purifier that uses the technique of Mass Transfer Vacuum Dehydration". This purifier is designed for use with medium to large oil systems, particularly where high viscosity fluids are employed. It uses vacuum dehydration to remove 100 % free water and as much as 90 % of dissolved water at minimum cost and ease of use.

Fig. 4.20- HNP075 Series Oil Purifier (Courtesy of Pall Corporation)

Figure 4.21 shows how removal of water to levels below the saturation curve ensures that free water will not be reformed after cooling the hydraulic fluid as follows:

1. Initial water content is above saturation (free water).
2. Maximum water removal capability of "free water removal" devices such as filter separators and centrifuges, etc.
3. Water content achieved with mass transfer dehydration is significantly below the oil's saturation point.
4. Water content achieved with mass transfer dehydration remains below the oil's saturation point even after oil is cooled by the system heat exchanger. This prevents the formation of free water which is detrimental to fluid system components and the fluid.
5. If only free water is removed at initial temperature, when oil is cooled the amount of free water in the oil can increase significantly.

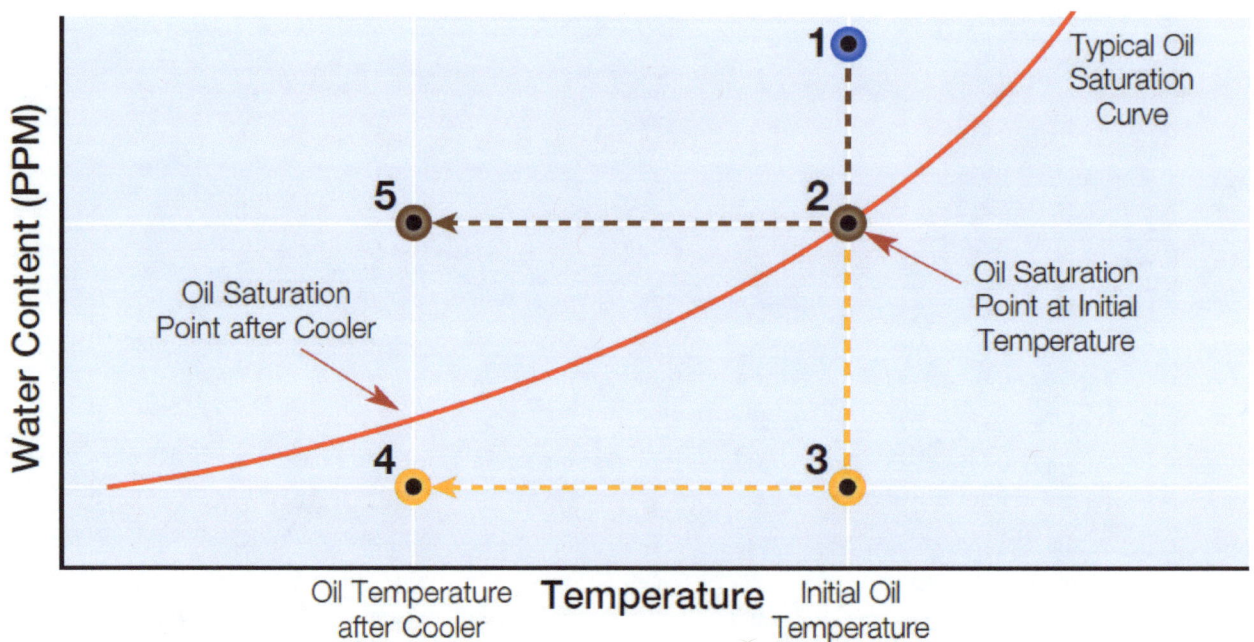

Fig. 4.21- Principle of Vacuum Dehydrator Performance (Courtesy of Pall Corporation)

Figures 4.22A through 4.22P explain, in steps, the procedure of water removal and purifying hydraulic fluid using Mass Transfer Vacuum Dehydrators.

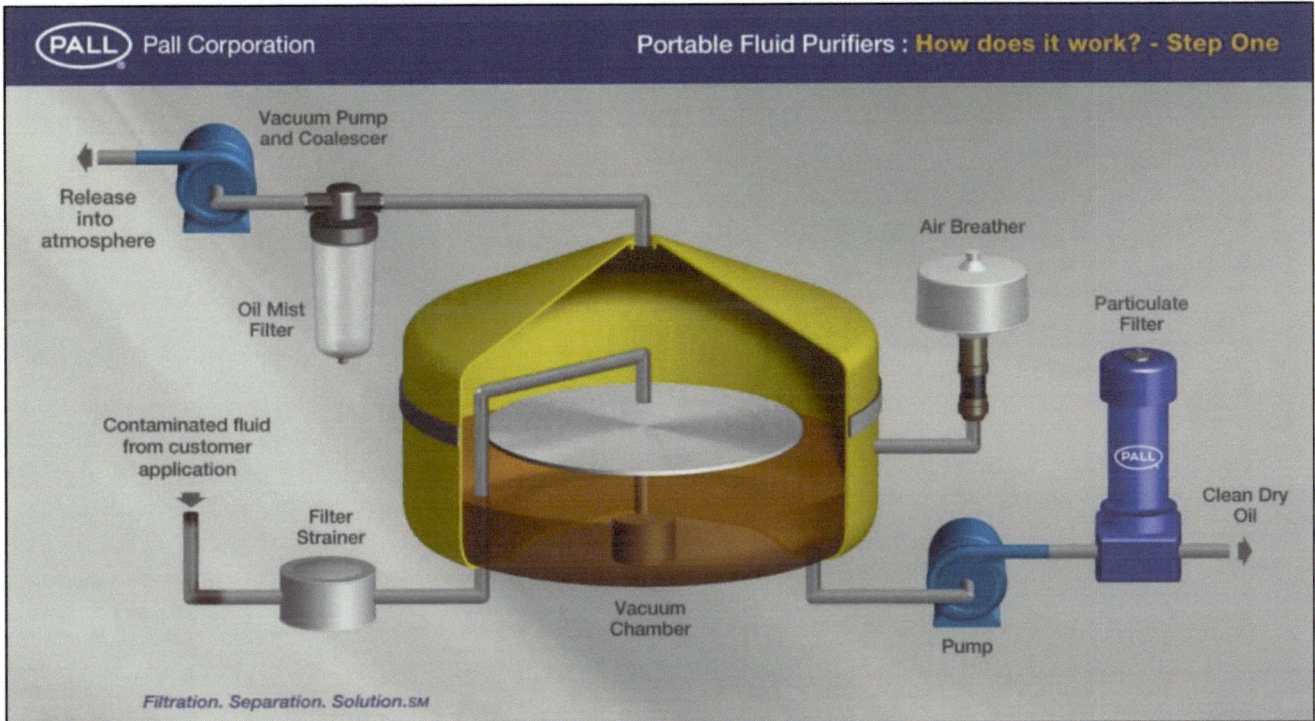

Fig. 4.22-A- Water Separator Operating Principle (Courtesy of Pall Corporation)

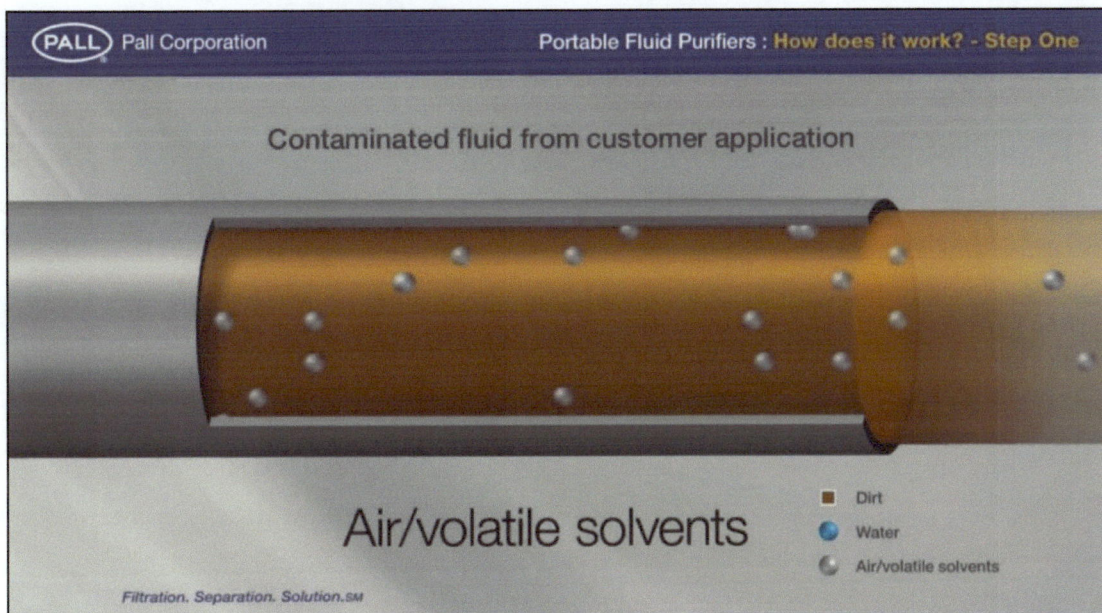

Fig. 4.22-B- Water Separator Operating Principle (Courtesy of Pall Corporation)

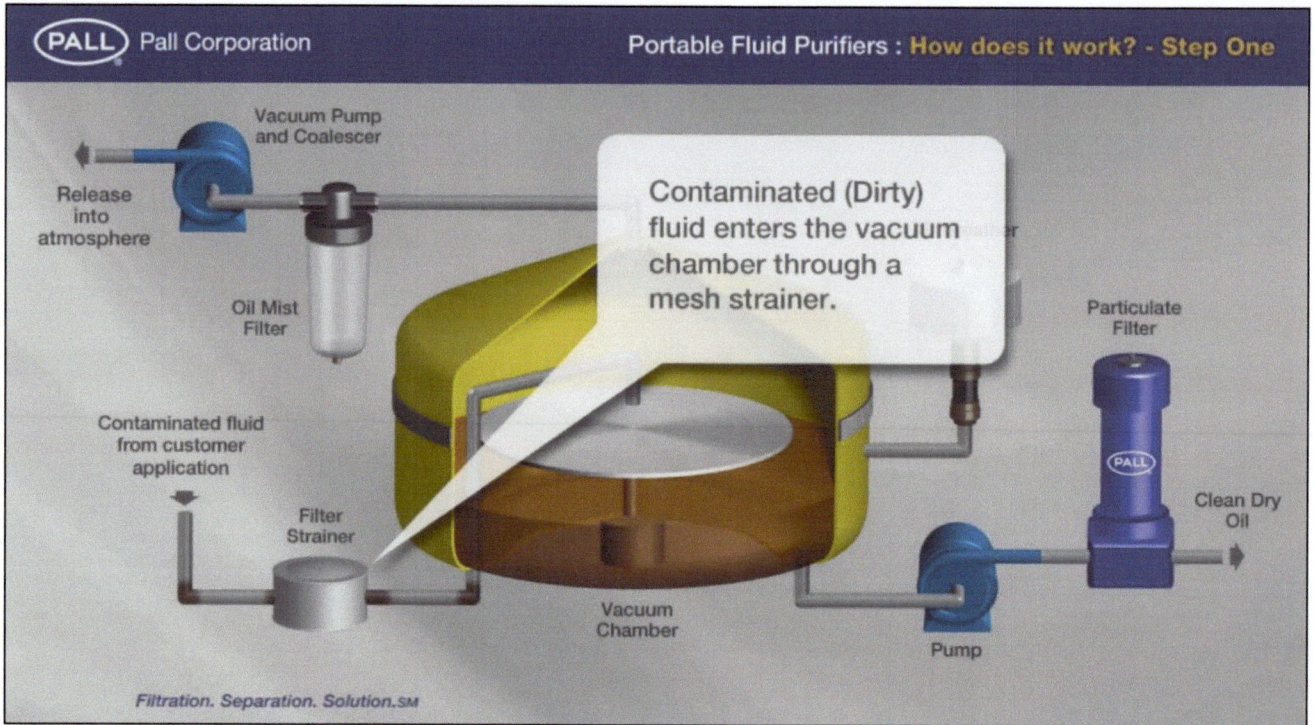

Fig. 4.22-C- Water Separator Operating Principle (Courtesy of Pall Corporation)

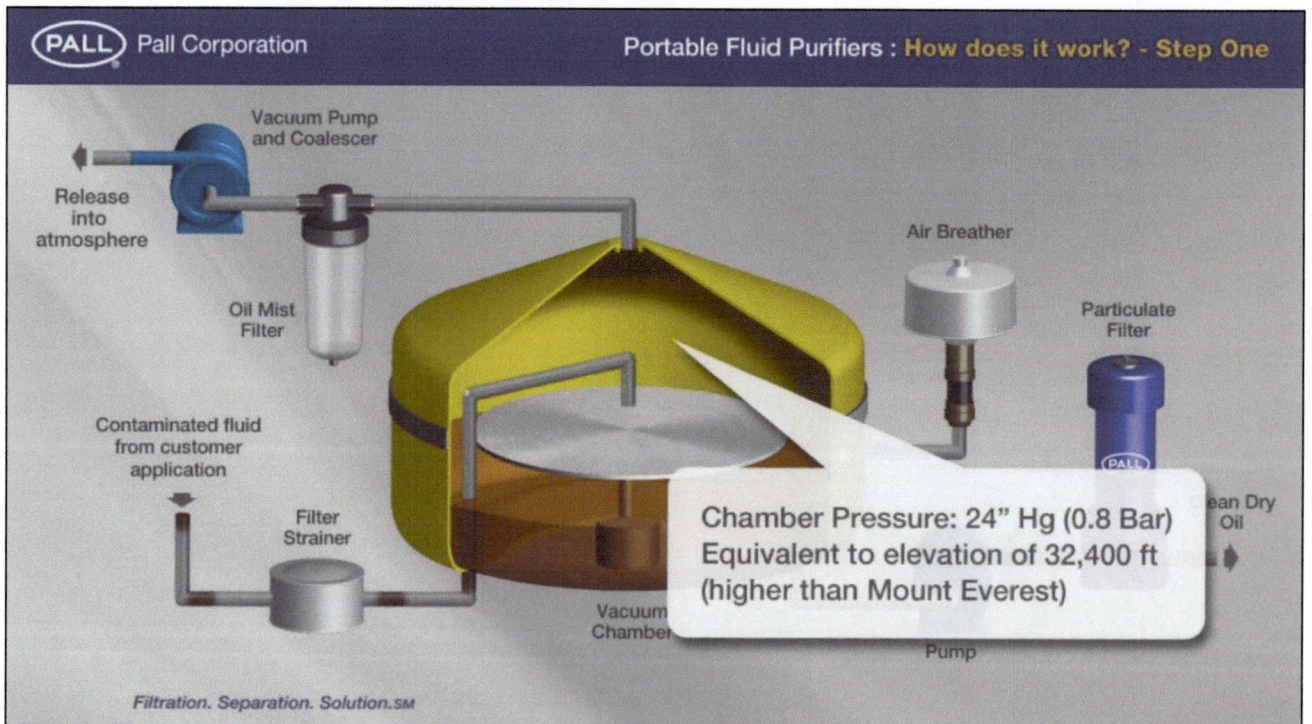

Fig. 4.22-D- Water Separator Operating Principle (Courtesy of Pall Corporation)

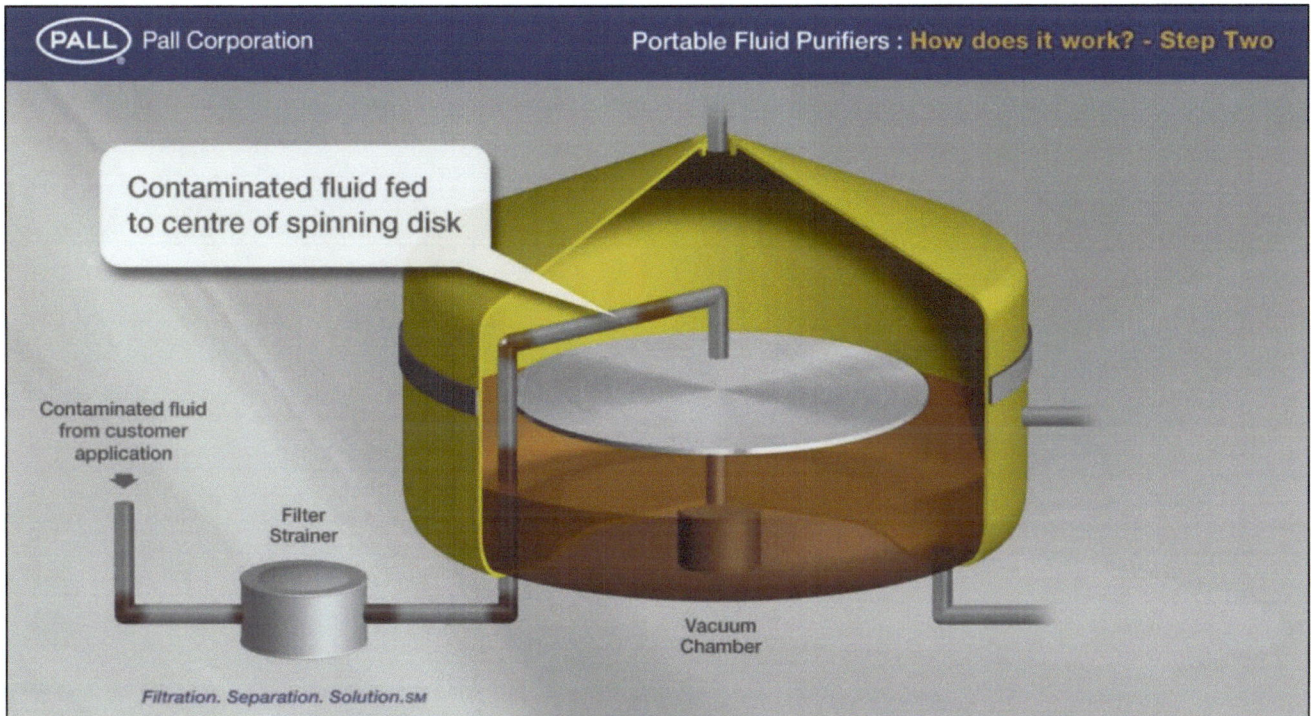

Fig. 4.22-E- Water Separator Operating Principle (Courtesy of Pall Corporation)

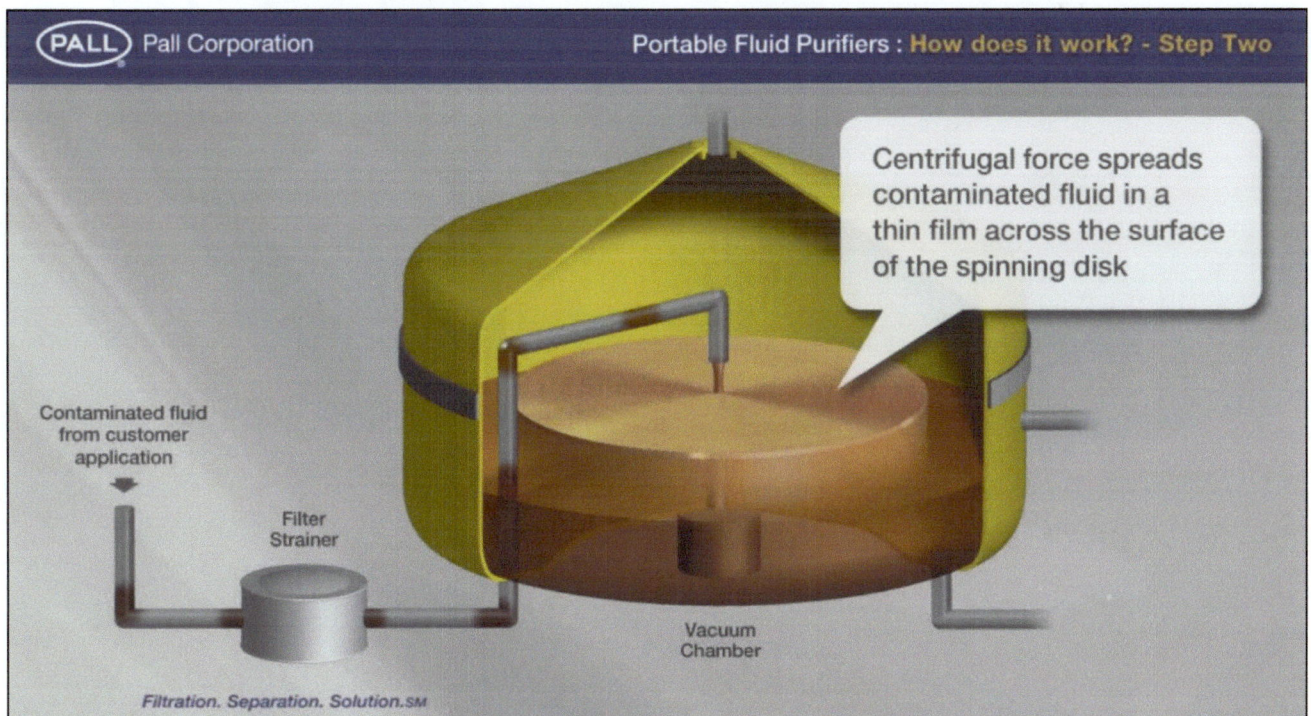

Fig. 4.22-F- Water Separator Operating Principle (Courtesy of Pall Corporation)

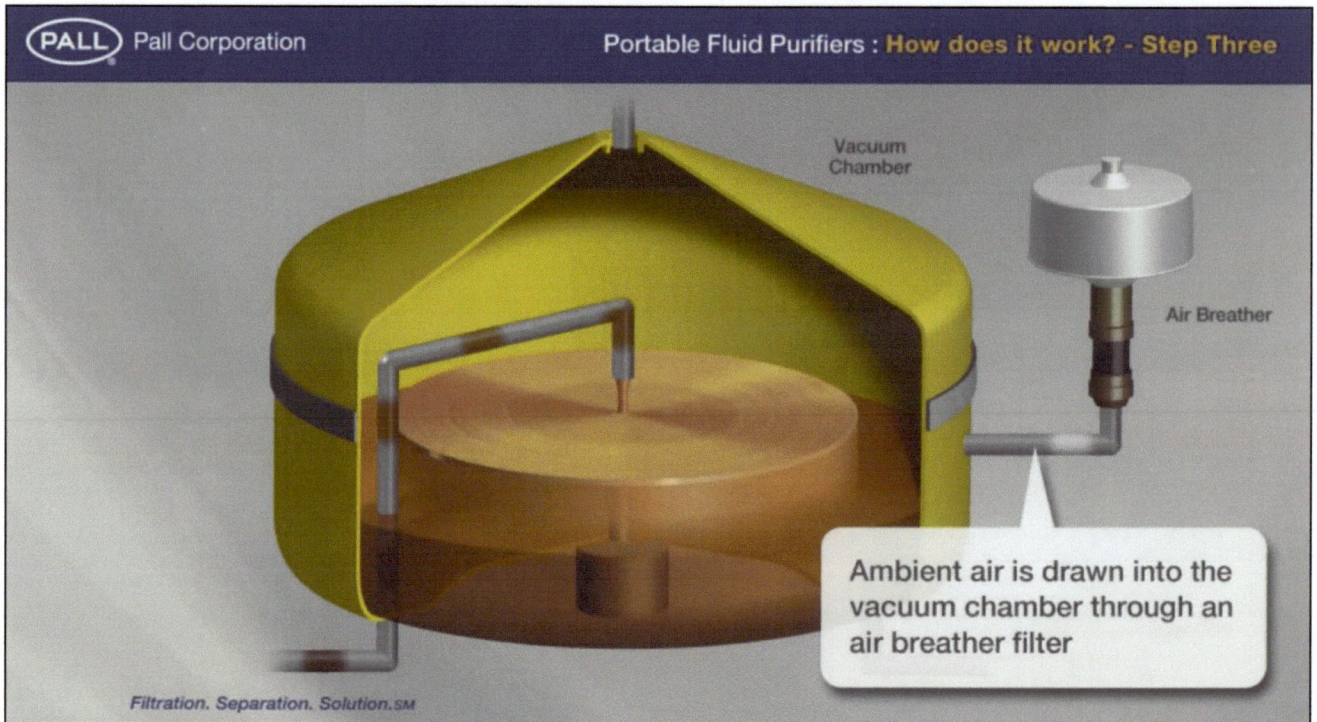

Fig. 4.22-G- Water Separator Operating Principle (Courtesy of Pall Corporation)

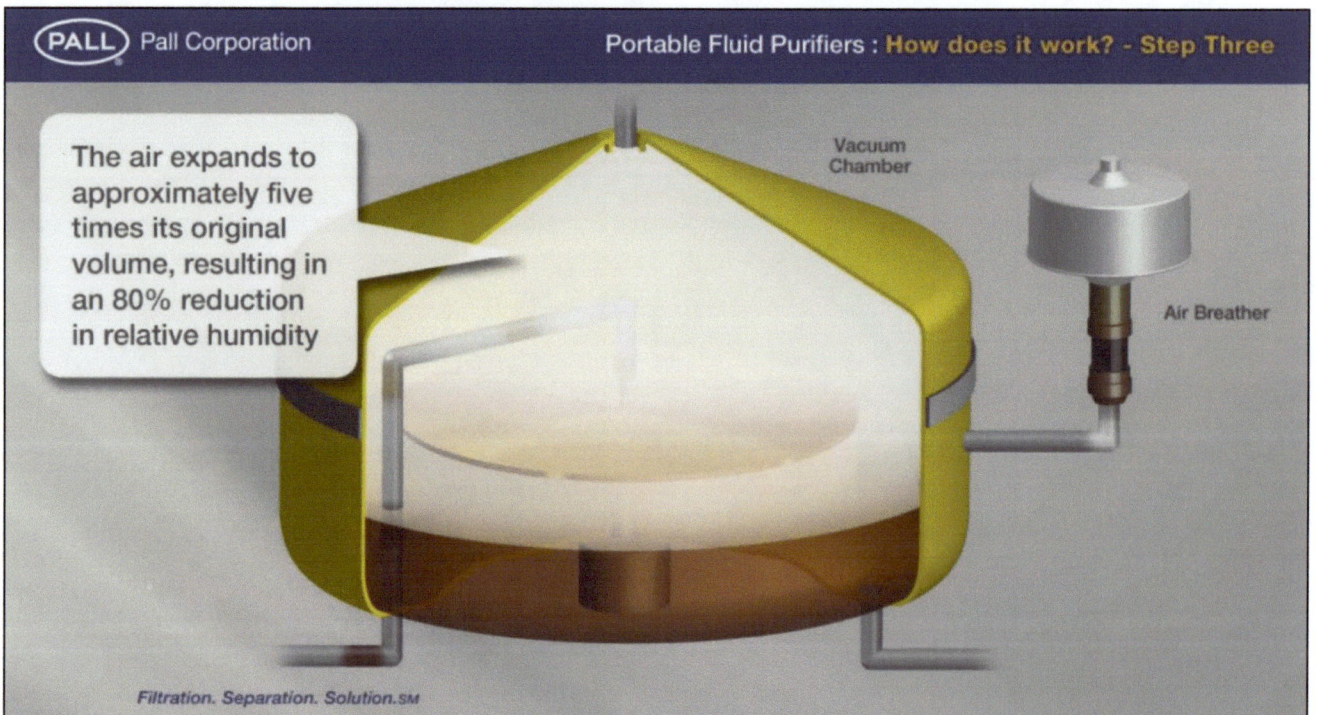

Fig. 4.22-H- Water Separator Operating Principle (Courtesy of Pall Corporation)

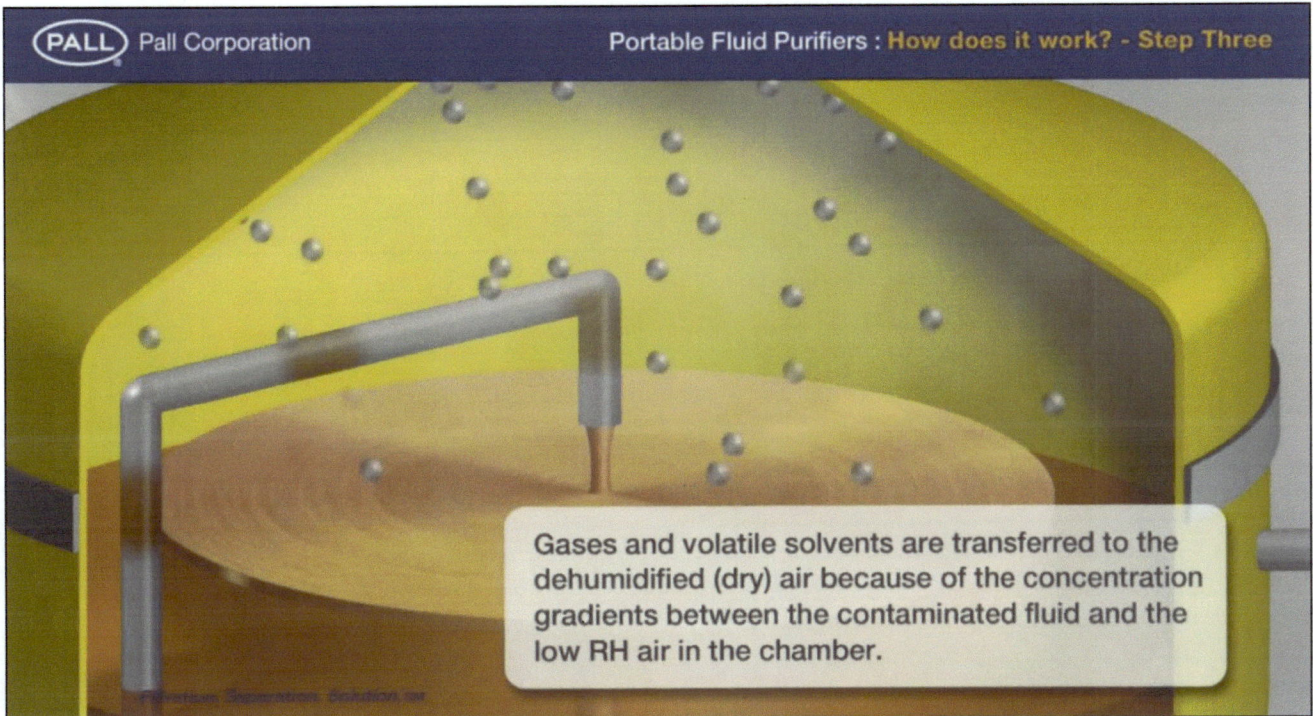

Fig. 4.22-I- Water Separator Operating Principle (Courtesy of Pall Corporation)

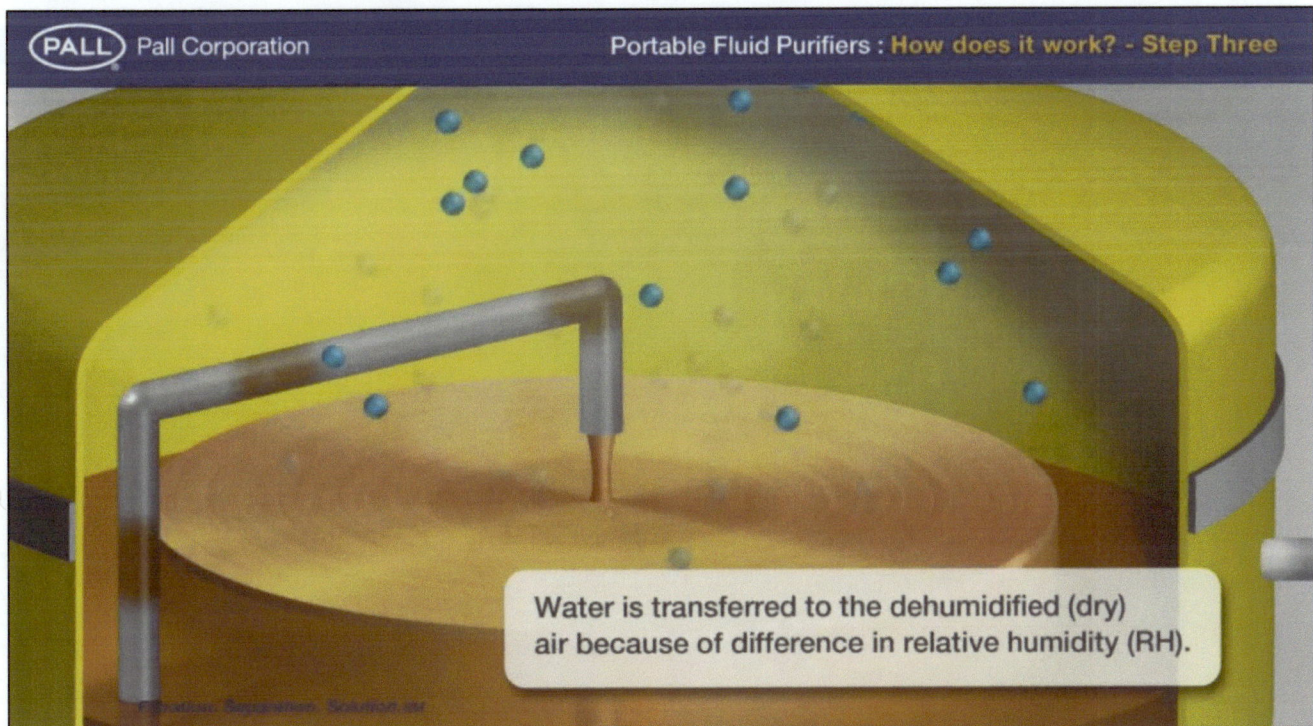

Fig. 4.22-J- Water Separator Operating Principle (Courtesy of Pall Corporation)

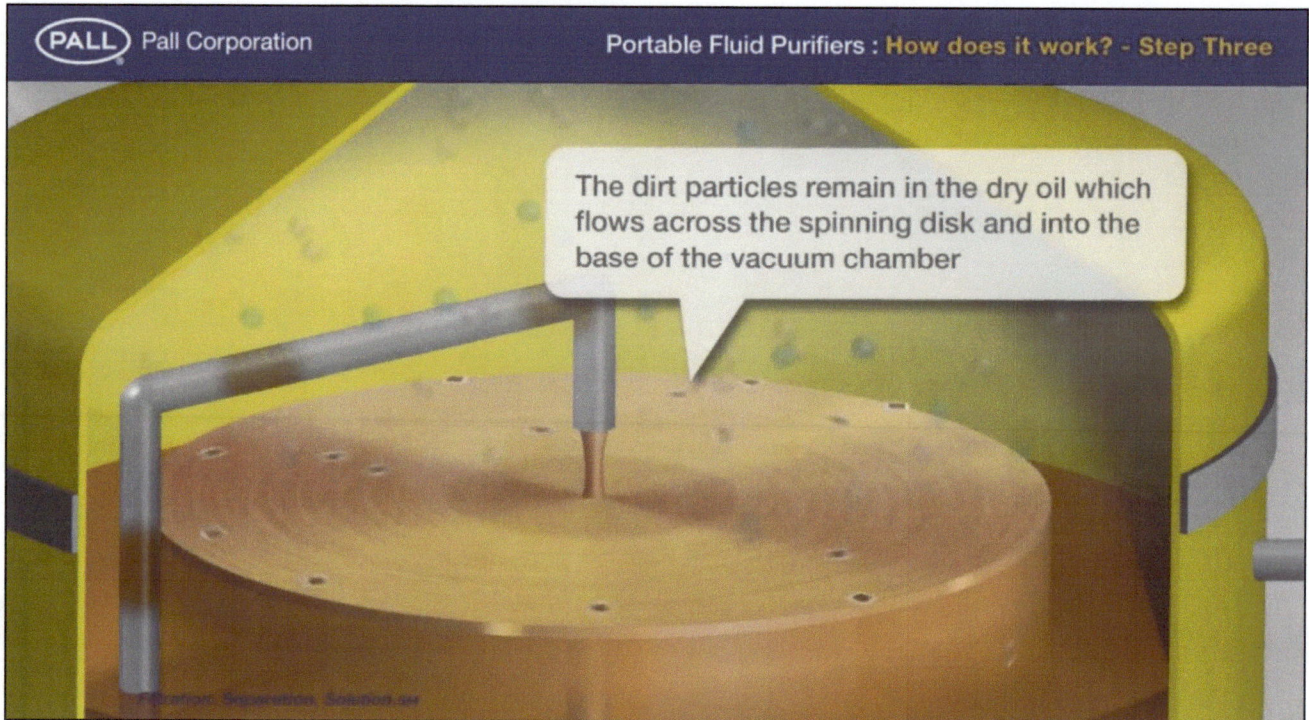

Fig. 4.22-K- Water Separator Operating Principle (Courtesy of Pall Corporation)

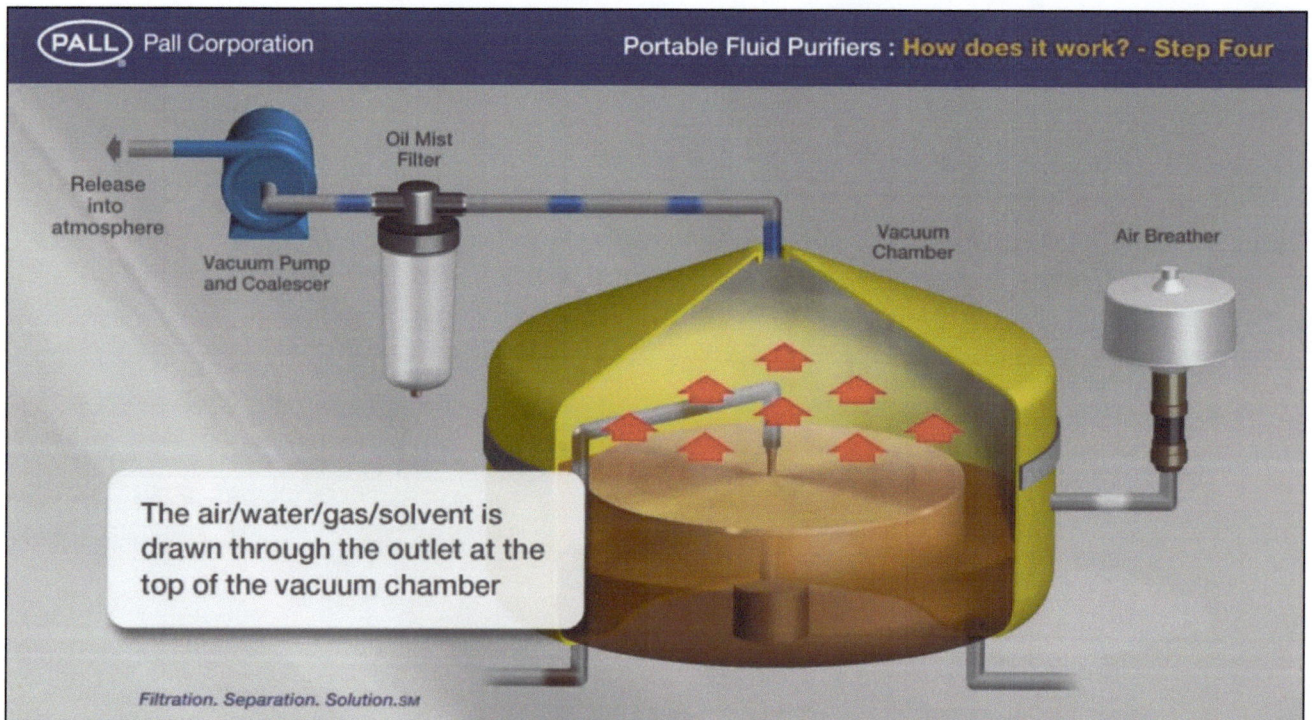

Fig. 4.22-L- Water Separator Operating Principle (Courtesy of Pall Corporation)

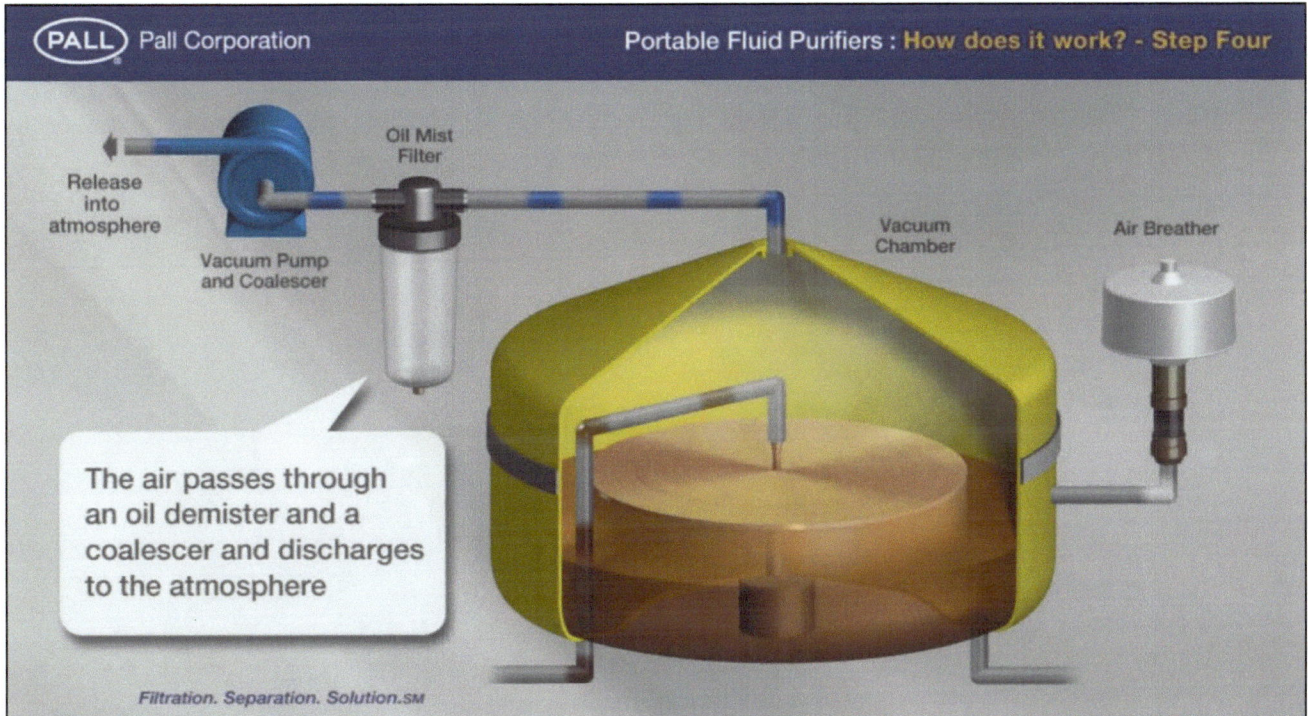

Fig. 4.22-M- Water Separator Operating Principle (Courtesy of Pall Corporation)

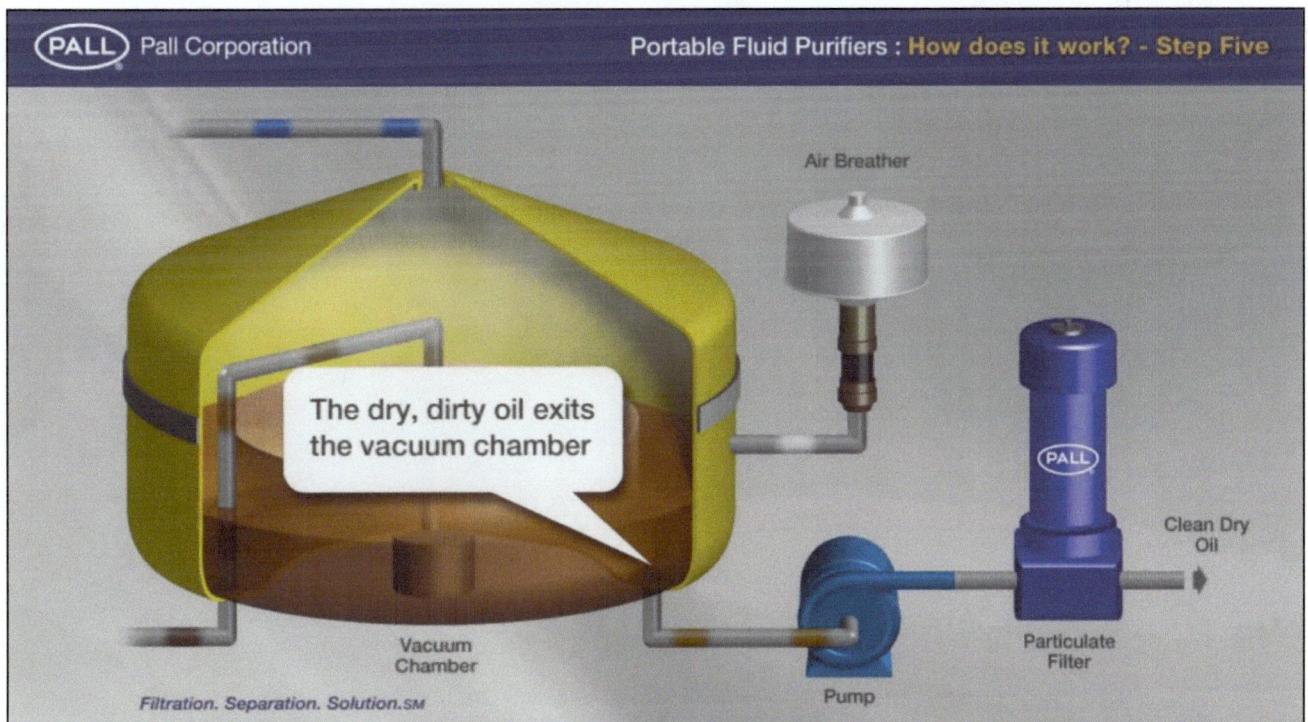

Fig. 4.22-N- Water Separator Operating Principle (Courtesy of Pall Corporation)

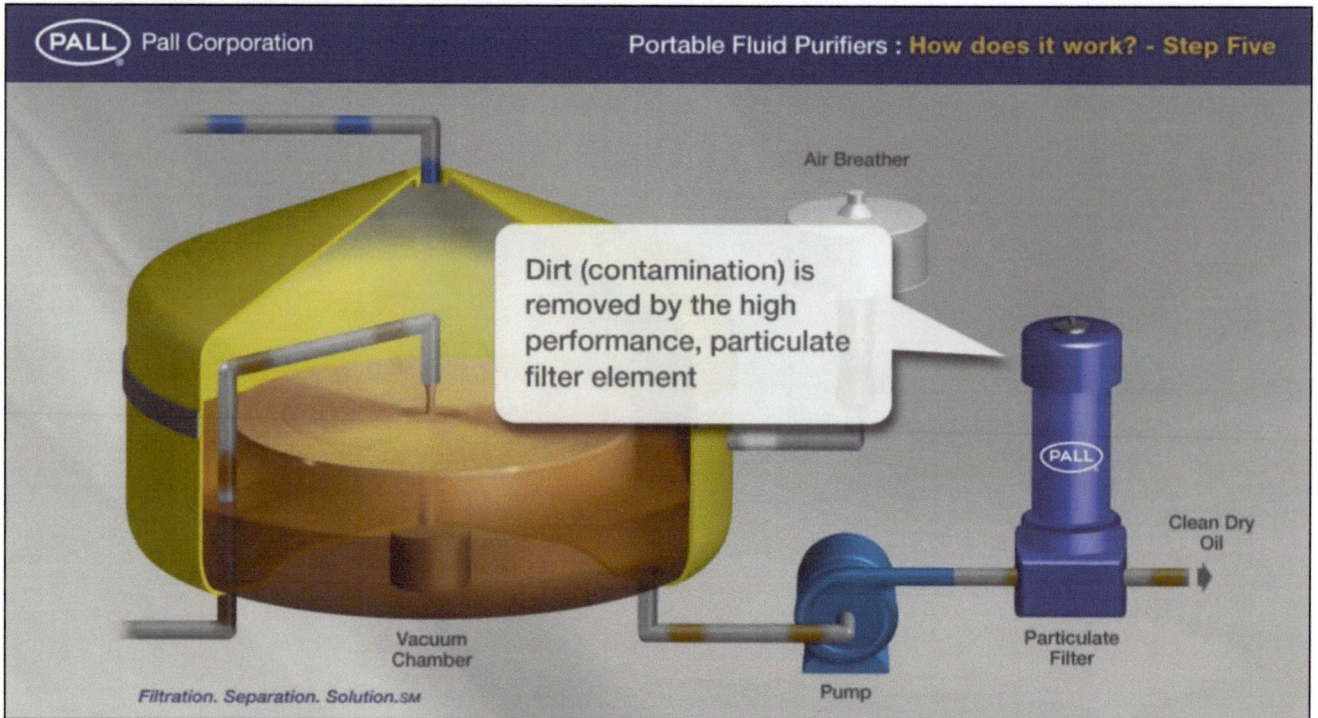

Fig. 4.22-O- Water Separator Operating Principle (Courtesy of Pall Corporation)

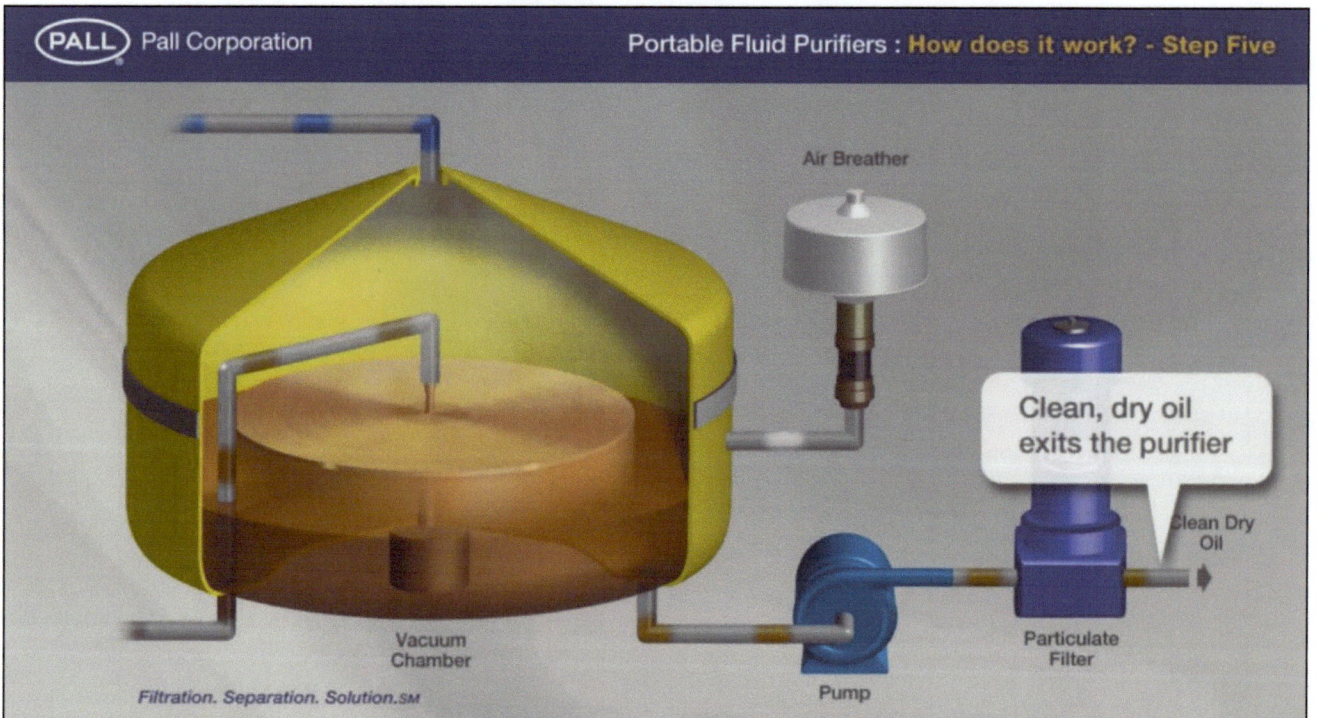

Fig. 4.22-P- Water Separator Operating Principle (Courtesy of Pall Corporation)

Chapter 5

Chemical Contamination

Objectives

This chapter presents the sources of chemical contamination. For each source, the chapter explains how the system performance will be affected and possible recommendations to minimize such consequences.

Brief Contents

5.1- Sources of Chemical Contamination
5.2- Products of Hydraulic Fluid Degradation
5.3- Effects of Chemical Contamination
5.4- Standard Test Methods for Measuring Oil Degradation
5.5- Best Practices to Minimize Chemical Contamination

Chapter 5 – Chemical Contamination

5.1- Sources of Chemical Contamination

Combination of gaseous, fluidic, and thermal contamination results in oil degrading that is a common problem both in lubrication and hydraulic systems. The main sources of oil degradation are typically one or combination of the four catalysts shown in Fig. 5.1. As a result, three different forms of oil degradation occur: *Oxidation*, *Hydrolysis*, and *Thermal Degradation*.

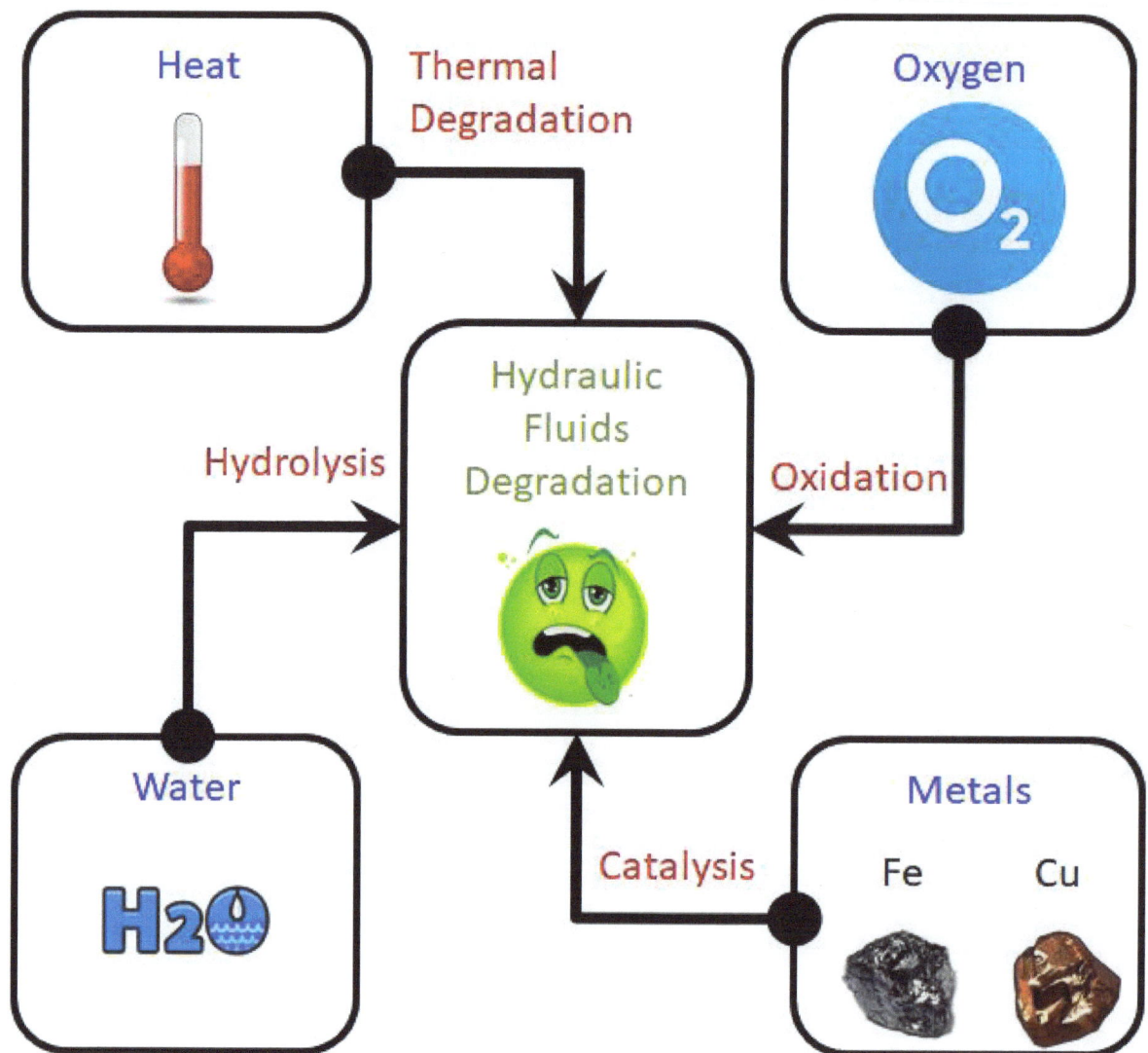

Fig. 5.1- Sources of Chemical Contamination

As shown in Fig. 5.2, four products of oil degradation are Rust, Varnish, Acids, and Sludge.

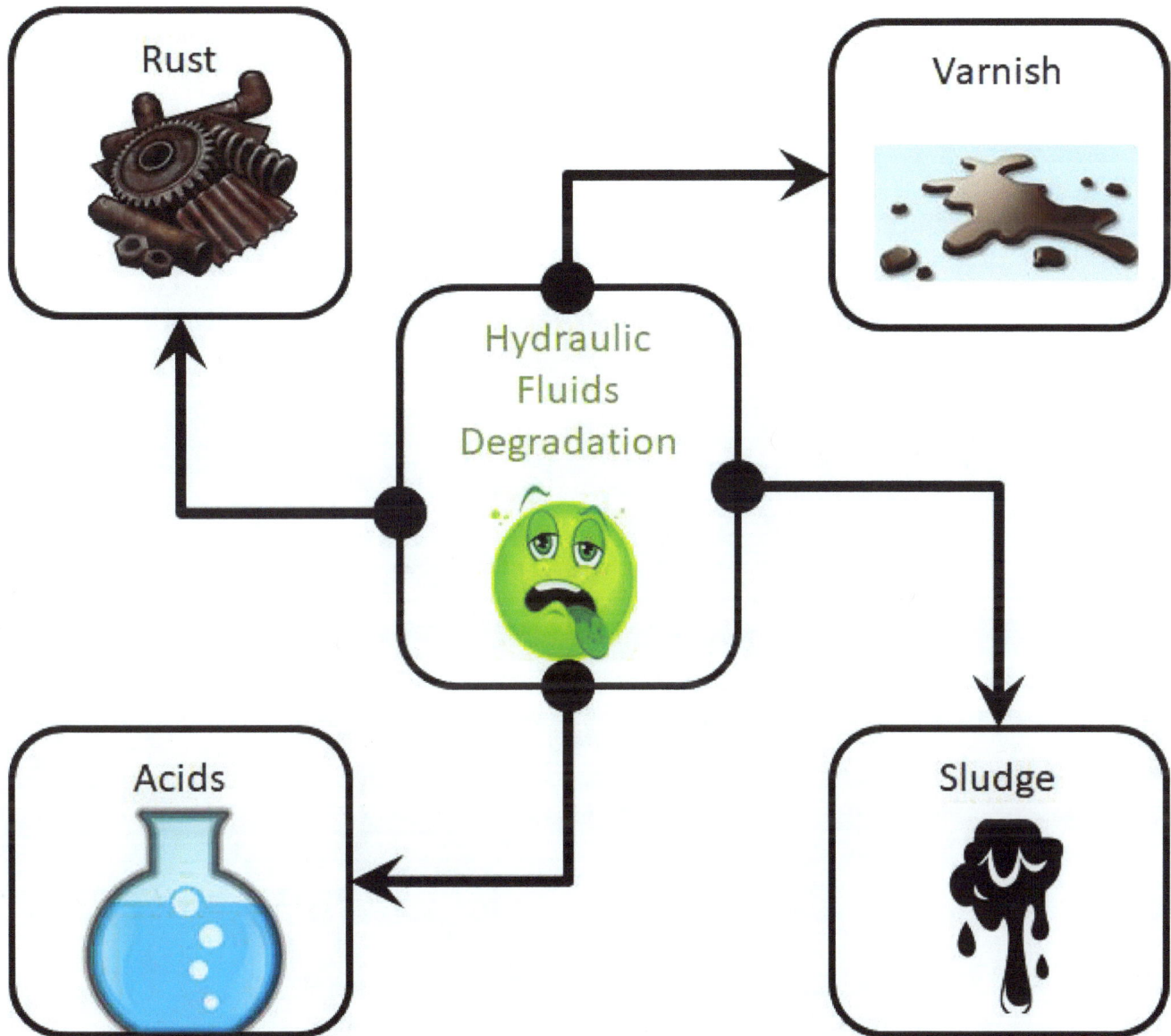

Fig. 5.2- Products of Hydraulic Fluids Chemical Degradation

5.2- Products of Hydraulic Fluid Degradation

As shown in the previous figure, when oil degrades, the composition and functional properties of the oil are changed resulting in formation of the following products such as *Rust, Acids, Sludge*, and *Varnish*.

5.2.1- Rust

Rust is a surface degradation of iron metallic components due to oxidation. As shown in Fig 5.3, water moisture expedites oil oxidation and consequently rust formation within the fluid. Consequently, corrosion and metal fatigue are seen within the components. Heat is one the major contributors to oxidation. In general oxidation can occur twice as fast for every 10 °C (18°F) temperature rise. Oxidation also is accelerated in the presence of copper or iron particles, in conjunction with water. As shown in Fig. 5.4, rust is also formed on the outer surfaces of the components and affect hydraulic lines (1), valves (2), pumps (3), cylinder rods (4), etc.

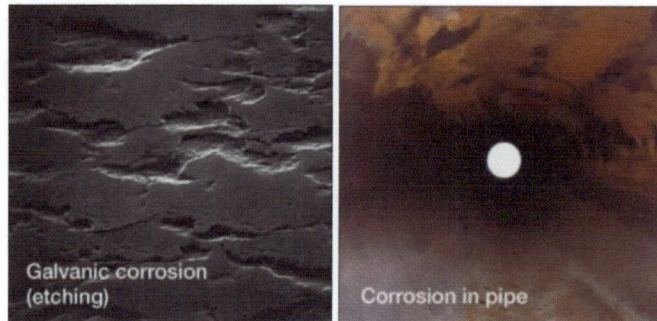

Fig. 5.3- Effect of Rust on Hydraulic Pipes (Courtesy of Pall Corporation)

Fig. 5.4- Effects of Rust on Hydraulic Components

5.2.2- Acids

Oil degradation due to hydrolysis of ester-based fluids, or the reaction of additives like zinc and calcium sulfonate, results in *acid* formation. As shown in Fig. 5.5, increased acidity promotes corrosion, shortens fluid and components service life, and leads to increased wear in the internal surfaces of machine.

Fig. 5.5- Corrosion in a Machine Component due to Acid Formation

5.2.3- Sludge

If the hydraulic fluid is exposed to high temperatures, many fluids will break-down and release resinous materials. When combined with other contaminates, sludge is formed, which tends to plug small openings and orifices and interfere with heat transfer.

As shown in Fig. 5.6, *Sludge* is thick polymerized compounds dissolved in warm oil. Sludge is a strong source of clogging filters, strainers, and control orifices causing sudden system failure.

Fig. 5.6- Sludge in Hydraulic Fluids

5.2.4- Varnish

Figure 5.7 shows the process of *Varnish* formation. Varnish is a thin, insoluble, non-wipeable, gummy, and sticky film deposit on metal surfaces. The figure shows examples of varnish formation on various machine components within a hydraulic system.

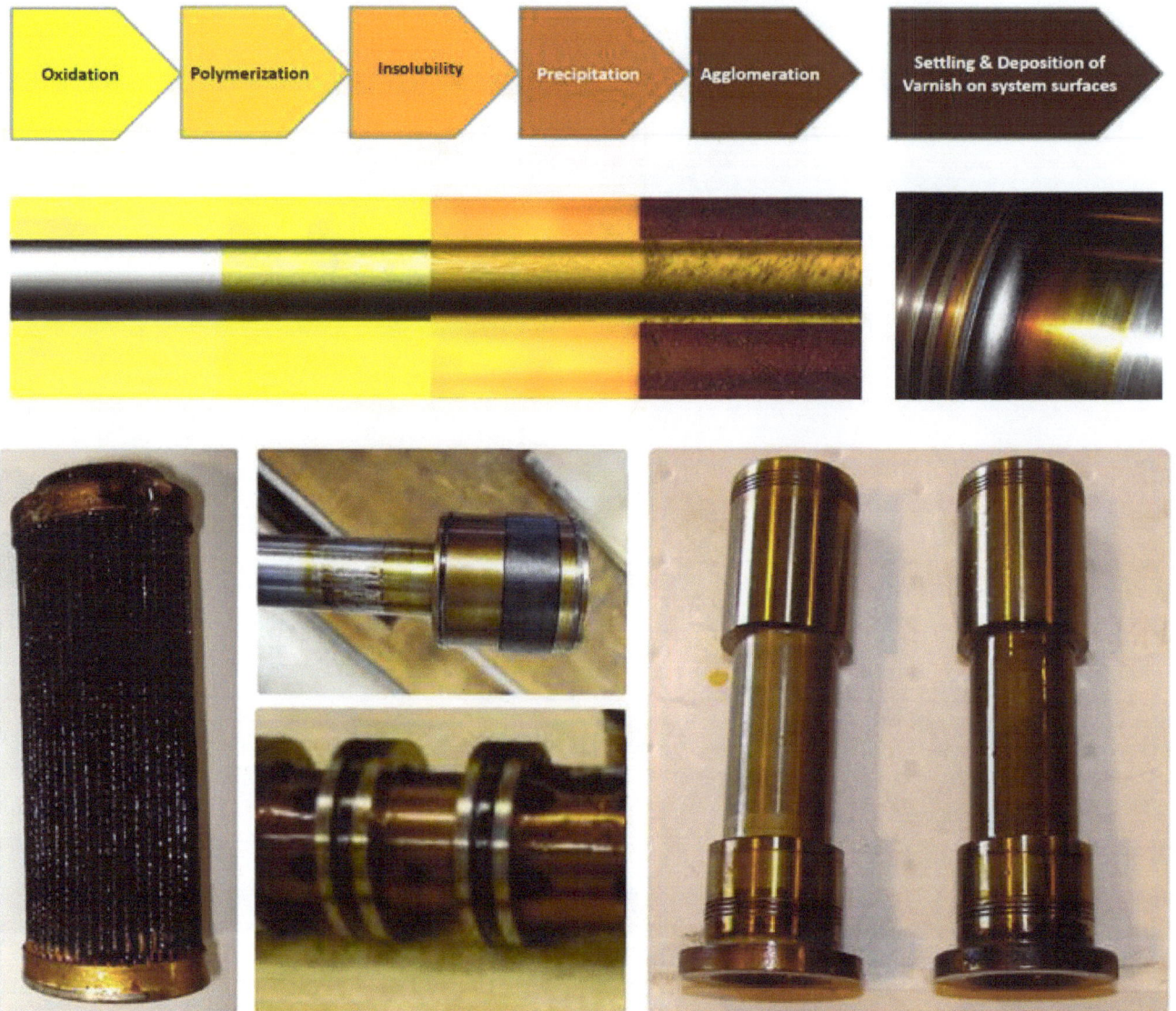

Fig. 5.7- Varnish Formation within Hydraulic Systems (Courtesy of C.C. Jensen Inc.)

As shown in Fig. 5.8, varnish creates a sticky layer. This layer attracts the abrasive particles of all sizes creating a sand-paper grinding surface which radically speeds up machine wear. As shown in Fig. 5.9, varnish can easily block fine tolerances, making spool valves (e.g. directional control valves) seize. As shown in Fig. 5.10, varnish clogs filters. Furthermore, varnish acts as an insulator reducing the effect of the heat exchangers.

Fig. 5.8- Varnish Sticky Layer Attracts Abrasive Particles (Courtesy of C.C. Jensen Inc.)

Fig. 5.9- Varnish Sticky Layer Seizes Valve Spools (Courtesy of C.C. Jensen Inc.)

Fig. 5.10- Varnish Sticky Layer Clogs Filters (Courtesy of C.C. Jensen Inc.)

5.3- Effects of Chemical Contamination

Oil degradation products are a widespread problem in most industries and cause problems in both hydraulic and lube oil systems. Oil degradation results in the following common problems.

Fluid Appearance: As shown in Fig. 5.11, as compared to a sample of new oil, oil that is degraded has dark color.

Fluid Odor As shown in Fig. 5.12, as compared to a sample of new oil, oil that is degraded has sour, putrid, and acidic smell.

Increased Oil Viscosity: As shown in Fig. 5.13, increasing oil viscosity results in a higher pressure drop in valves and transmission lines.

Reduced System Performance: Varnishes build-up on surfaces affects movement of valves. And results in stick-slip actuators motion.

Decreased Additive Performance: Some additives react with the degrading products and consequently lose their effect, and instead accelerate the deterioration process.

Shorter Oil Life: Oil life is significantly reduced.

Filter Life: Reduce filter life because of sludge formation.

Component Life: Reduced components service life because of corrosive wear.

Reduced Productivity: Productivity reduced due to increased downtime and filter change frequency.

Increased Maintenance Costs: Maintenance cost increased due to shorter oil and component service life.

Environmental Pollution Consequences: Environmental pollution increased due to frequent disposals and possible leakage.

Fig. 5.11- Fluid Appearance (Courtesy of Noria Corp.)

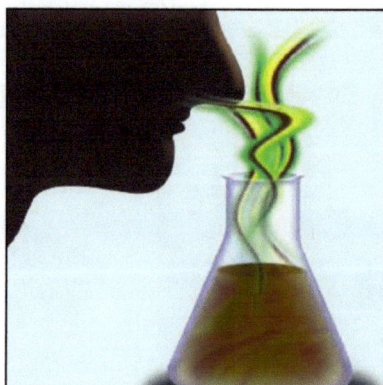

Fig. 5.12- Fluid Odor (Courtesy of C.C. Jensen Inc.)

Fig. 5.13- Fluid Viscosity (Courtesy of C.C. Jensen Inc.)

5.4- Standard Test Methods for Measuring Oil Degradation

As shown in Fig. 5.14, there are several ways for measuring and evaluating hydraulic fluids degradation. Some of these methods are just quantitative and some are qualitative:

1- **Ultracentrifuge Test:** *Ultracentrifuge Test* uses the centrifugal forces to extract and settle the contaminants of the oil. The sediments are compared with a sedimentation rating system to determine the degradation of the oil.

2- **Fourier Transformation Infrared Spectroscopy Analysis (FTIR):** *FTIR* analysis is same as used in water content measuring.

3- **Membrane Patch Colorimetry (MPC):** *Member Patch Colorimetry (MPC)* analysis is an indication that the oil contains degradation products. The varnish is captured in the white MPC membrane (0.45-micron cellulose membrane), and shows as a yellow, brownish or dark color depending on the amount of varnish present in the oil. A microscopic magnification shows if the color comes from varnish or hard particles.

4- **QSA Test:** *QSA* method identifies the varnish potential rating and is based on colorimetric analysis. By comparing the result to a large database of QSA tests, a 1 to 100 severity rating scale indicates the tendency of the lubricant to form sludge and varnish.

5- **Gravimetric Analysis:** *Gravimetric Analysis* can determine the level of oil degradation by measuring the weight of residual components.

6- **Viscosity Test:** *Viscosity Test* can be used as an indicator of oil degradation.

7- **Remaining Useful Life Evaluation Routine (RULER) Test:** *Remaining Useful Life Evaluation Routine (RULER) Test* measures the remaining amount of anti-oxidants (oil additives). When the additives get depleted due to oil degradation, RULER number decreases.

8- **Total Acid Number (TAN):** *Total Acid Analysis (TAN)* analysis measures the level of acidic compounds. It can also be used as an indicator of oil degradation, since acidity is a product of degradation.

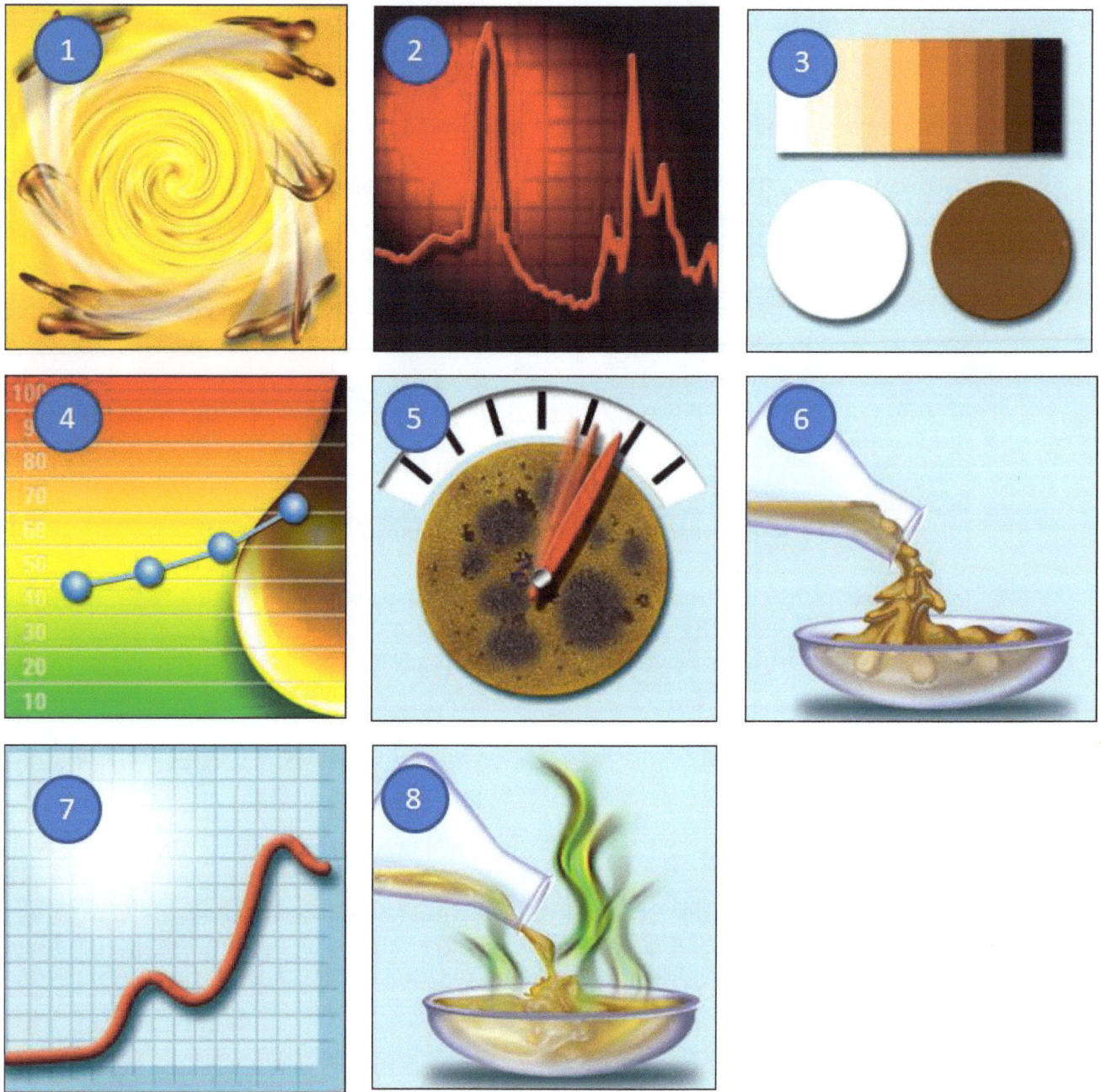

Fig. 5.14- Standard Test Methods for Measuring Oil Degradation (Courtesy of C.C. Jensen Inc.)

5.5- Best Practices to Minimize Chemical Contamination

5.5.1- Preventive Practices to Minimize Chemical Contamination

Oil degradation is a common problem in both lubrication and hydraulic systems. Therefore, considering some of the preventive practices limits the consequences of such a contamination.

Water Control: Since water is one of three elements that expedite chemical degradation of hydraulic fluids, all preventive methods that has been listed in water control are also applicable here.

Temperature Control: Since heat is one of three elements that expedite chemical degradation of hydraulic fluids, working temperature must be properly controlled and monitored on a continuous basis.

Gaseous Contamination Control: Since oxygen is one of three elements that expedite chemical degradation of hydraulic fluids, every action must be taken to eliminate gaseous contamination.

Acids Control: The amount of acidity in oil should be limited, since acidity will cause chemical corrosion of machine components and shorten the life of the oil. Acid numbers should not be allowed to increase more than +0.5 TAN higher than that of new oil. If +1 TAN is measured, an immediate action is required (i.e. if new oil has 0.5 TAN, then 1.0 TAN is alert and 1.5 TAN is alarming value).

Hydraulic Fluid Analysis: Periodic testing for measuring fluid conditions such as TAN, varnish and sludge formation, oil viscosity, etc. are key information for predictive maintenance.

Hydraulic Fluid Additives: Use of proper additive package such as anti-oxidation, rust inhibitors, emulsifiers, and foam suppressors.

As an example of new technology, Figs. 5.15 and 5.16 show the performance of a patented type of hydraulic fluid called (Parker DuraClean™). DuraClean™ is an ultra-premium multi-grade hydraulic oil provided exclusively by Parker. The fluid has a unique additive chemistry designed to maximize oil life while providing optimum anti-wear protection for the components of today's advanced hydraulic systems. The following are

- ISO 46, all season, multi-grade hydraulic fluid.
- Replaces ISO 32, 46, and 68 mono-grades.
- High viscosity index for wide operating temperature ranges.
- Outstanding oxidation life to maximize component life.
- Formulated to help extend the life of hoses and seals.

- Prevents varnish formation.
- Clean, as packaged, to ISO 17/15/12 cleanliness level.
- Special formulation that allows for rapid air release and water separation.
- Excellent filterability to minimize filter blockage.
- Outstanding acrylate anti-foam agent contains no silicones, which can lead to inaccurate particle counts.
- Excellent shear stability for stable viscosity over time.
- Superior thermal stability for uncompromised performance at high temperatures.

Without DuraClean

With DuraClean

Fig. 5.15- Effect of using DuraClean Fluid on Varnish Formation (Courtesy of Parker)

DuraClean™	Product B	Product C
ISO 15/14/12	ISO 22/20/14	ISO 25/24/21
100X	100X	100X

Initial samples taken directly from a 5 gallon pail.

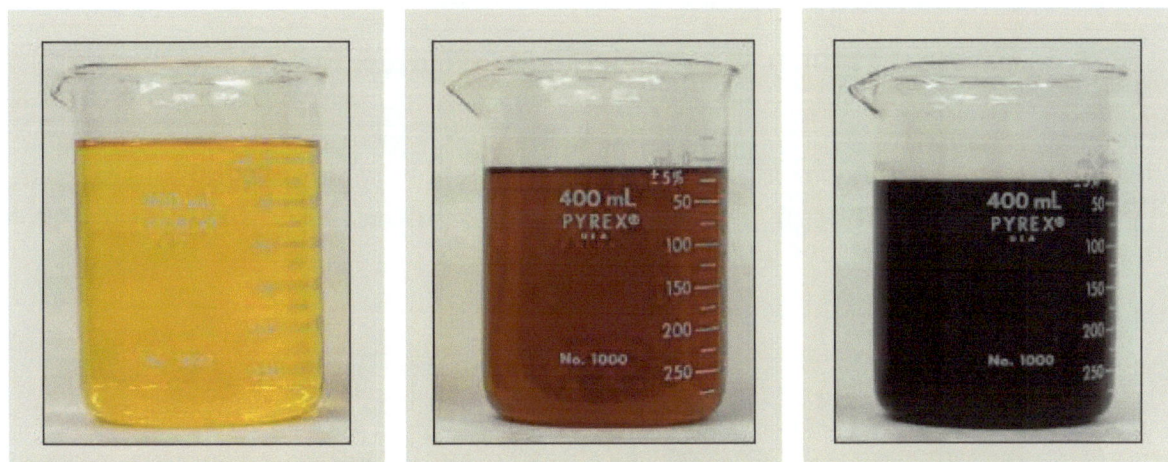

Same samples after 1,300 hours of exposure to 93 °C (200 °F)

Fig. 5.16- Effect of using DuraClean Fluid on Oxidation after 1300 Working Hours (Courtesy of Parker)

5.5.2- Curative Practices to Remove Chemical Contamination

If hydraulic fluid degrades, serious and immediate actions are required. The following sections provide most common methods of water removal.

Fluid Replacement: As it has been previously mentioned, if the quantity of the contaminated oil is small (< 500 gallon =2000 liter), it is recommended to replace it and flush the system.

Acidity Neutralization: The alkalinity of the oil is supposed to neutralize incoming acidity. Acid number 3-5 times higher than that of new oil results in severe acidic corrosion of system components. In such fluids the acid number can be lowered and maintained by changing the fluid. Acidity can be neutralized or removed from oil in different other ways.

Varnish and Sludge Removal: Oil degradation products, shown in Fig. 5.17, cannot be removed with conventional mechanical filters because they are submicron particles and a fluid in a fluid, like when sugar is dissolved in water. These degradation products can be removed by fine filters through a combination of adsorption and absorption processes.

Fig. 5.17- Oil Degradation Products (Courtesy of C.C. Jensen Inc.)

As shown in Fig. 5.18, Adsorption is the physical or chemical binding of molecules to a surface (like getting a cake thrown into your face). In contrast with absorption, molecules are absorbed into the media. See illustrations.

Fig. 5.18- Difference between Absorption and Adsorption (Courtesy of C.C. Jensen Inc.)

Figure 5.19 shows a specialized varnish removal unit. The figure shows the filter element before and after passing the contaminated oil through it. Table 5.1 shows the technical data of the unit.

Fig. 5.19- Varnish Removal Unit (Courtesy of C.C. Jensen Inc.)

TECHNICAL DATA		
Varnish Removal Unit		**VRU 27/108**
		380 - 420V @ 50 Hz & 440 - 480V @ 60 Hz
Pump inlet pressure max.	bar/psi	0.5/7
Power consumption aver.	kW	2
Full load current max.	A	4
Filter Insert VRi 27/27	pcs.	4
Oil reservoir volume max. *)	ltr/gal	45,000/11,900
Oil viscosity **)		<ISO VG68
Oil temperature max *)	°C/°F	105/221
Varnish holding capacity up to	kg/lb	8/18
Total weight	kg/lb	290
Design pressure, filter	bar/psi	4/58
Dimensions lxwxh incl. + free height	mm inches	1600x650x1598+575 63x25.6x62.9+22.6

*) For more than 45,000 L or higher temperatures, please contact us
**) For viscosities higher than ISO VG68, please contact us

Table 5.1- Technical Data of the Varnish Removal Unit (Courtesy of C.C. Jensen Inc.)

As an example of new technology, CJC™ Filter Inserts, made of *Cellulose Fibers*, have a high surface area and are effective as adsorbents and absorbents. In addition, due to their chemical nature, they are highly suited to pick-up oxygenated organic molecules, such as oil degradation products.

As shown in Fig. 5.20, each cellulose fiber consists of millions of cellulose molecules. Each strand of cellulose molecule has a diameter of 10-30 microns. Degradation products are adsorbed and absorbed into the cellulose material.

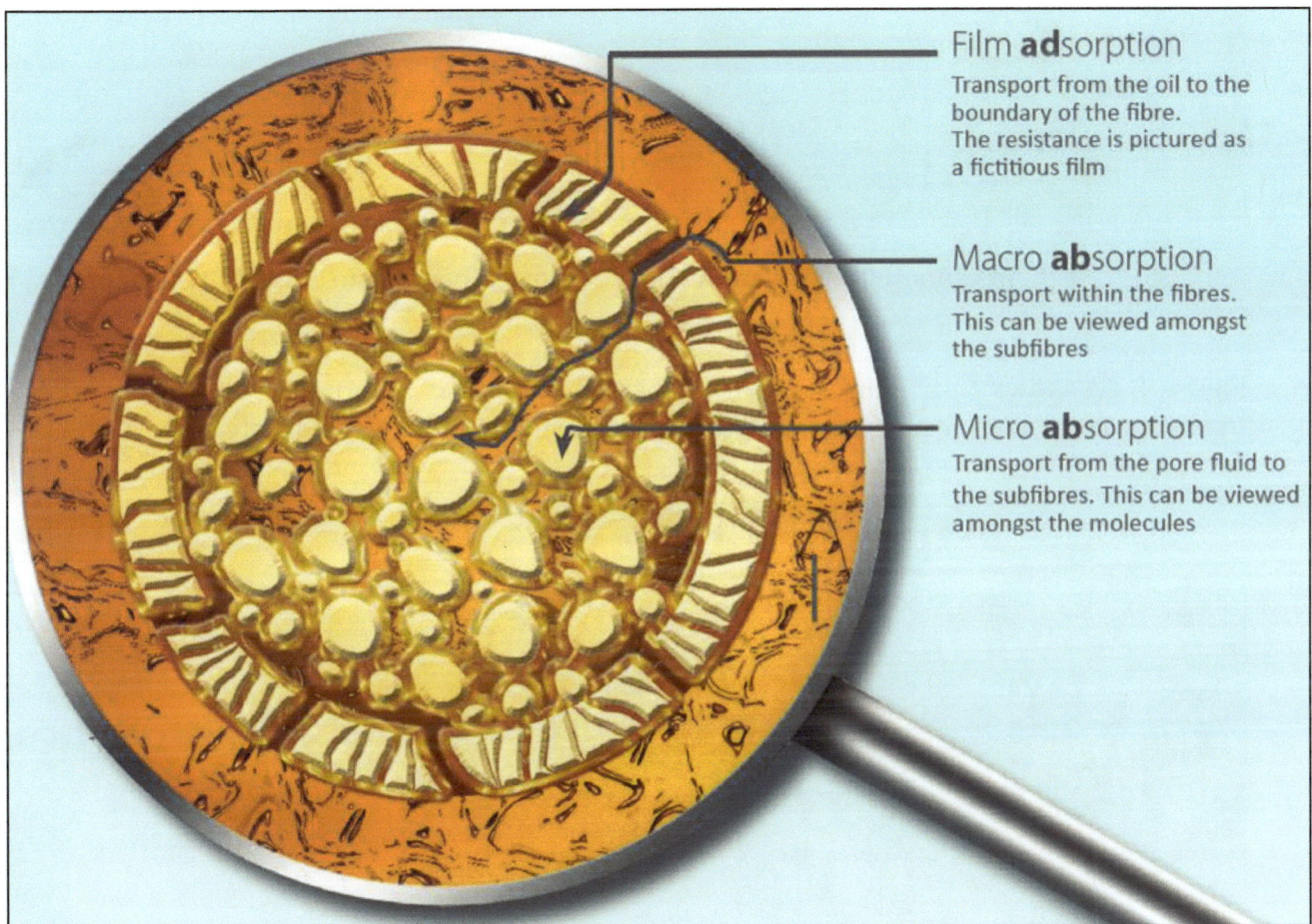

Film **ad**sorption
Transport from the oil to the boundary of the fibre. The resistance is pictured as a fictitious film

Macro **ab**sorption
Transport within the fibres. This can be viewed amongst the subfibres

Micro **ab**sorption
Transport from the pore fluid to the subfibres. This can be viewed amongst the molecules

Fig. 5.20- Cross-section of a Cellulose Fiber (Courtesy of C.C. Jensen Inc.)

Figure 5.21 shows the contaminated oil approaching the cellulose fibers in an almost new Filter Insert.

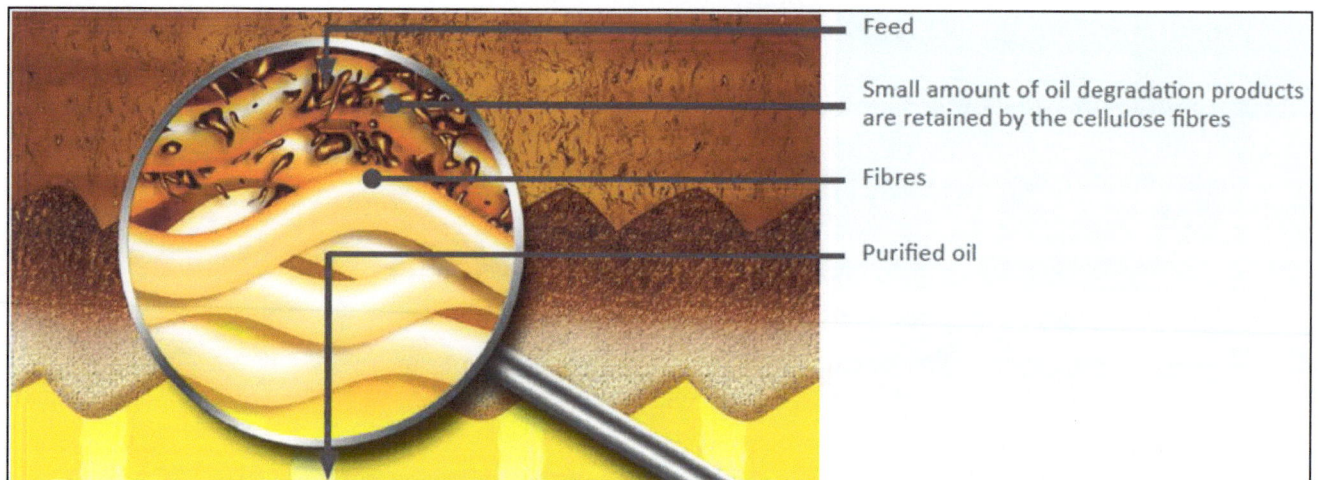

Fig. 5.21- Contaminated Oil Approaching Cellulose Fibers (Courtesy of C.C. Jensen Inc.)

Figure 5.22 shows CJC™ Filter Insert near saturation. This illustration shows that the Filter Insert is still delivering clean oil even though the cellulose fibers are nearly saturated.

Fig. 5.22- Filter Inserts Near Saturation (Courtesy of C.C. Jensen Inc.)

Chapter 6

Particulate Contamination

Objectives

This chapters presents the sources of particulate contamination. For each source, the chapter explains how the system performance will be affected and possible recommendations to minimize such consequences.

Brief Contents

6.1- Forms of Particulate Contamination
6.2- Sources of Particulate Contamination
6.3- Contamination Particle Sizes
6.4- Critical Clearances in Hydraulic Components
6.5- Effects of Particulate Contamination
6.6- Best Practices for Controlling Particulate Contamination

Chapter 6 – Particulate Contamination

6.1- Forms of Particulate Contamination

Particulate Contaminants are extraneous material that do dissolve in the hydraulic fluid. As shown in Fig. 6.1, particulate contaminants can take one of the following forms:

Abrasive Particles: *Abrasive* particles are either hard particles with rounded shape or extremely hard particles with sharp edges. Most of these abrasive particles are metallic due to component wear such as aluminum, chromium, copper, iron, lead, tin, silicon, sodium, zinc, barium and phosphorous. Some other abrasive particles are nonmetallic such as sand.

Silt: *Silt* is defined as very fine particulate, under 5 μm in size. Most of the silt is from dust and dirt.

Nonabrasive Particles: These are soft particles but are not dissolvable in the hydraulic fluids. Most of these particles are elastomeric due to seal wear such as rubber, fibers, paint chips, sealants. Some others nonabrasive particles are gelatinous particles or microorganisms.

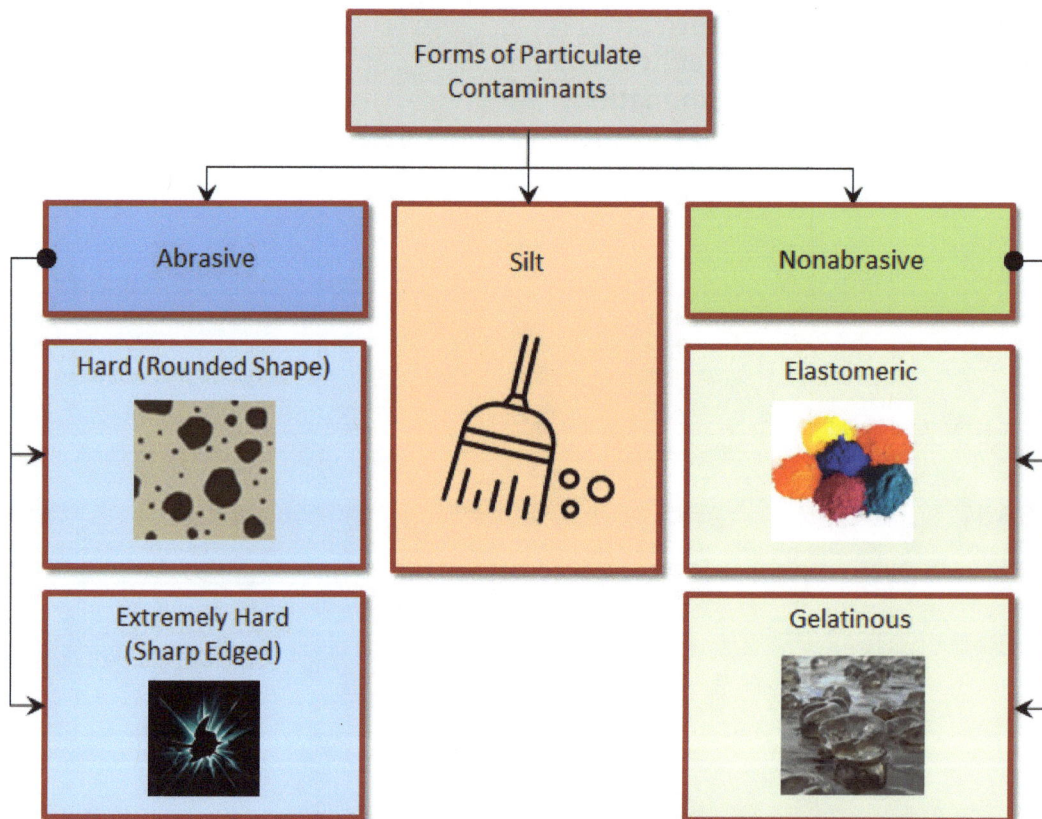

Fig. 6.1- Forms of Particulate Contaminants

6.2- Sources of Particulate Contamination

As shown in Fig. 6.2, particulate contaminants find their way into the hydraulic system through different sources as follows:

- **Built-in:** during manufacturing, assembly, and storage.
- **Introduced (Ingested):** from the environment.
- **Introduced (Induced):** during system servicing, make up fluid, and cleaning.
- **Generated:** due component wear and system normal operation.

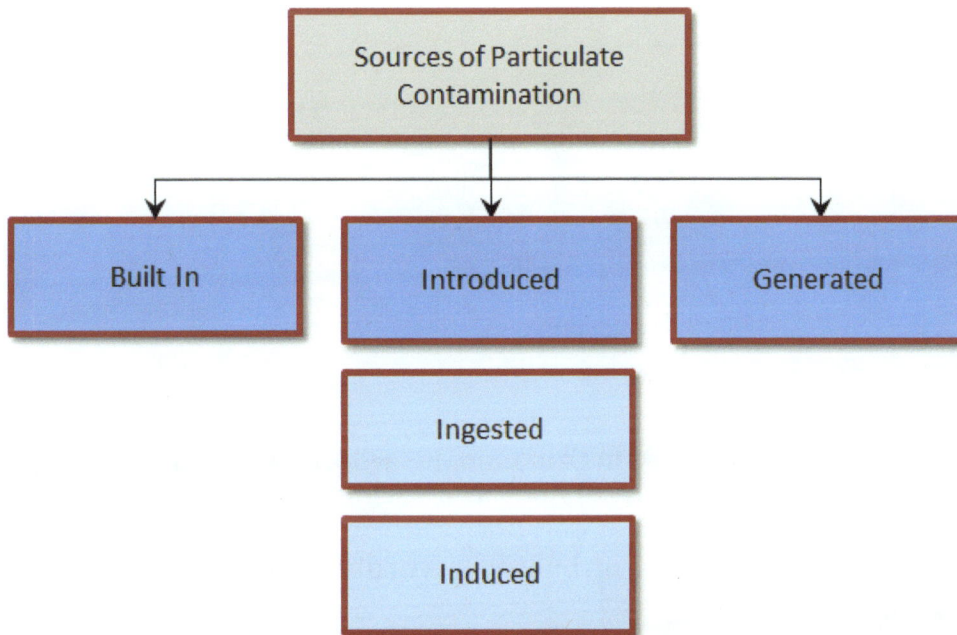

Fig. 6.2- Sources of Particulate Contaminants

6.2.1-Built-in Particulate Contamination

Built-in Particulate Contamination is defined as the particles remaining in the system following initial construction of the hydraulic components and system. Built-in contamination is also called *Primary Contamination*.

It is normally the responsibility of the machine builder to remove these contaminants before shipment. However, the end user should not assume that new components and systems are 100% clean. It is wise to pre-clean all hydraulic system components prior to assembly and utilize "good housekeeping" techniques in the assembly area.

As shown in Fig. 6.3, Built-in particulate contamination is a result of, but not limited to:

- **Foundry Operations:** Core sand and dust
- **Machining Operations:** metal chips and weld splatter.
- **Painting:** Paint flakes and overspray particulates.
- **Assembly:** Lubricants, Teflon tape, and other sealing materials.
- **Plumbing:** Hose cutting, tube bending and flaring, pipe threading, and fittings tightening.
- **Testing:** particles from testing fluid and environment.
- **Initial Cleaning:** Sands from sandblasting, fibers and lint from rags.
- **Storage and Handling:** Dust, insects, rust, scale from pipes, and airborne contaminants.
- **Shipping:** Packaging materials.

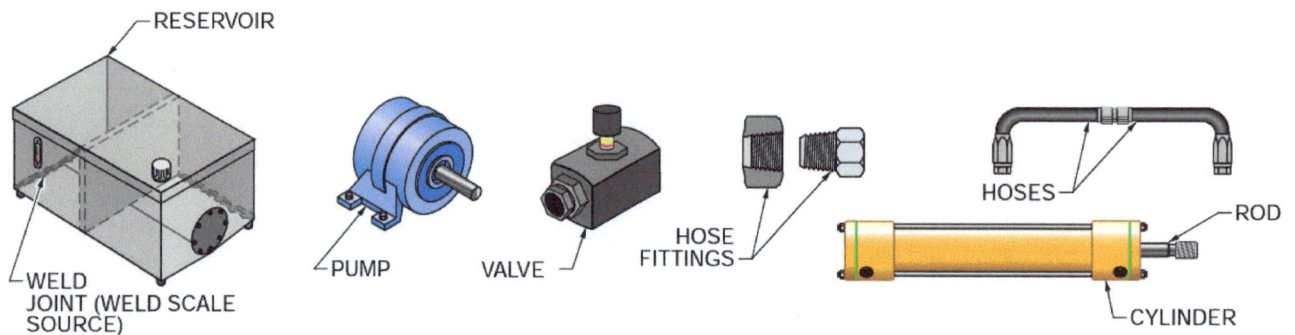

**Fig. 6.3- Built-in Contaminants within New Components
(Courtesy of American Technical Publishers)**

Figure 6.4 shows various types of built-in particulate contamination during production of hydraulic components.

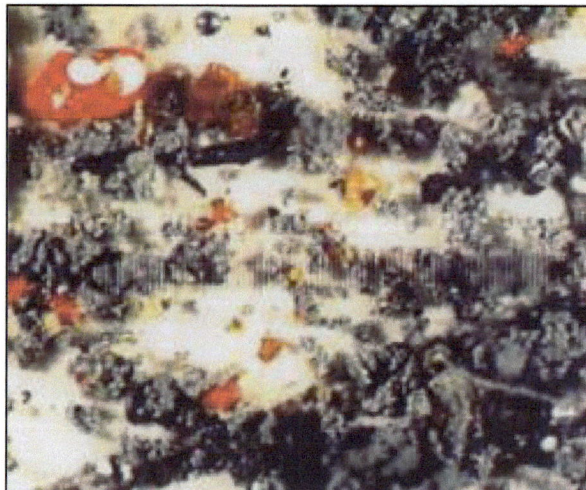

Fig. 6.4- Examples of Built-in Contaminants (Courtesy of Bosch Rexroth)

6.2.2-Ingested Particulate Contamination

Ingested (Ingressed) Particulate Contamination is defined as particles introduced to the system from the surrounding environment during system operation.

As shown in Fig. 6.5, ingested particulate contamination is a result of, but not limited to:

- **System Openings (1):** Most dirt particulates are introduced into hydraulic systems through vents, breather caps, filler tubes, and other system openings.
- **Lack of Cleaning (2):** Accumulated dirt on the surface of hydraulic reservoirs and components find their way into the system.
- **Cylinder Wipers (3):** Cylinder rods and seal systems are major contributors to contaminant ingression. The extended rod, coated with an oil film, will capture particulate contamination from the surrounding atmosphere. When the rod re-enters the cylinder housing, system fluid rinses the particles from the rod into system hydraulic oil.
- **Shaft Seals (4):** Particles are introduced into hydraulic systems through air leak around failed shaft seals.

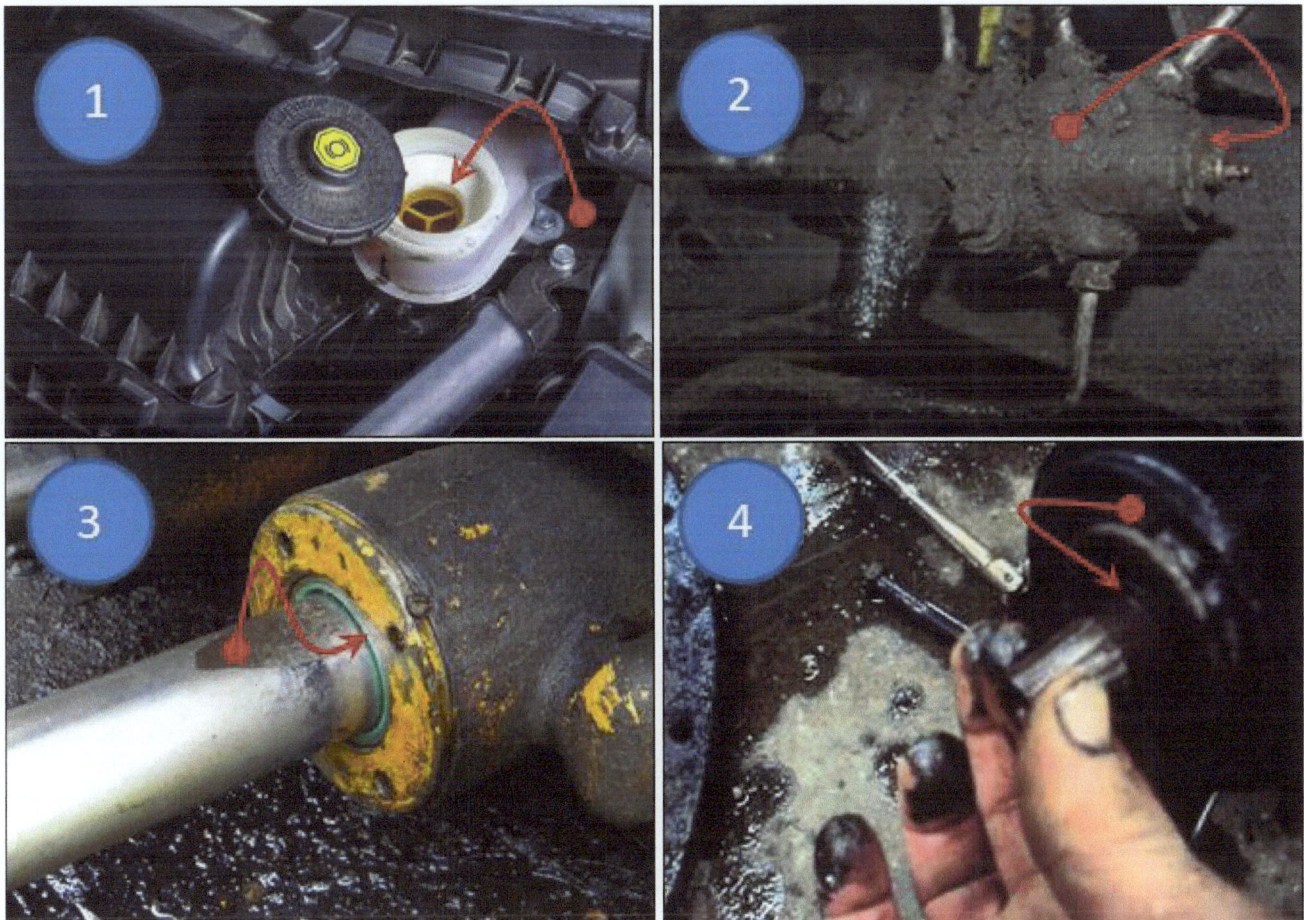

Fig. 6.5- Ingested Contamination

6.2.3-Induced Particulate Contamination

Induced Particulate Contamination is defined as the particles introduced into the system during maintenance, repair, and troubleshooting. Whenever a hydraulic system is "opened up" contaminates may be induced into the system.

Induced particulate contamination is a result of, but not limited to:

- **Make Up Hydraulic Fluids:** As shown in Fig. 6.6, new hydraulic fluid, as delivered from the drum, is not necessarily clean, even though it may appear clean. The smallest particles human eyes can see is about 40 μm (0.00158 inch). Therefore, someone might say they can see that a fluid sample is dirty; however, they cannot claim to see that a fluid sample is clean or acceptable. Oil out of shipping containers is usually contaminated to a level above what is acceptable for most hydraulic systems:

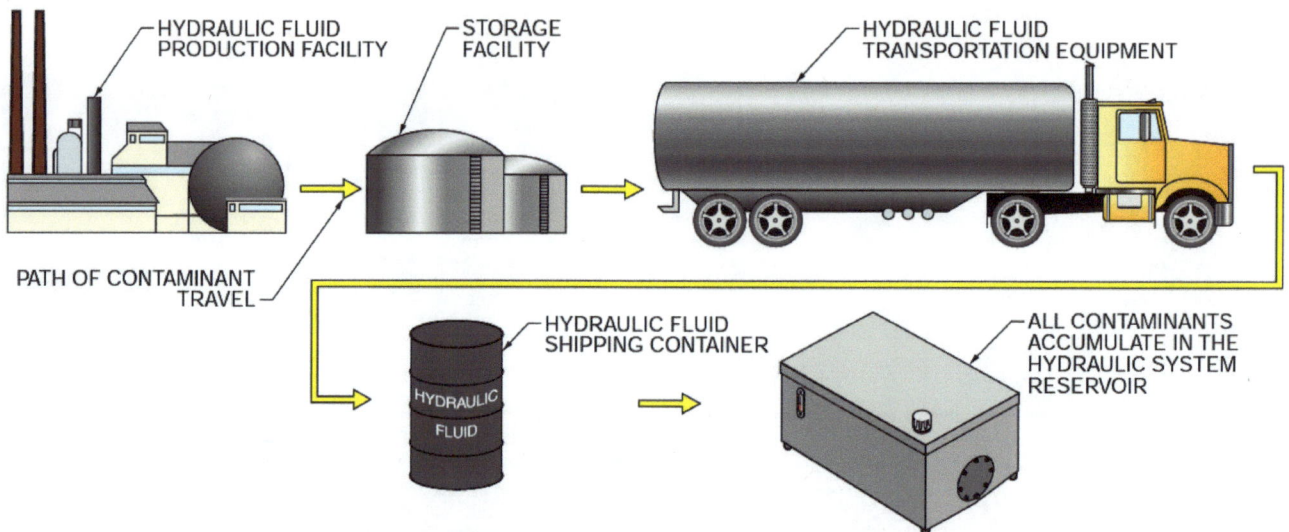

**Fig. 6.6- Introduced Contaminants During Hydraulic Fluid Handling
(Courtesy of American Technical Publishers)**

- **Filter Change:** Changing a filter element requires opening the filter housing and replacing the current element with a new one that was taken out of package. This process can induce particles into the system

- **Component Rebuilding:** Component overhauling process requires system dissembling, cleaning, possibly machining, reassembling, cleaning, lubrication, testing, packaging, and storage. All these steps can be accompanied by introducing some amount of dirt into the component.

- **Reservoir Clean-out:** Cleaning a reservoir may result in inducing lint from rags or sand if sandblasting is used.

- **Hydraulic Line Replacement:** Cutting hoses by saw blade, bending and flaring tubes, threading and welding pipes, and fitting tightening are accompanied by induced particulate contamination.

- **Open Ports of Components:** Leaving ports of hydraulic components open (such as pump intake and discharge ports) during servicing a hydraulic system provides continuous ingression of particles into the system.

6.2.4-Generated Particulate Contamination

Generated Particulate Contamination is defined as the particles internally generated during normal system operation. As shown in Fig. 6.7, Generated particulate contamination is a result of, but not limited to:

- Metallic particles due to components wear or loss of metal due to other reasons.
- Sludge products due to oxidation.
- Rubber compounds and elastomers degradation due to aging, temperature and high velocity fluid streams. During degradation, the elastomers will start releasing particles into the hydraulic system. Sources include hoses, accumulator bladders and seals, particularly dynamic seals, such as cylinder pistons and shaft seals.

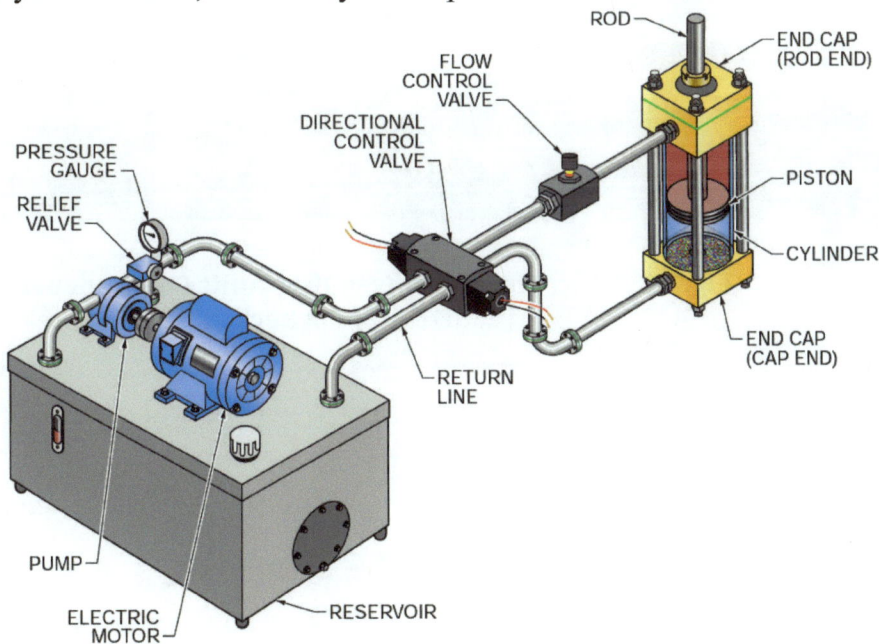

**Fig. 6.7- Particulate Contamination Generated During Normal System Operation
(Courtesy of American Technical Publishers)**

6.2.5- Wear Mechanisms in Hydraulic Components

As shown in Fig. 6.8, wear and loss of material due to abrasive particles is caused by different mechanisms depending on the existing combination of factors causing the wear.

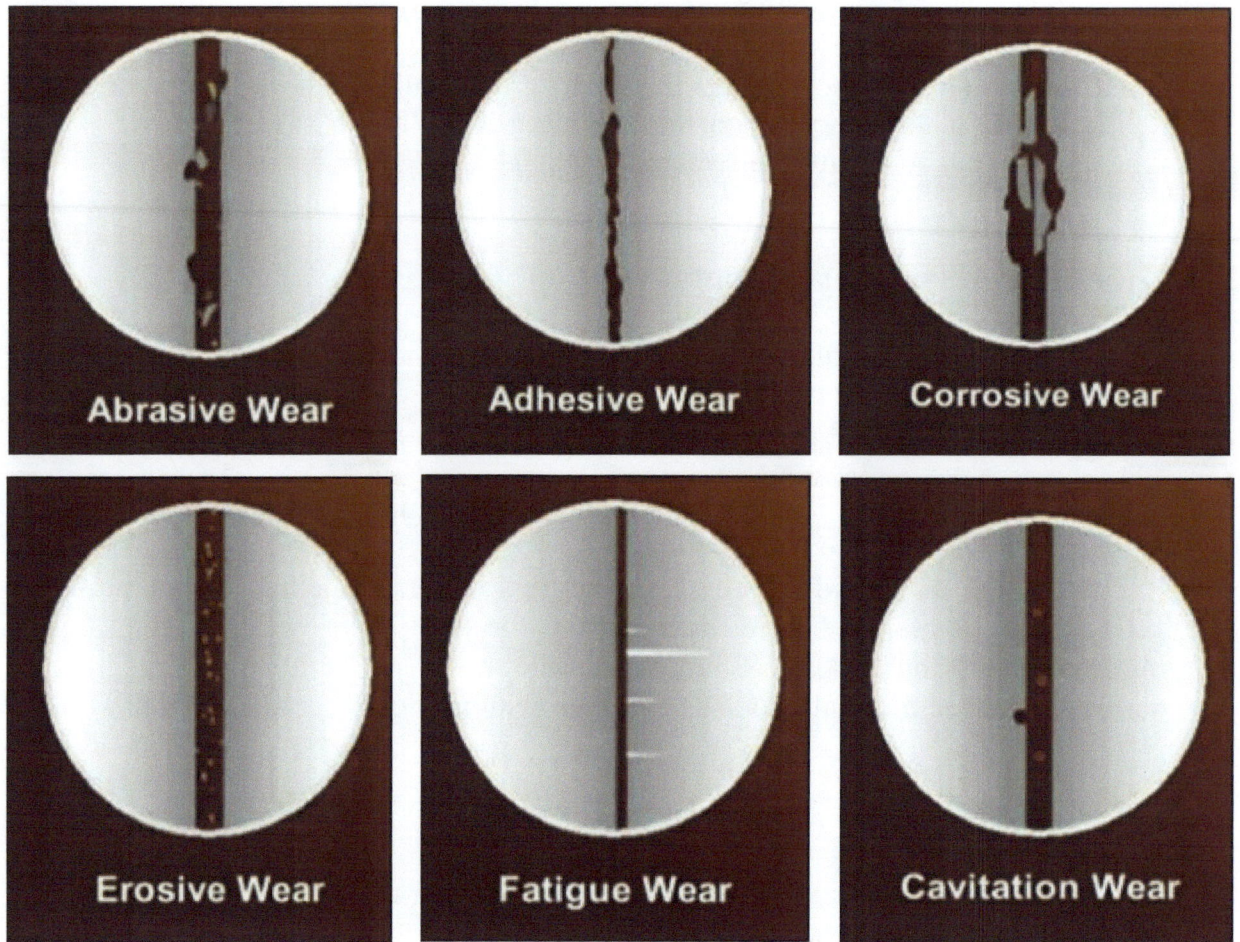

Fig. 6.8- Wear Mechanisms in Hydraulic Components
(Courtesy of Parker)

6.2.5.1- Abrasive Wear Mechanism

Abrasive wear is caused when hard particles bridge two moving surfaces, scraping one or both.

As shown in Fig. 6.9, hydraulic fluid is expected to create a lubricating film to separate moving surfaces, prevent metal-to-metal contacts, and allow the silt (small) particles to pass through causing no damage. Ideally, the lubricating film is thick enough to completely fill the clearance between moving surfaces. When the wear rate is low, a component is likely to reach its intended life expectancy, which may be millions of pressurization cycles.

It is to be noted that, *Operating (Dynamic) Clearance* and consequently the actual thickness of a lubricating film depends on:

- F, Applied load.
- v, Relative speed of the two surfaces.
- ν, Fluid viscosity.

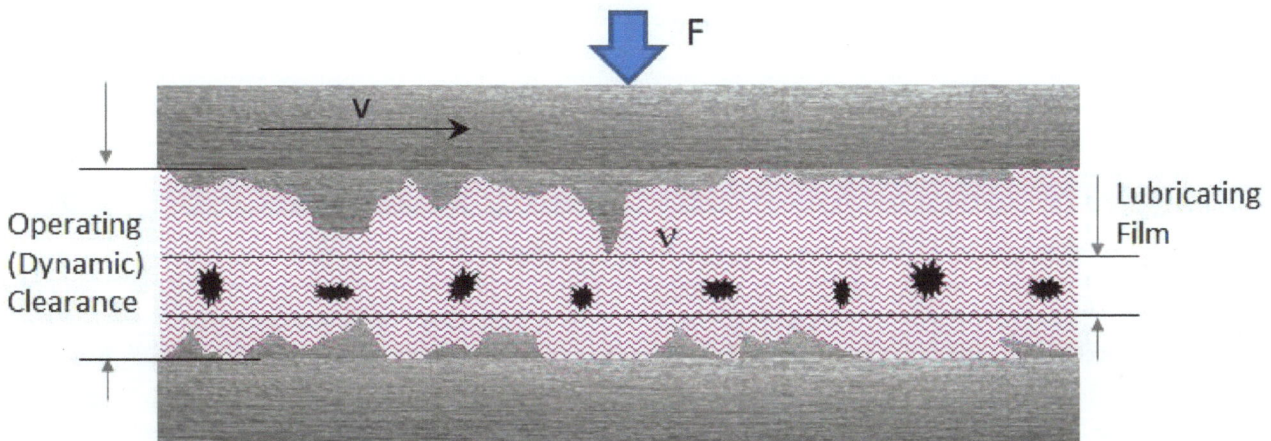

Fig. 6.9- In Normal Conditions, Silt Particles Pass Through Causing No Damage

As shown in Fig. 6.10, operating (dynamic) clearance in bearings is not equal to the machine clearance of the bearing but depends upon the load, speed, and lubricant viscosity.

Table 6.1 shows typical *Dynamic Oil Film Thickness* in various hydraulic components.

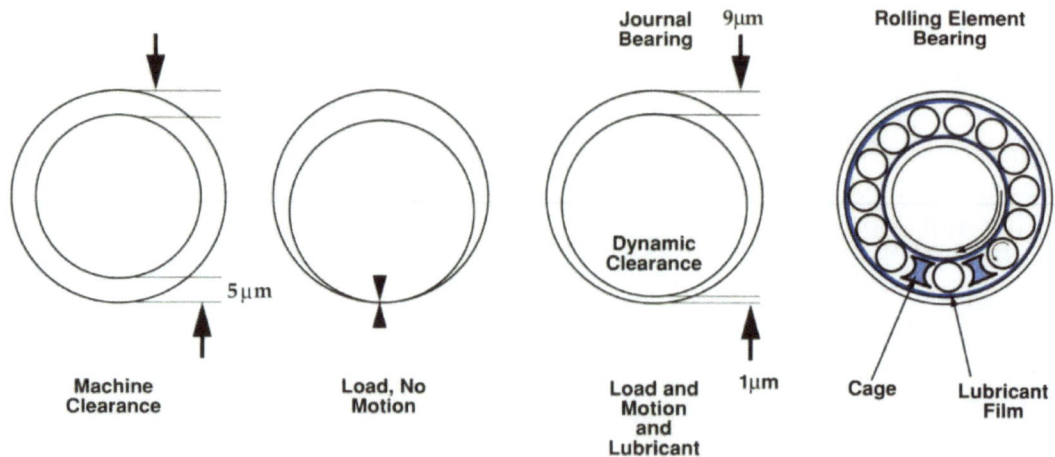

Fig. 6.10- Dynamic Clearances in Bearings (Courtesy of Pall)

Dynamic oil film	
Component	**Oil film thickness in micron (μm)**
Journal, slide and sleeve bearings	0.5-100
Hydraulic cylinders	5-50
Engines, ring/cylinder	0.3-7
Servo and proportional valves	1-3
Gear pumps	0.5-5
Piston pumps	0.5-5
Rolling element bearings / ball bearings	0.1-3
Gears	0.1-1
Dynamic seals	0.05-0.5

**Table 6.1- Typical Dynamic Oil Film Thickness in Various Hydraulic Components
(Courtesy of Noria Corporation)**

As shown in Fig. 6.11, the particle size causing the most damage is the one that is equal to or slightly larger than the dynamic clearance. When a particle enters the clearance space between two moving surfaces, it acts like a grinding tool to remove material from the opposing surface. Figure 6.12 shows grooving caused by hard abrasive particles.

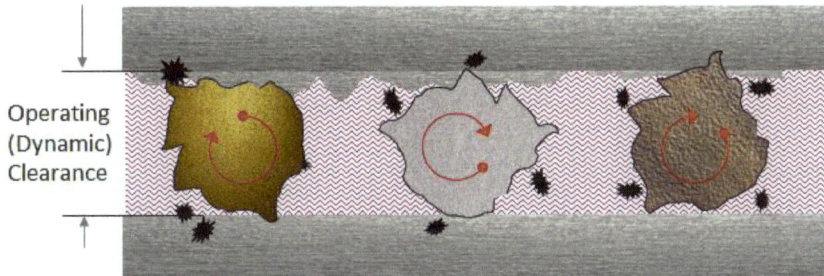

Fig. 6.11- Abrasive Wear Mechanism

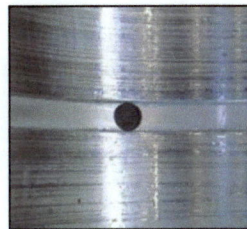

Fig. 6.12- Abrasive Wear Damage

6.2.5.2- Adhesive Wear Mechanism

Adhesive wear results when moving surfaces tend to stick together because of lubricating oil film collapse that allows metal-to-metal contact.

In many components, mechanical loads are such an extreme that they squeeze the lubricant into a very thin film, less than 1 micrometer thick. As shown in Fig. 6.13, due to excessive load, low speed and/or reduction in fluid viscosity:
- Lubrication film is squeezed and penetrated by the moving parts surface asperities.
- Metal-to-metal contact occurs and moving surfaces are "cold welded" together.
- Particles are generated as the surface asperities are sheared off.

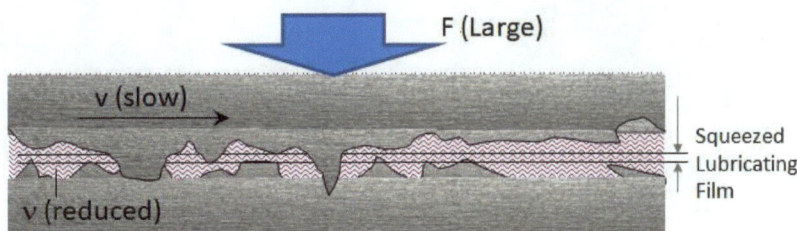

Fig. 6.13- Adhesive Wear Mechanism

6.2.5.3- Corrosive Wear Mechanism

Corrosive wear is the loss of material over a large area typically caused by water, chemical, or microbial contamination in the fluid. As shown in Fig. 6.14, rust due to oxidation on a cylinder rod is one form of corrosive wear. Rust entered the cylinder through failed rod seals causing more wear due to surface abrasion.

Fig. 6.14- Corrosive Wear due to Rust (www.gallagherseals.com)

6.2.5.4- Erosive Wear Mechanism

Erosive wear occurs when fine particles (silt) in a high-speed stream of fluid eat away metering edges or critical surfaces. As shown in Fig. 6.15, particles already found in the fluid are flowing at high-speed eroding spool lands, metering orifices, and component surfaces. As pressure rises, even the smallest particles contribute to the erosion process.

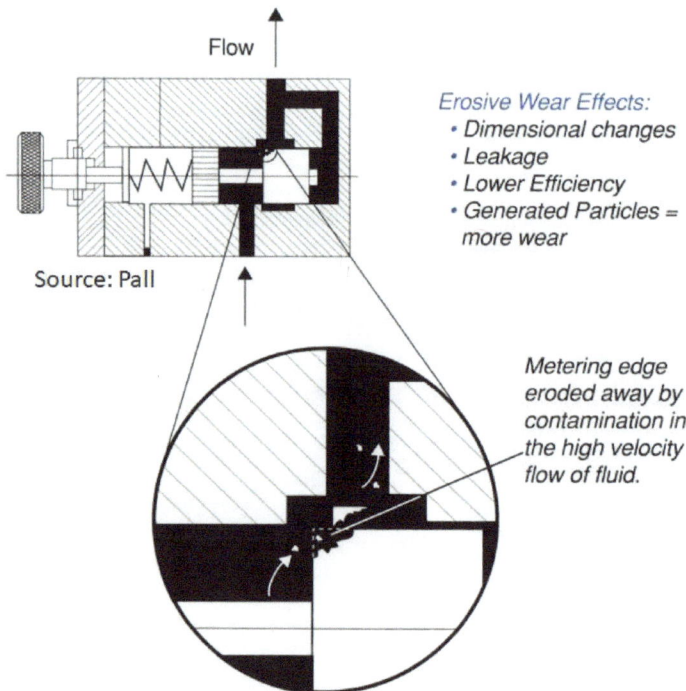

Flow

Erosive Wear Effects:
• *Dimensional changes*
• *Leakage*
• *Lower Efficiency*
• *Generated Particles = more wear*

Source: Pall

Metering edge eroded away by contamination in the high velocity flow of fluid.

Source: Ultra Clean

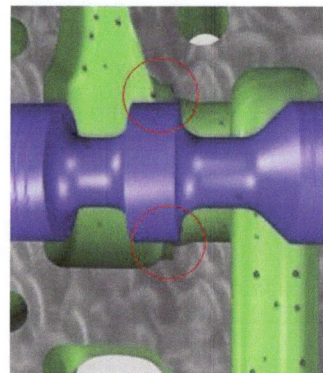

Source: MSOE

Fig. 6.15- Erosive Wear Mechanism

6.2.5.5- Fatigue Wear Mechanism

Fatigue wear is surface degradation due to periodic or reversable loads.

Figure 6.16 explains the fatigue wear mechanism. In the beginning, tiny abrasive particles are wedged in the fine clearances of rotating hydraulic components. the bearing surfaces within these components are microcracked. These cracks spread and propagate under the effect of periodic load. Eventually, even without additional particulate contaminates, the surface fails producing additional particles.

Source:
Ultra clean

1- Particle Trapped   2- Microcracks Initiated

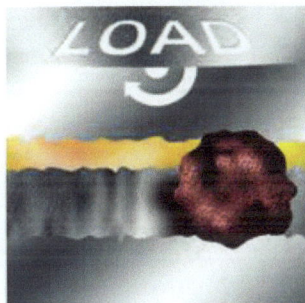 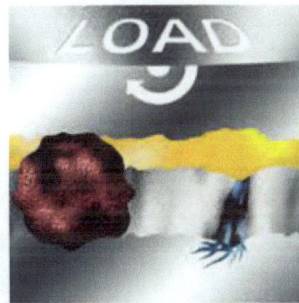

Source: CJC

3- Cracks Propagated   4- Surface Degrades

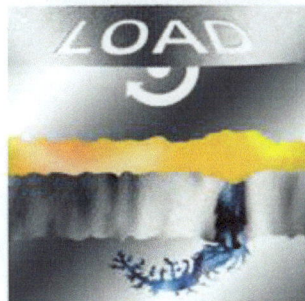 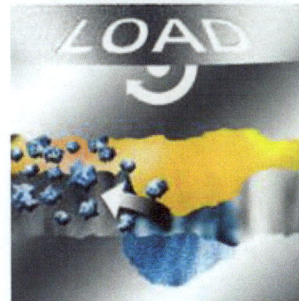

Fig. 6.16- Fatigue Wear Mechanism (Courtesy of C.C. Jensen Inc)

6.2.5.6- Cavitation Wear Mechanism

Cavitation wear is due to surface pitting caused by implosion of air bubbles putting shock loads on a small surface area.

The mechanism of cavitation wear can be described as follows: Hydraulic fluids contain (7-10) % by volume air. This amount of air, under normal operating temperature and atmospheric pressure, is homogeneously dissolved on the molecular level within the fluid. This amount of dissolved air does not affect the fluid properties or performance. When the fluid passes through a negative pressure zone, dissolved air separates from the fluid in form of bubbles. It is commonly known that cavitation starts when the hydraulic fluid is subjected to negative pressure. However, the formation of bubbles within the liquid could begin even in the presence of positive pressure that are equal to or close to the vapor pressure of the fluid at the given temperature. Bubbles increase rapidly in size and in numbers. Subsequently as shown in Fig. 6.17, when the bubbles enter a zone of high pressure, they are condensed (imploded). Implosion of bubbles is accompanied by a microjet shock load, destruction of material bonds, sound emission, and other undesirable effects.

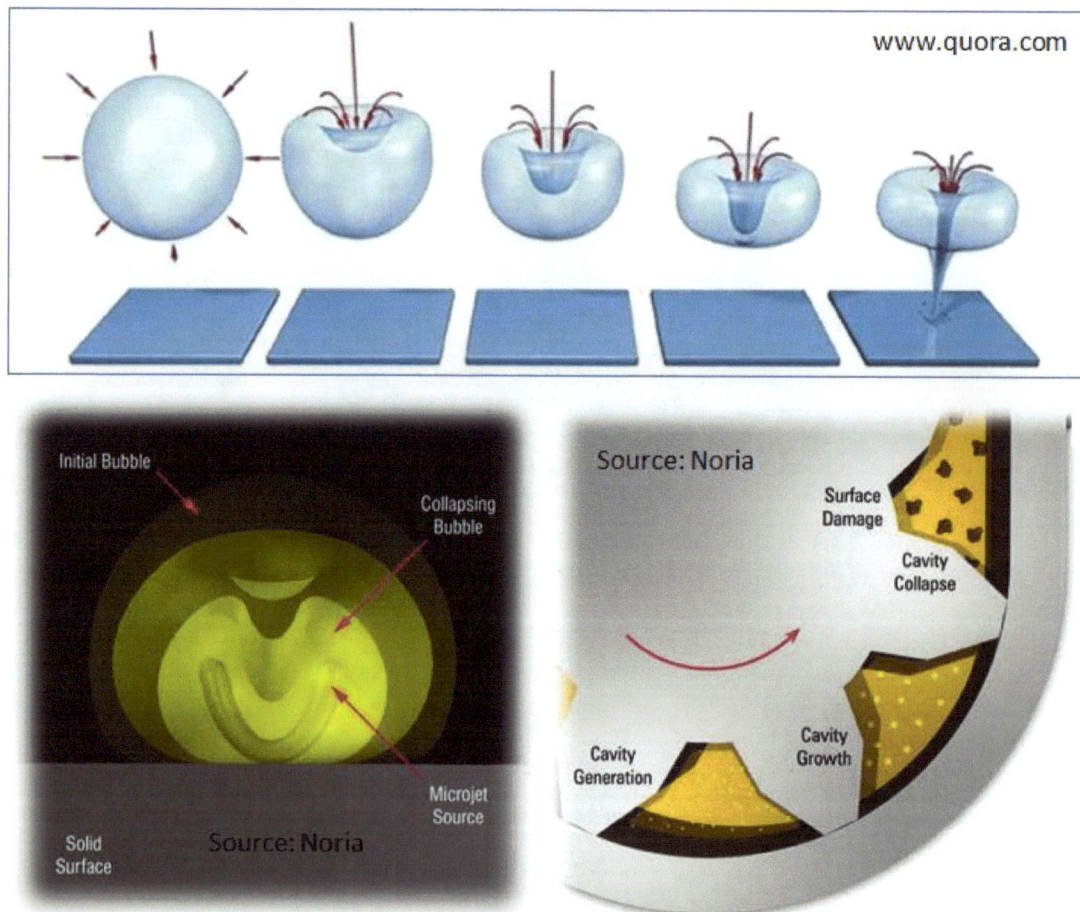

Fig. 6.17- Cavitation Wear Mechanism

6.3- Contamination Particle Sizes

Particle sizes are generally measured on the micrometer scale. One micrometer (or "micron") is one millionth of a meter or 39 millionth of an Inch. To get a better sense of the particle size, Fig. 6.18 shows relative sizes of different substances.

The limit of human visibility is approximately 40 micrometers. Keep in mind that most damage-causing particles in hydraulic or lubrication systems are smaller than 40 micrometers. Particulates in the (1 – 20) micron range are typically the most damaging to the hydraulic system. Therefore, they are microscopic and cannot be seen by the unaided eye.

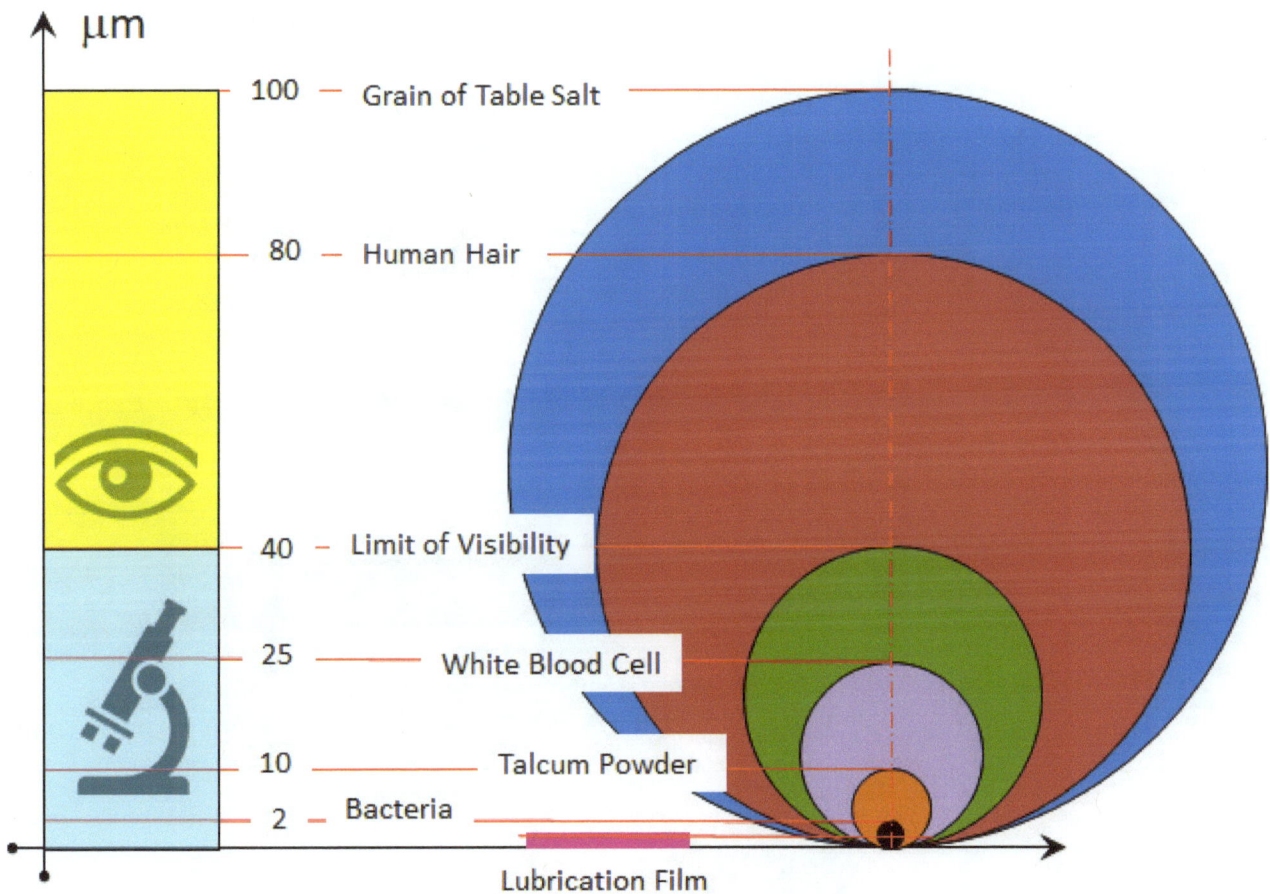

Fig. 6.18- Relative Particles Size

Figure 6.19 shows, in a demonstrative way, the range of particle sizes that affects hydraulic systems.

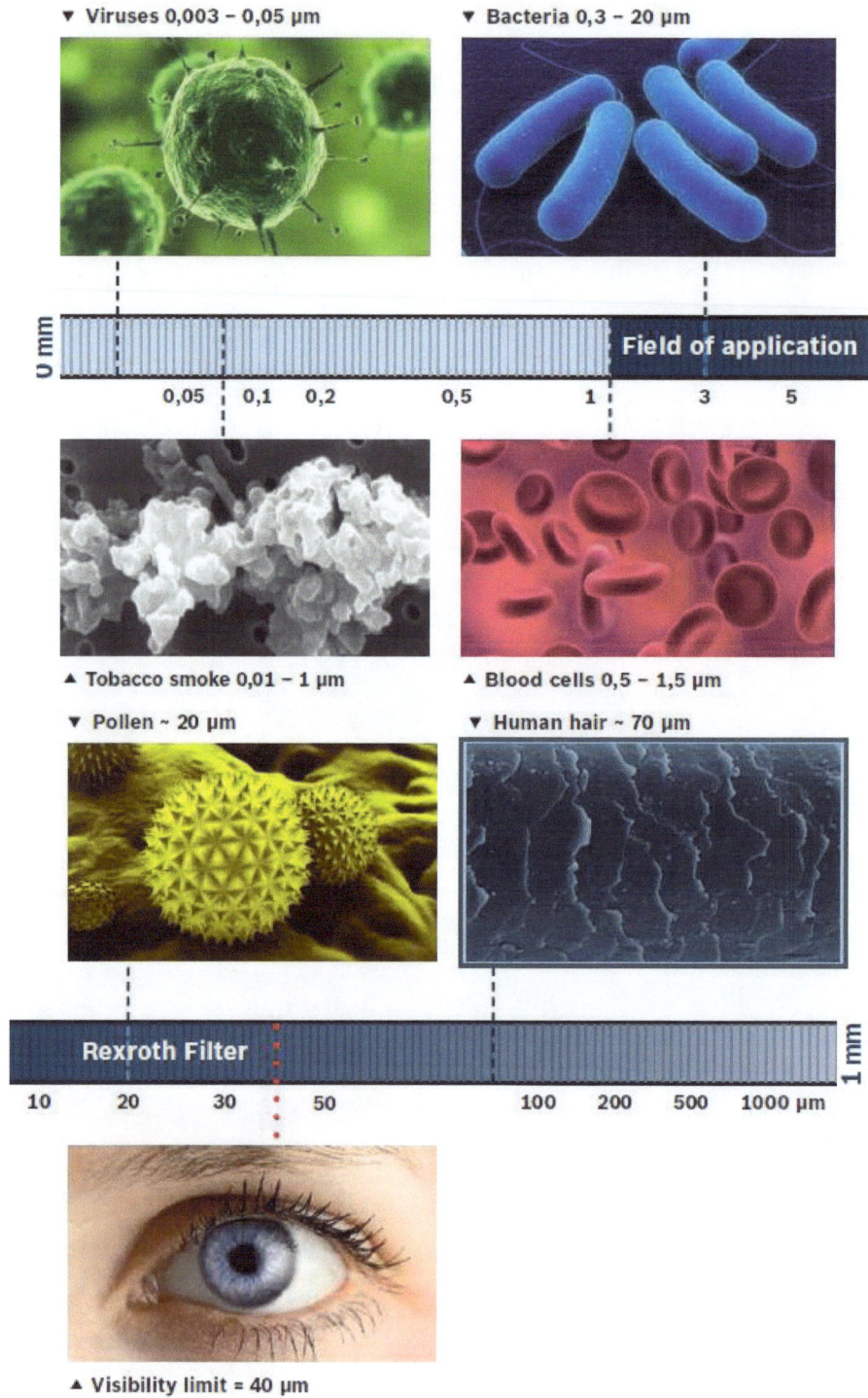

Fig. 6.19- Range of Particle Sizes that affects Hydraulic Systems (Courtesy of Bosch Rexroth)

Figure 6.20 shows typical particle sizes in a contaminated sample of a hydraulic fluid.

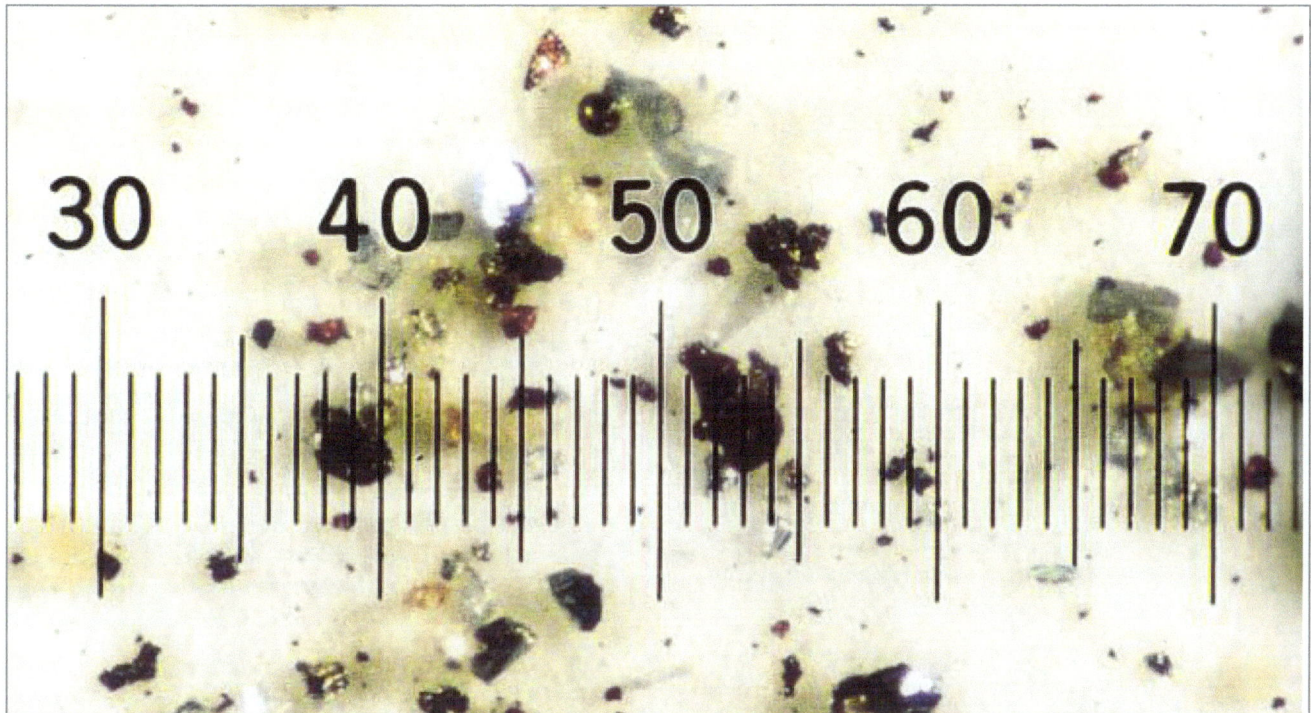

Actual photomicrograph of particulate contamination - (Magnified 100x Scale: 1 division = 20 microns)

**Fig. 6.20- Typical Particle Sizes in a Contaminated Sample of a Hydraulic Fluid
(Courtesy of Parker)**

6.4- Critical Clearances in Hydraulic Components

Table 6.2 shows typical clearances in bearings

Component	Clearance (µm)
Roller Element Bearings	0.1-1
Journal Bearings	0.5-100
Hydrostatic Bearings	1-25

Table 6.2- Typical Bearing Clearances (Courtesy of Pall)

Figure 6.21 shows typical clearances in hydraulic pumps and valves. These clearances justify the importance of keeping the oil free of particulate contaminants.

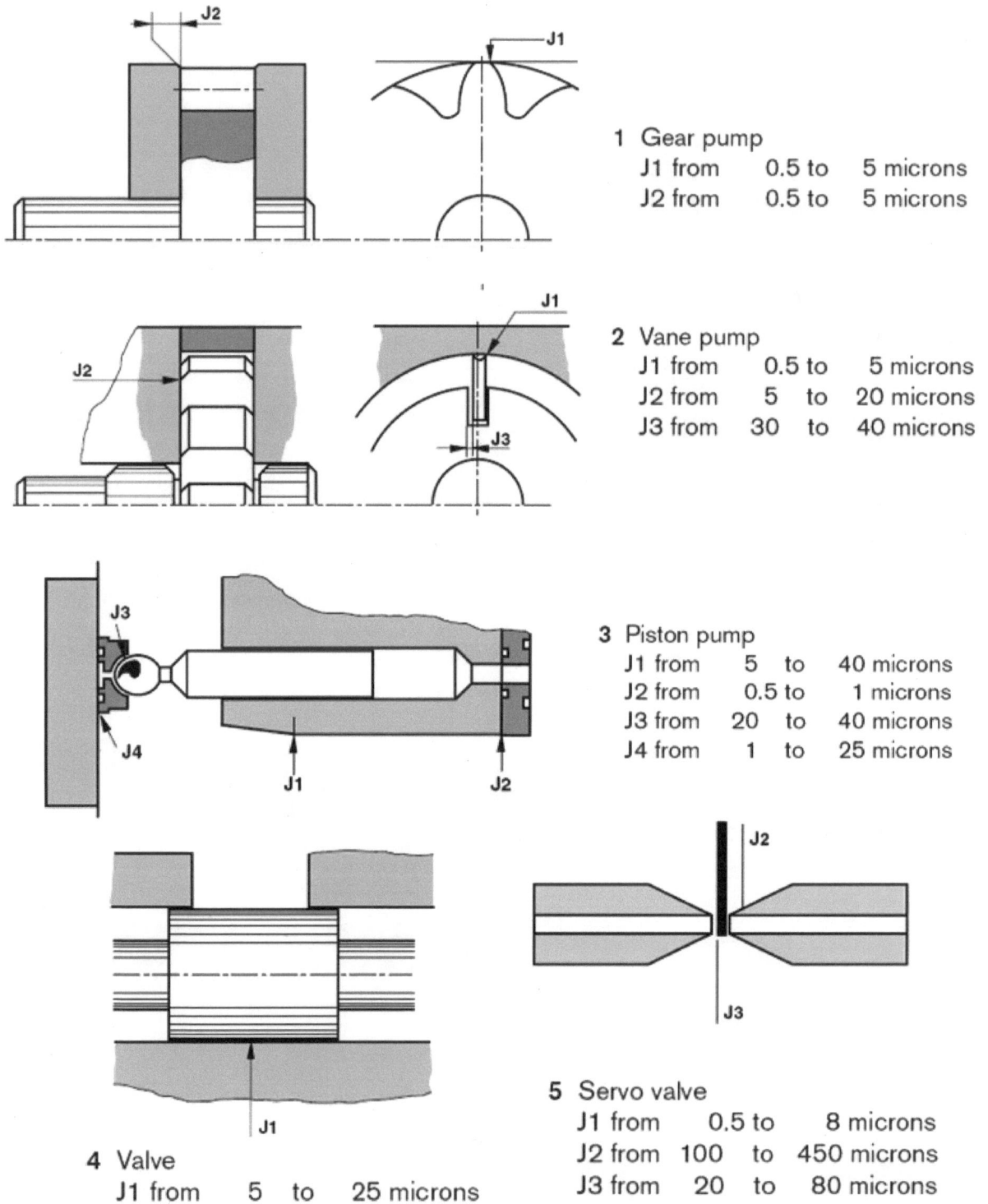

1 Gear pump
   J1 from   0.5 to   5 microns
   J2 from   0.5 to   5 microns

2 Vane pump
   J1 from   0.5 to   5 microns
   J2 from   5   to   20 microns
   J3 from   30 to   40 microns

3 Piston pump
   J1 from   5   to   40 microns
   J2 from   0.5 to   1 microns
   J3 from   20 to   40 microns
   J4 from   1   to   25 microns

4 Valve
   J1 from   5   to   25 microns

5 Servo valve
   J1 from   0.5 to   8 microns
   J2 from   100 to   450 microns
   J3 from   20 to   80 microns

Fig. 6.21- Typical Clearances in Hydraulic Components (Courtesy of Bosch Rexroth)

6.5- Effects of Particulate Contamination

Particulate contamination directly affects the reliability of the hydraulic system and longevity of components.

6.5.1- Replication of Particulate Contamination

Particulate contaminants circulating in hydraulic systems cause surface degradation through various mechanical wear mechanisms (abrasion, adhesion, corrosion, erosion, fatigue, and cavitation).

As shown in Fig. 6.22, regardless the wear mechanism, each abrasive dirt particle acts like an "Abrasive Seed" that produces additional dirt particles in a *"Chain Action"*. So, solid particles are replicating.

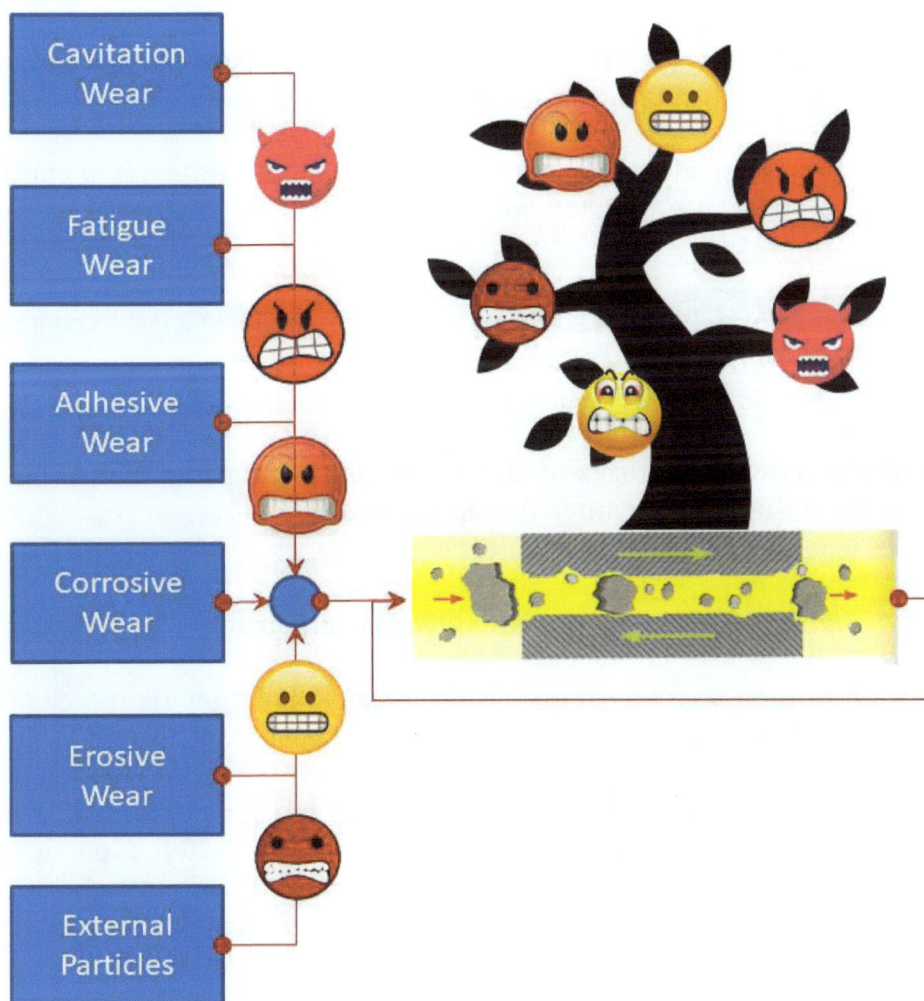

Fig. 6.22- Abrasive Particles Replication

6.5.2- Factors Affecting Level of Damage due to Particulate Contamination

Abrasive particles are mainly responsible for the wear of components. The level of damage due to abrasive particles depends on:

- Particle Size.
- Particle Shape.
- Particle Material.

6.5.2.1- Effect of Particle Size

As it has been discussed, each component of an oil hydraulic system has critical dynamic clearances between sliding or sealing surfaces. With respect to the dynamic clearance, particle size has the following effect on the level of damage:

Very Small Particles: Very fine particles are called *Silt*. These particles usually pass through most clearances slowly causing surface degradation.

Degradation Failure is usually a long-term failure which occurs when a component is worn to the extent that the system no longer operates at the speed, precision or force defined in its original specifications.

Less than Clearance Size Particles: They can pass through the clearance causing a low rate of wear that leads to *Normal Life Failure* that is failure at the end of a component expected lifetime.

Clearance Size Particles: Particles of about the same size as the operating clearance are the most dangerous because they remove material increasing the clearance which generates even more particles leading to catastrophic failures.

Catastrophic Failures occur rapidly or suddenly such as seizure or breakage of component.

Larger than Clearance Size Particles: If the particle is larger than a clearance or orifice, it may not enter the clearance causing wear, but it may cause Intermittent Failures.

Intermittent Failures do not occur frequently such as blocking a control orifice or a seat of a poppet valve. Usually in a control orifice or a valve seat, when a particle is removed, the component will continue to function normally. However, if the particle restricts or prevents lubrication flow into a bearing or other critical area, serious failure can occur almost immediately.

6.5.2.2- Effect of Particle Shape

As shown in Fig. 6.23, particles with irregular shape and sharp edges (left side) cause deep scratches and are more dangerous than spherical particles (right side).

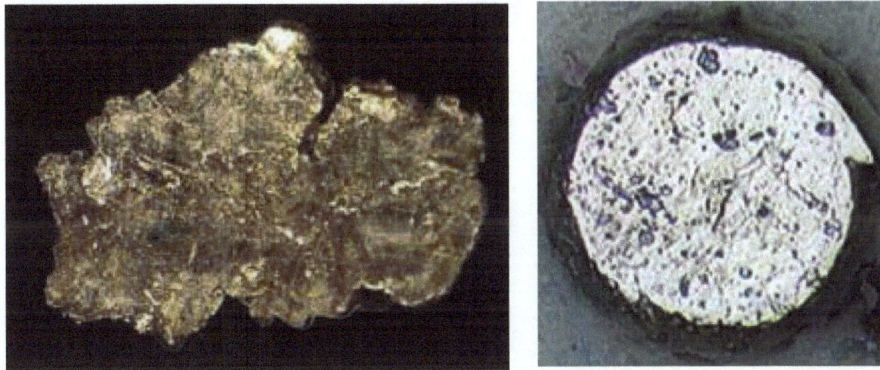

Fig. 6.23- Shapes of Particulate Contamination (Courtesy of Noria Corporation)

6.5.2.3- Effect of Particle Material

Very Severe Damage results from particles of:
- Rust.
- Scale.
- Carbide steel.
- Iron.
- Silica (sand) and other very hard materials.

Severe Damage results from particles of:
- Brass.
- Aluminum.
- Bronze.
- Calcium and Sulphur products.

Slight Damage results from particles of:
- Packaging plastics.
- Laminated fabrics.
- Elastomeric and rubber particles from seal residues.
- Paint chips or overspray.
- Gelatinous particles.

6.5.3- Typical Failures due to Particulate Contamination

The ingress of contaminants not only can cause preliminary damage to system components but also premature failure as well. Particulate contamination in a hydraulic system can lead to one or a combination of the following consequences:

Mechanical Efficiency:
Increased friction between surfaces can decrease the efficiency of hydraulic components.

Volumetric Efficiency:
Internal clearances grow larger increasing internal leakage and decreasing pump and motor volumetric efficiency.

Lubrication:
Blocked lubrication passages can cause catastrophic component failure.

Damage to Rotating Components:
Under high friction and temperature, seizure of rotating parts in pumps and motors can occur.

Damage to Valves:
Under high contamination conditions seizure or breakage of shifting elements, such as valve spools, could occur. Also, higher internal leakage lowers efficiency and increases heat generation.

Damage to Cylinders:
- Hydraulic cylinder barrel scratching resulting in load creeping or drifting.
- Hydraulic cylinder rod scratching and rod seal failure resulting in external leakage.

Filter Clogging:
- Pressure and return filter clogging add back pressure to the system.
- Suction filter clogging causes pump cavitation.
- Frequent filter replacement increases operating cost of the machine and disposal cost of spent filters.

System Efficiency:
- Loss of components efficiency reduces the overall system performance.
- Machine efficiency loss is gradual.
- Hydraulic systems efficiency can drop as much as 20% before the operator will notice.

System Performance:
- Increased orifices dimensions results in loss of component controllability.
- Stick-Slip motion of sliding parts in valves resulting in jerkiness of actuator motion.
- Internal leakage results in slower system performance.

System Productivity: Increased machine down time reduces productivity.

Silt Lock: *Silt Lock* (also known as *Contamination Lock*) is an accumulation of silt causing seizure or jamming of components. It is a type of failure that usually doesn't involve wear or permanent internal damage to components, it is rather sudden and unpredictable.

Because of its lack of warning or predictability, silt lock is responsible for a significant number of catastrophic failures in mechanical machinery including even loss of human life. Silt lock has been found to be the root cause of countless failures related to aircraft, spacecraft, passenger cars, elevators, turbine generators, tower cranes, etc.

Silt Lock usually occurs in control valves preventing spool movement from neutral to a shifted position and vice versa. This results in unpredicted actuator movement or failure of the actuator to stop moving.

Electrohydraulic spool valves such as solenoid, pulse-width modulated (PWM), proportional control, and servo valves are sensitive to silt lock. As shown in Fig. 6.24, silt particles can enter the clearances between the spool and bore in the leakage path. This increases the static friction of the spool when the valve is actuated. This reduces the valve dynamic response and cause a *stick-slip* movement, which is also known as a *hard-over* condition.

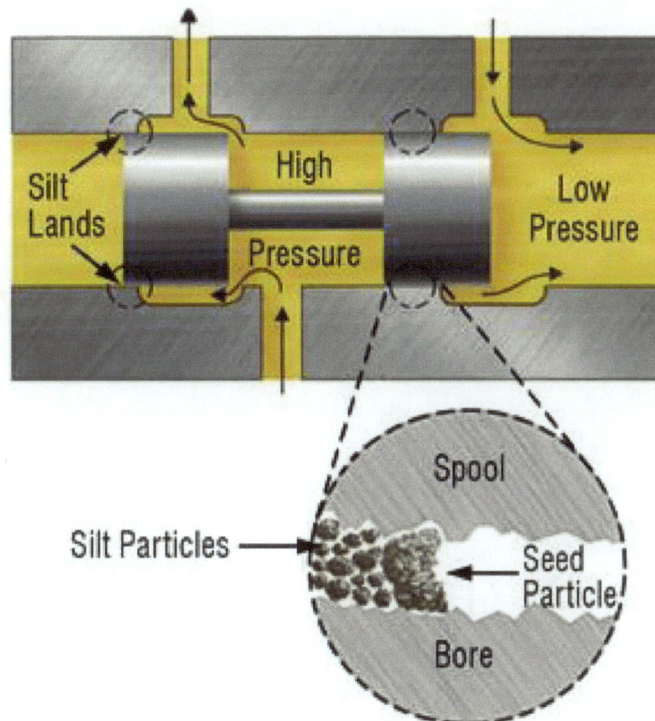

Fig. 6.24- Silt Lock in Spool Valves (Courtesy of Noria Corporation)

6.5.4- Examples of Failed Components due to Particulate Contamination

6.5.4.1- Pump Failure due to Particulate Contamination

Figure 6.25 shows opposing moving surfaces within hydraulic pumps that are commonly affected by abrasive wear as follows:

- **In Gear Pumps and Motors:**
  - o The radial clearance between opposite teeth of a gear pump or motor and between the tip of the teeth and the housing.
  - o The side clearance between the face of the gears and the bearing plates.

- **In Vane Pumps and Motors:**
  - o The radial clearance between the tip of each vane and the cam ring.
  - o The side clearance between the body of the vane and the rotor.

- **In Piston Pumps and Motors:**
  - o The clearance between the cylinder block and the valve plate.
  - o The clearance between each piston and its piston chamber.
  - o The clearance between the spherical head of each piston and its slipper pad.
  - o The clearance between the slipper pads and the swash plate.

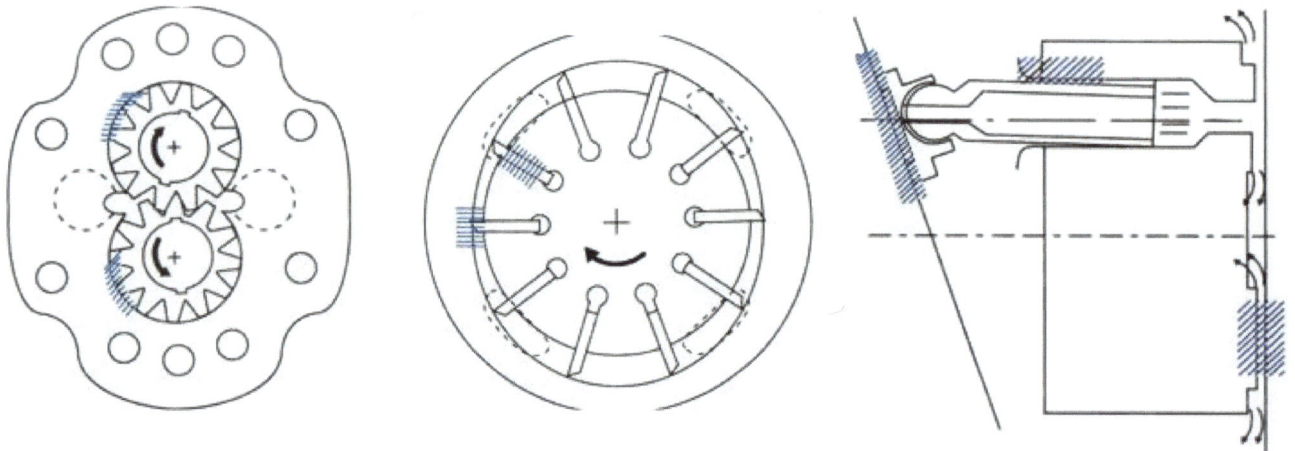

Fig. 6.25- Commonly Worn Areas within Hydraulic Pumps and Motors (Courtesy of Pall)

Figure 6.26 shows examples of piston pump failures due to particulate contamination. Damage happens when particulate contamination level (ISO 4406) exceeds manufacturers recommendations. The figure shows worn/broken slipper pads, retaining plates, and valve plate.

Fig. 6.26- Examples of Piston Pumps Failure due to Particulate Contamination

6.5.4.2- Valve Failure due to Particulate Contamination

Figure 6.27 shows how spool valves affected by abrasive wear.

- **In Spool Valves:** The clearance between the spool and the sleeve.
- **In Poppet Valves:** The poppet/seat of the valve.

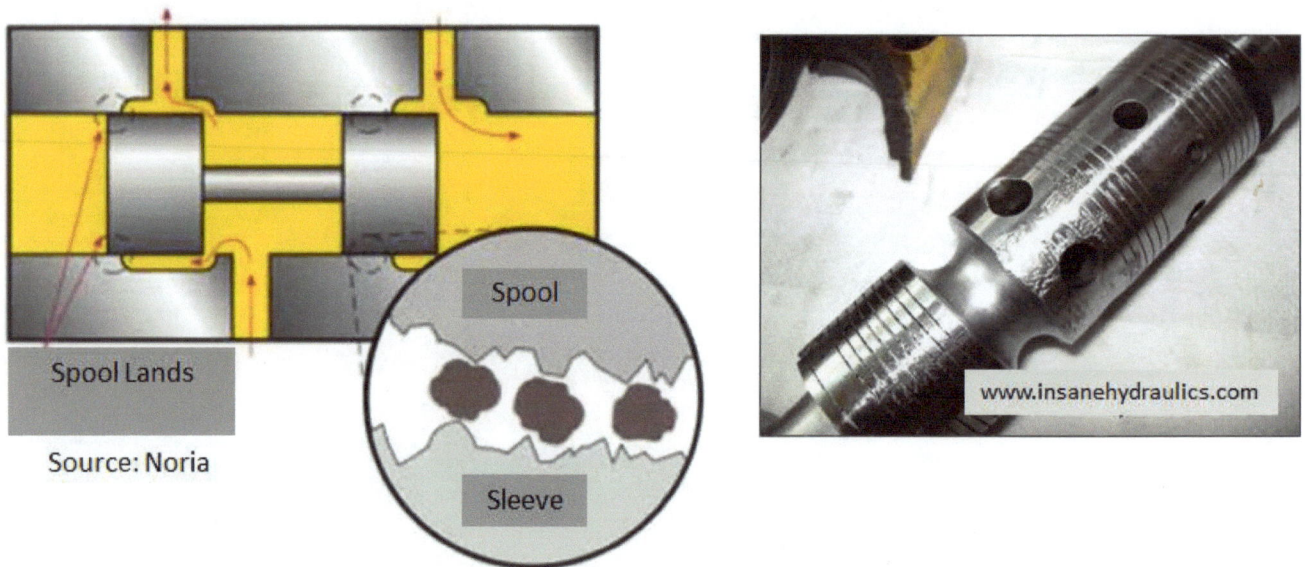

Fig. 6.27- Commonly Worn Surfaces in Spool Valves

Figure 6.28 shows how poppet valves affected by abrasive wear.

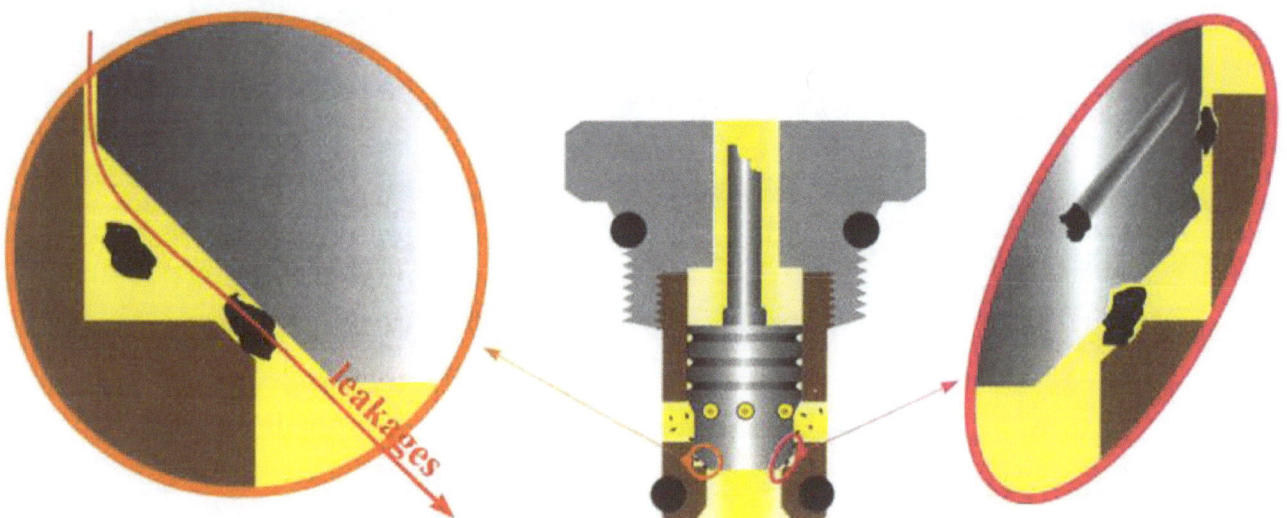

Fig. 6.28- Commonly Worn Surfaces in Poppet Valves (Courtesy of ASSOFLUID)

6.5.4.3- Cylinder Failure due to Particulate Contamination

Figure 6.29 shows the opposing moving surfaces within hydraulic cylinders that are commonly affected by abrasives.

- **At Rod Seals:** The clearance between the cylinder rod, rod seals and wipers.
- **At Piston Seals:** The clearance between the piston seal package and the cylinder barrel.

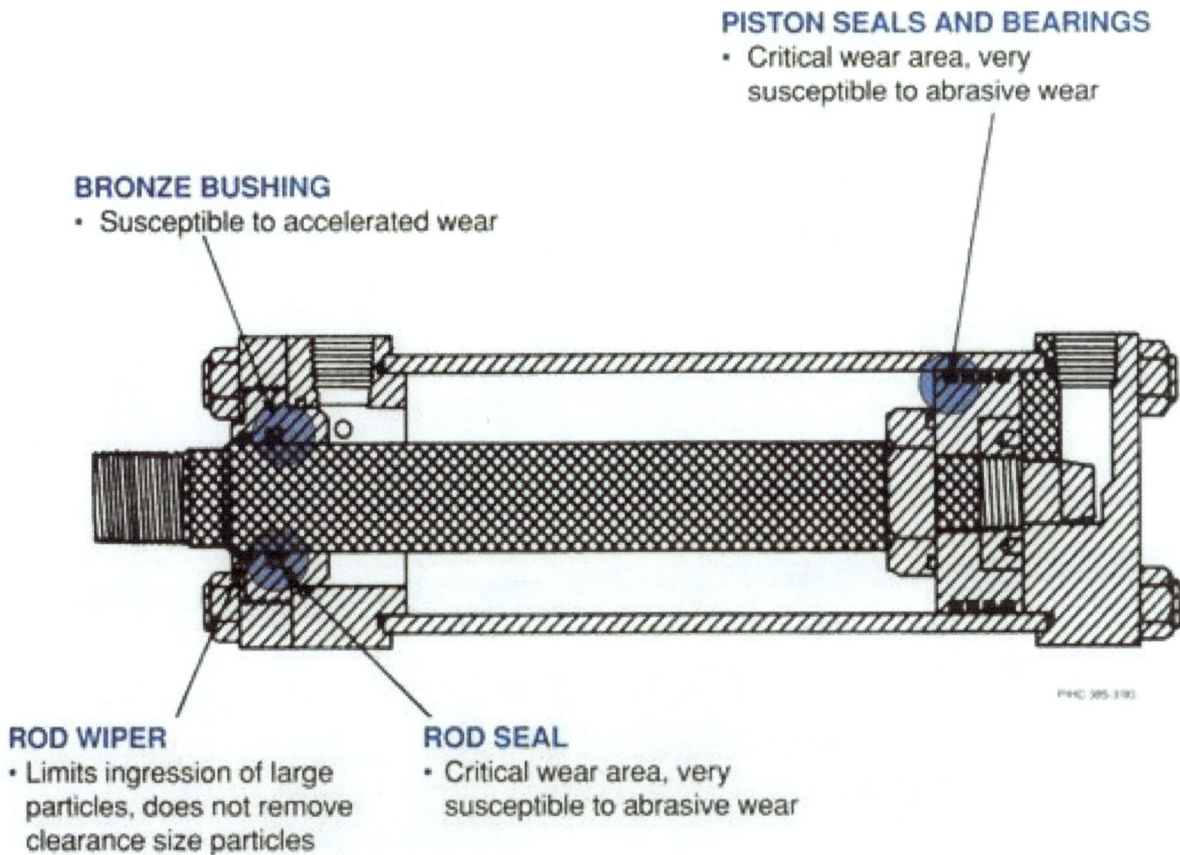

PISTON SEALS AND BEARINGS
- Critical wear area, very susceptible to abrasive wear

BRONZE BUSHING
- Susceptible to accelerated wear

ROD WIPER
- Limits ingression of large particles, does not remove clearance size particles

ROD SEAL
- Critical wear area, very susceptible to abrasive wear

Fig. 6.29- Commonly Worn Areas within Hydraulic Cylinders (Courtesy of Pall)

Figure 6.30 shows examples of hydraulic cylinder failure due to particulate contamination. The upper part of the figure shows visible leakage due to seal failure caused by abrasive particulate contamination. The figure shows (on lower left) piston rings that were eaten away by contaminants. The figure shows (on lower right) a scored cushion plunger resulting in a loss of cushioning effect.

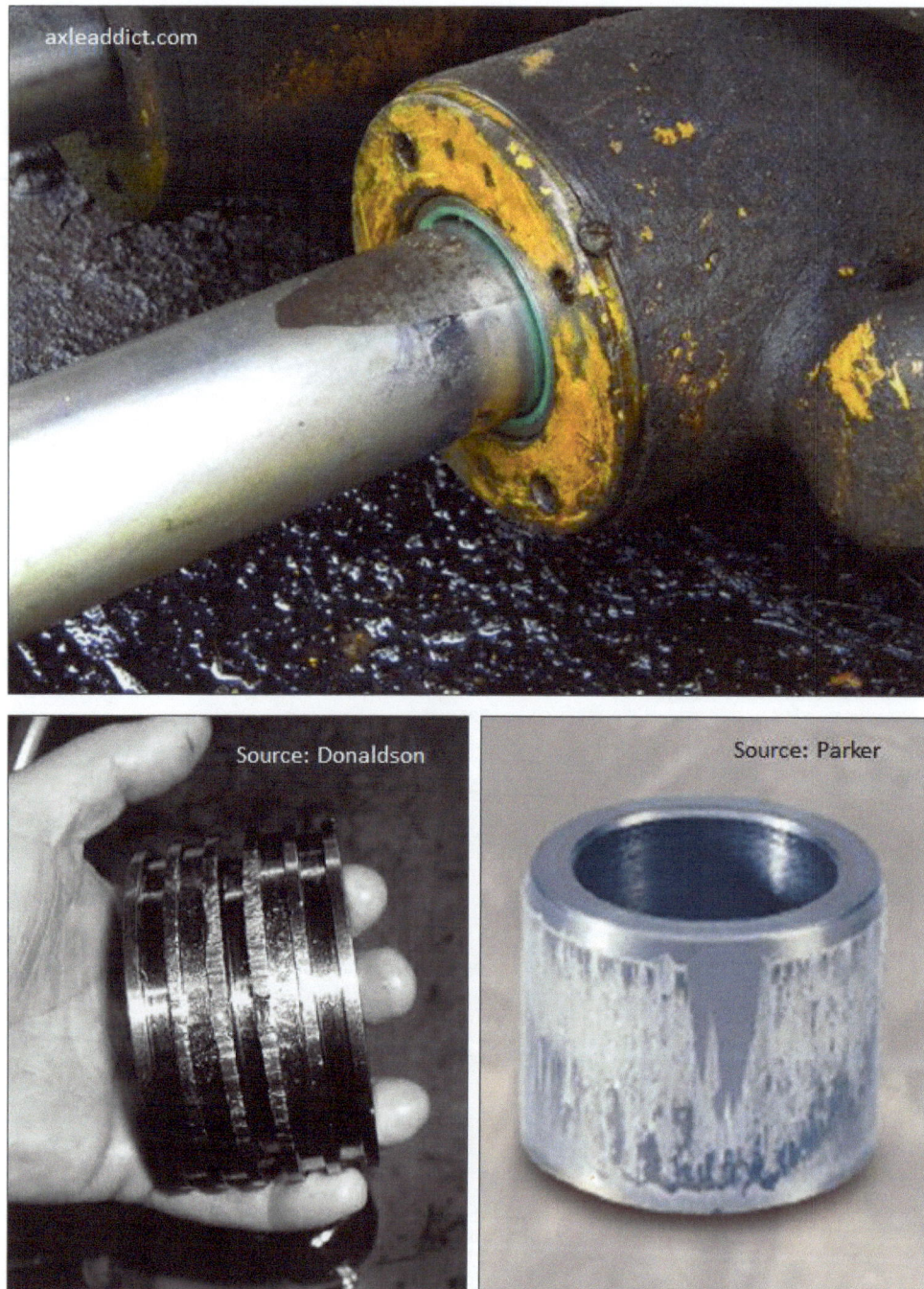

Fig. 6.30- Examples of Hydraulic Cylinder Failures due to Particulate Contamination

6.5.4.4- Bearing Failure due to Particulate Contamination

As shown in Fig. 6.31, a positive displacement pump is an unbalanced pump because it has high pressure at the discharge side and low pressure at the suction side. Therefore, bearing load is not evenly distributed along the bearing circumference, and rather concentrated on one side. Severe bearing wear occurs in presence of particulate contamination.

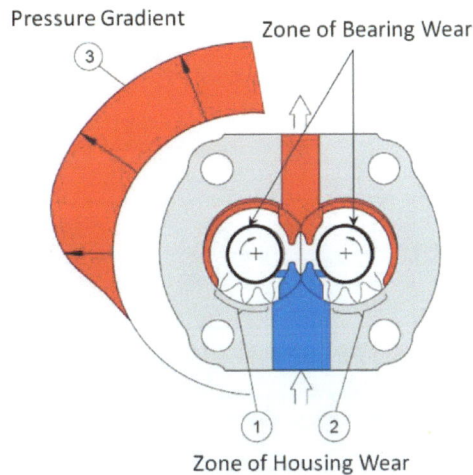

Fig. 6.31- Wear Zones in Gear Pump and Motor Bearings

Figure 6.32 shows examples of bearing failure due to particulate contamination. The figure shows a destroyed raceway (1) of a ball bearing caused by particulate contamination, a chip (2) embedded in a surface of an anti-friction bearing and a destroyed roller bearing (3) in a piston pump.

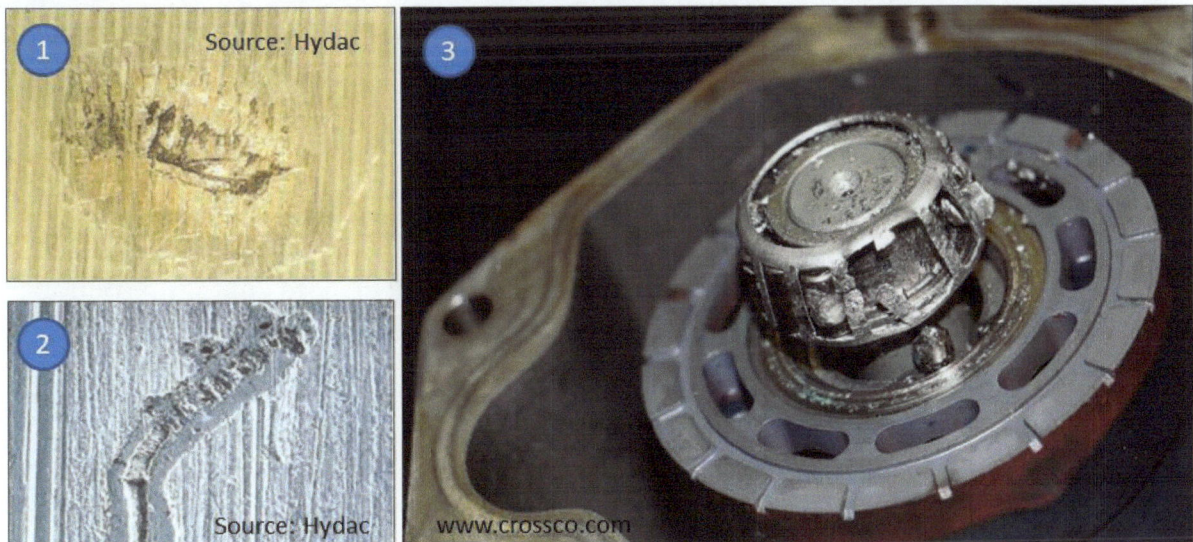

Fig. 6.32- Examples of Bearing Failures due to Particulate Contamination

6.5.4.5- Filter Clogging due to Particulate Contamination

Figure 6.33 shows an example of a filter that has been clogged by dirt. The filter appears normal but the particles clogging it are smaller than the limit of vision. In operation, this filter will by-pass due to high differential pressure, thus a pressure indicator is needed to detect when the filter has reached maximum dirt holding capacity.

**Fig. 6.33- Example of Filters Blockage due to Particulate Contamination
(Courtesy of Noria Corporation)**

6.6- Best Practices for Controlling Particulate Contamination

6.6.1- Preventive Practices to Control Particulate Contamination

Particulate contamination can't be 100% avoided. However, Table 6.3 shows the preventive practices for controlling the different forms of particulate contamination.

Form of Particulate Contamination	General Preventive Actions
Built-in	▪ **Contamination Limits** for new components should be verified. ▪ **Hydraulic Transmission Lines** should be cleaned before and after assembly. ▪ **System Flushing** before first use and after major maintenance.
Introduced (Ingested and Induced)	▪ **Service and Maintenance** proper procedures help in minimizing ingested and induced contamination.
Generated	▪ **Filtration System Design** based on the system requirements to maintain recommended fluid cleanliness level. ▪ **Hydraulic Fluid Analysis** in order to predict possible future causes of failure and the required action that should be taken to prevent it. ▪ **Hydraulic Reservoir Design and Maintenance** is an important preventive action for controlling generated contamination.

Table 6.3- General Preventive Actions for Controlling Particulate Contamination

The following sections presents some best practices for controlling particulate contamination.

Hydraulic Fluid Analysis (Fig. 6.34):
▪ Cleanliness level of the operating hydraulic fluids must be checked frequently to make sure the system complies with the standard cleanliness level recommended by the system manufacturer. Frequency and methods of *Hydraulic Fluid Analysis* will be discussed in Chapter 8. However, it can be done by intermittent oil sampling and analysis offline or by installing online particle counter.
▪ Always observe maximum cleanliness and accuracy during sampling.
▪ Always use independent analysis resources with high quality control and repeatability.
▪ If the system is sensitive, use online particle counters or contamination sensors for continuous monitoring of contamination.
▪ Check the oil after machine malfunctions or other incidents which might affect the oil.

- When replacing seals, compatibility with the oil must be checked.
- Never apply new additives without consulting the oil supplier/consultant. Ask for written confirmation of the measures to be taken.

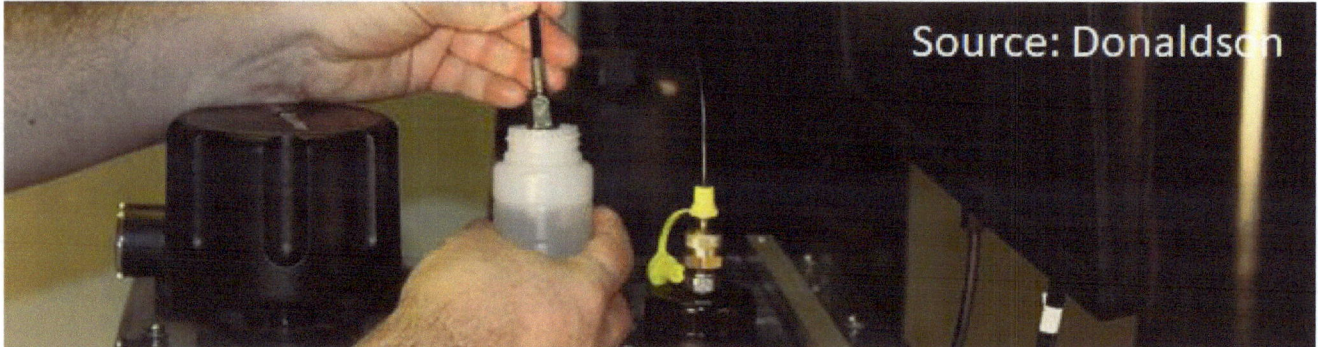

Fig. 6.34- Hydraulic Fluid Analysis (Courtesy of Donaldson)

Filtration System Design:

Proper design of the filtration system for a hydraulic-driven machine is a crucial factor in machine reliability. *Hydraulic Filters Performance Ratings* will be discussed in Chapter 9. Filtration system design depends on many factors, the important of which is the components in the system. For example, Servo and proportional valves, in particular, are extremely sensitive to particulate contamination. For this reason, systems containing these types of valves need non-bypass filters placed directly upstream of the component.

As shown in Fig. 6.35, the filter should be equipped with a visual or electronic differential pressure indicator. When the bypass valve in a filter opens due to a clogged filter element, all filtration ceases, and contaminants are free to enter the system. Therefore, regardless of what type of alarm or indicator is chosen, the device should be activated at a differential pressure below the bypass valve cracking pressure. This gives time to service the element, before the bypass valve open.

Fig. 6.35- Hydraulic Filter Differential Pressure Indicator

Hydraulic Fluid - New Oil is not Clean!! (Fig. 6.36):

- New hydraulic fluid added to the system from a drum or any storage container can be a source of contamination.
- Therefore, new fluid should be considered contaminated until a sample has been analyzed.
- New oil should always be introduced to the system through an appropriate filter.
- Never transfer fluid using buckets, containers, funnels, etc.

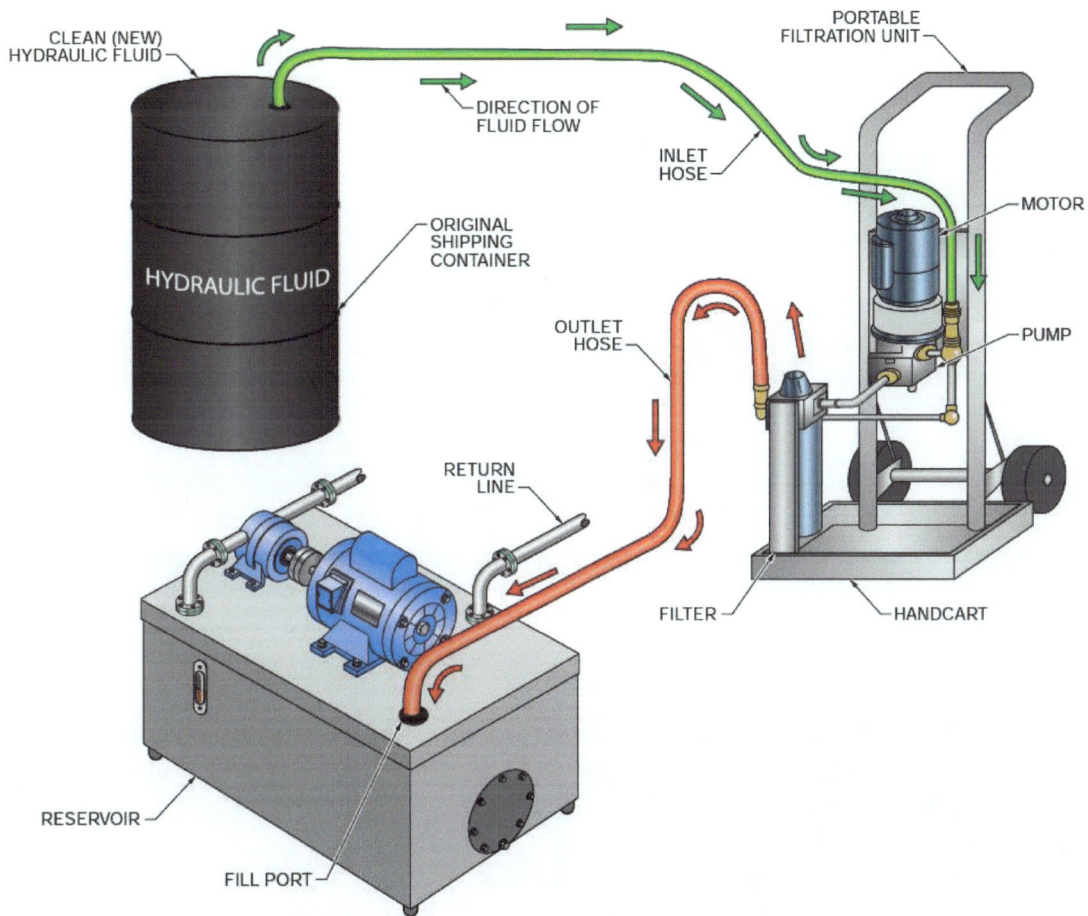

**Fig. 6.36 – Hydraulic Fluid Filtration before Filling a Reservoir
(Courtesy of American Technical Publishers)**

Hydraulic Transmission Lines - New Transmission Lines are not Clean!!:

New hydraulic transmission lines contain built-in contaminants from manufacturing and the storage process. Particulate contaminants find their way inside the transmission lines during packaging, shipping, storage, pipe welding, tube bending, hose cutting, and fitting crimping. Therefore, such transmission lines must be cleaned before and after assembly. *Contamination Control in Hydraulic Transmission Lines* will be discussed in Chapter 10.

Contamination Limit in New Components (New Components are not Clean!!):
Do not assume that the new components are 100% clean. It is wise to pre-clean all hydraulic system components prior to assembly.

The Fluid Power Institute at Milwaukee School of engineering (MSOE) recently evaluated the contamination level of more than 100 new hydraulic components. As shown in Fig. 6.37, the study included hoses, tubes, fittings, valves, cylinders, pumps and reservoirs. The results show that one-third of the new components has particulate contaminants exceeded 8 mg. Abrasive dirt and debris from these components will attack the rest of the hydraulic system as soon as the machine is powered up.

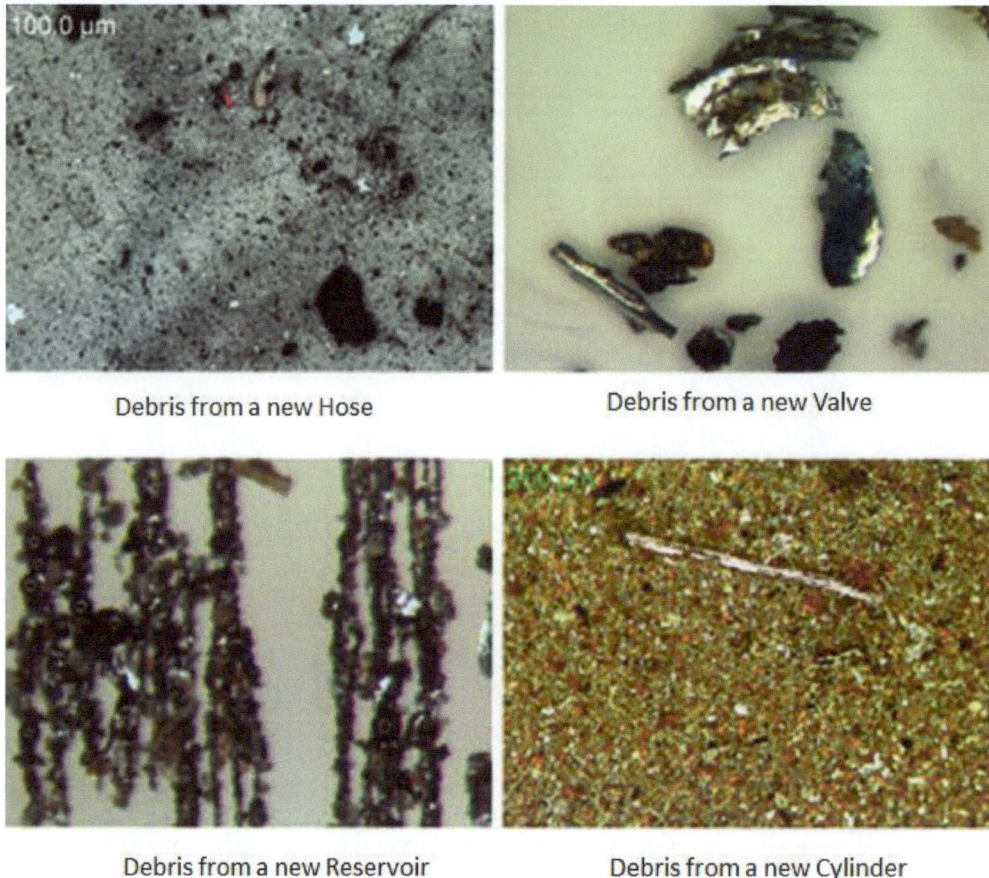

Debris from a new Hose   Debris from a new Valve

Debris from a new Reservoir   Debris from a new Cylinder

Fig. 6.37- Debris from New Components (Courtesy of MSOE)

As it has been discussed, in hydraulic fluid power systems, power is transmitted and controlled through a pressurized liquid within an enclosed circuit. Contaminants present in circuiting working liquid may degrade system performance. One method of reducing the amount of these contaminants within the system is to manufacturer, package, ship, store, and install components in ways that achieve and control the desired component cleanliness level.

As a general principle, the manufacturer is responsible for providing components that meet the requirements agreed upon with the purchaser.

For more information about cleanliness requirement of new components, refer to the following ISO Standards:
- **ISO 18413:** Hydraulic fluid power - Cleanliness of components - Inspection document and principles related to contaminant extraction, analysis, and data reporting.
- **ISO 12669:** Hydraulic fluid power - Method for determining the required cleanliness level (RCL) of a system.
- **ISO/TR 10949:** Hydraulic fluid power – Component cleaning – Guidelines for achieving and controlling cleanliness of components from manufacture to installation

A method has been developed to set a maximum allowable *Contamination Limit* in new components. Component manufactures should take the required steps to comply with this method. Establishing the contamination limits for components, like most other engineering decisions, involves a cost/benefit analysis. In this case the cost associated with achieving a given level of cleanliness versus possible damage are subject to analysis. The method of setting contamination limits in new components is based on the *Volume-to-Area Ratio* of hydraulic components. Table 6.4 shows, in order of magnitude, the volume-to-area ratio of hydraulic components. In this case, the area applies to wetted surfaces that are in direct contact with the hydraulic fluid.

COMPONENT	VOLUME-TO-AREA RATIO
Reservoirs	1 to 5
Hoses and Tubes	0.2
Cylinders	0.5 to 0.6
Pumps and Motors	0.001 to 0.05
Valves	0.001
Complete Systems	0.2 to 4

Table 6.4- Volume-to-Area Ratio of Hydraulic Components (Courtesy of MSOE)

Generally, particulate contamination limits for components are specified in units of milligrams (mg) of contamination and the length (longest chord) of the largest particle. In order to account for differences in volume and wetted surface area, different units of measure are used to define built-in contamination levels as follows:

- Mass per unit volume (mg/liter). This unit is used for components that have a high volume-to-area ratio.
- Mass per unit weight (mg/kg). This unit is used for components that have a low volume-to-area ratio.
- Mass per unit area (mg/m^2). This unit is also used for components that have a low volume-to-area ratio
- Mass per unit length (mg/m). This unit is used for hydraulic transmission lines.

Table 6.5 provides a list of contamination limits for new components expressed in common units of measure.

UNIT OF MEASURE	TYPICAL RANGE
Mass per unit volume (mg/liter)	3 to 10
Mass per unit weight (mg/Kg)	0.5 to 5
Mass per unit area (mg/M^2)	25 to 1,000
Mass per unit length (mg/M)	6 to 12

**Table 6.5- Contamination Limits for New Hydraulic Components
(Courtesy of MSOE)**

The following are examples on how to use the previous data:

- Example 1: A hydraulic reservoir that has a volume-to-area ratio of 5 has a contamination limit range from 3-10 mg/liter.
- Example 2: A hydraulic cylinder that has a volume-to-area ratio of 0.5 has a contamination limit range from 0.5-5 mg/kg or 25-1000 mg/m^2.
- Example 3: A hydraulic pipe has a contamination limit range from 6-12 mg/m.

Service and Maintenance:
Service and maintenance technicians and their work affect the cleanliness of the hydraulic system. Each time a system is "opened-up" for repair or maintenance, particulate contamination can be induced into the system. The following best practices help minimize particulate contamination during service and maintenance:

- As shown in Fig. 6.38, maintain organized and clean housekeeping. Repair work should be performed in a dust-free environment and on clean work benches.
- Parts and seals should be kept in sealed plastic bags until needed.
- When parts washing and solvent flushing, only pre-filtered solvents should be used. Make sure solvents are compatible with seals and other component parts.
- When replacing or cleaning filters, consult the service manual for best procedure.

Fig. 6.38- Organized, Dry and Clean Housekeeping

- As shown in Fig. 6.39, maintaining clean outside surfaces of the hydraulic components and their surrounding areas limits the amount of dirt particles that can find their way into the system. When cleaning, only lint-free wipes that contain no fibers should be used. Ordinary shop towels and waste rags should not be used.

Fig. 6.39- Keeping the Hydraulic System Clean is an Important Practice

- As shown in Fig. 6.40, covering cylinder rods minimizes ingression of particulate contaminants through the rod wipers.

Fig. 6.40- Covers for Hydraulic Cylinder Rods

- As shown in Fig. 6.41, methods that can be employed to help prevent contaminants from entering the system includes use of pipe plugs, tube caps, etc. during disassembly, assembly, shipping, and storage. Make sure installation and removal of caps and plugs does not generate contaminants in the threaded area of the component.

Fig. 6.41- Covers for Hydraulic Components and Parts (www.capsnplugs.com)

Hydraulic Reservoir Design and Maintenance (Fig. 6.42):
- A well-designed reservoir helps hydraulic fluid to get rid of all contaminants (particulate, fluidic, gaseous, and thermal). It allows settling of particulate contaminants will also help in keeping particulates out of the mainstream fluid.
- Reservoir drain-plug or strainer magnets help capture ferrous particulates and rust.
- Since most of the reservoir has continuous exchange of air with the surrounding environment, leaving the system open during operation provides continuous ambient particle ingression through the reservoir cap or breather. Therefore, systems should be well-sealed, and all permanent openings should be equipped with venting filters (preferably desiccant breathers) with same micron rating as liquid side filters.
- When changing the oil, the tank and the system should be emptied completely, and the tank should be cleaned with an appropriate compatible solvent.

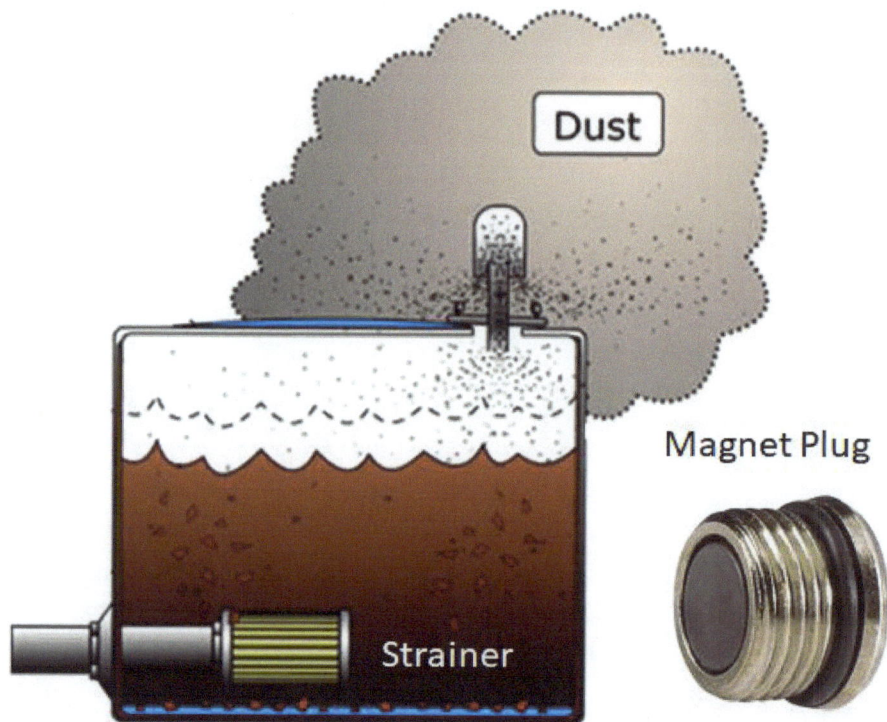

Fig. 6.42- Reservoir Design and Maintenance for Controlling Generated Particulate Contamination

Hydraulic System Flushing: In some cases, contamination is so severe that the cost of removing it exceeds the cost of replacing the fluid itself. In such a case, or after major maintenance, hydraulic system flushing is needed. This topic will be discussed in a separate chapter. *Hydraulic System Flushing* will be discussed in Chapter 11

6.6.2- Curative Practices to Remove Particulate Contamination

In case if hydraulic system cleanliness level was found not acceptable limit recommended by manufacturer, then additional filtration is needed.

Offline Filtration (Fig. 6.43): One very effective way to adjust the cleanliness level of a hydraulic fluid is by using *off-line* circulation loop, or "*kidney Loop*" filtration.

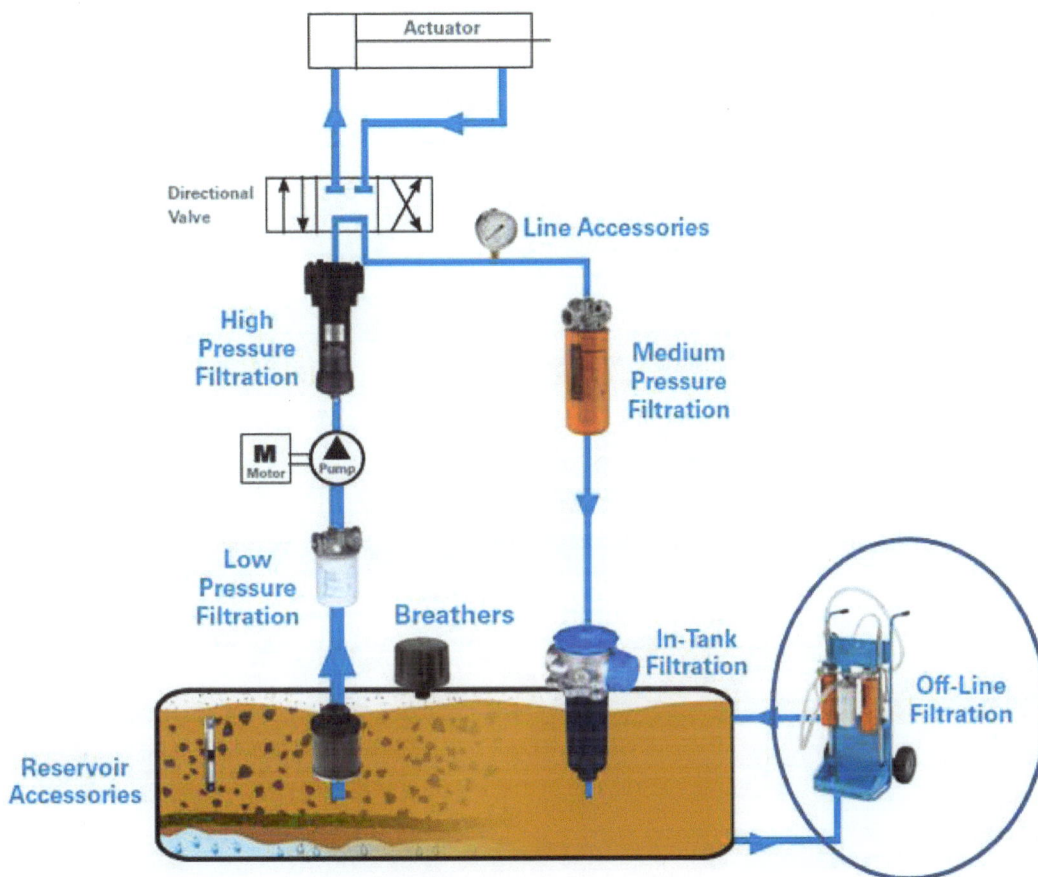

Fig. 6.43- Offline Filtration (Courtesy of Donaldson)

Fluid Purification Units: In addition to mechanical filters, other advanced *Fluid Purification Units* that use the concept of *Centrifuging* can be used to remove particulate contamination. Some of this equipment was presented in Chapters 5 and 6.

Oil Changing and System Flushing: If any of the previous methods didn't help in removing particulate contamination, the only remaining solution is to drain, clean the reservoir, and flush the complete system with appropriate fluid.

Chapter 7

Maintenance of Filters

Objectives

This chapter provides guidelines for **Filters** selection, replacement, maintenance scheduling, installation, testing, storage and transportation. This chapter is supported by examples and figures granted by leading fluid power manufacturers.

Brief Contents

7.1-BP-Filters-01-Selection and Replacement
7.2-BP-Filters-02-Maintenance Scheduling
7.3-BP-Filters-03-Installation and Maintenance
7.4-BP-Filters-04-Standard Tests and Calibration
7.5-BP-Filters-05-Transportation and Storage

Chapter 7: Maintenance of Filters

The following set of best practices provide general guidelines and may not be applicable for all cases. They are not intended to replace the instructions given by the component manufacturer. It is strongly advisable to adhere to instructions provided by the manufacturer.

7.1-BP-Filters-01-Selection and Replacement

Filters are originally specified based on cleanliness level specified by the system designer, placement in the circuit, size, dirt holding capacity, static and dynamic working conditions (pressure, temperature, and flow), and mechanical mounting method. A filter acts like a kidney in a human body. So, replacing an existing filter with a new one of different specifications impacts the hydraulic system reliability on short term operation. So, when replacing an existed filter, none of the originally specified design and operating specifications shall be changed.

The shown below example explores the importance of maintaining the specification of the filter. Table 7.1 shows the amount of the dirt that passes through a pump as function the oil cleanliness level based on conditions of (200 lit/min flow, 18 hours a day, and 340 working days per year). The figure also shows that, even new oil is typically contaminated with particles to ISO 19/17/14.

ISO Code	NAS 1638	Description	Suitable for	Dirt/year
ISO 14/12/10	NAS 3	Very clean oil	All oil systems	7.5 kg *
ISO 16/14/11	NAS 5	Clean oil	Servo & high pressure hydraulics	17 kg *
ISO 17/15/12	NAS 6	Light contaminated oil	Standard hydraulic & lube oil systems	36 kg *
ISO 19/17/14	NAS 8	New oil	Medium to low pressure systems	144 kg *
ISO 22/20/17	NAS 11	Very contaminated oil	Not suitable for oil systems	> 589 kg *

**Table 7.1- Amount of Dirt Pass through a Filter based on Oil Cleanliness Level
(Courtesy of C.C. Jensen Inc.)**

7.2-BP-Filters-02-Maintenance Scheduling

Unless otherwise is stated by components and systems manufacturer, Table 7.1 provides guidelines for *scheduling* preventive maintenance actions for hydraulic filters.

#	Preventive Maintenance Actions	Daily	Weekly	Monthly	Biannually	Annually
1	Clean the dust on outer surface		✔	✔	✔	✔
2	Check hydraulic connections			✔	✔	✔
3	Check status of the filter through clogging indicator **(Note 1)**		✔	✔	✔	✔
4	Check electrical connections (if found)			✔	✔	✔
5	Disassemble and inspect/clean/wash/replace filter element and filter housing **(Note 2)**			✔	✔	✔
6	Check valve performance through standard tests				✔	✔

Table 7.2- BP-Filters-02-Maintenance Scheduling

Note (1): Continuous monitoring of filter conditions in a hydraulic system provide good insight about how healthy the system is. All filter assemblies shall be equipped with a device that indicates when the filter requires servicing. The indication shall be readily visible to the operator or maintenance personnel. When this requirement is not available, scheduled filter element replacement shall be addressed routinely in the operator manual.

Note (2): A disassembled filter can be visually inspected by taking a good look at the filter and check if the filter has any of the following signs:
- Damage such as pleats have signs of cuts or bunched together.
- Filter elements aren't properly sealed on both end caps.
- Center tube is collapsed or buckled.
- Oil degradation products accumulated on the filter element or the end caps.
- High concentration of debris as a sign of metal wear.
- Nonmetallic particles such as paint chips, fibers, seal wear products, etc.

7.3-BP-Filters-03-Installation and Maintenance

One of the keys to consistent filtration performance is good maintenance practices. Remember, any contamination induced by a filter change goes directly into the system. Electro-hydraulic systems utilizing servo valves are the most sensitive and very susceptible to any contamination or air. Referring to Fig. 7.1, the following set of bullets provide common best practices for installation and maintenance of hydraulic filters considering a spin-on filter as an example:

Review available instruction (1): leading manufacturers provide step-by-step servicing guidelines. Where possible, follow the filter service instructions supplied by the original equipment manufacturer. Also, check if there are any servicing instructions that given by filter pictogram. Filter *Pictogram* is a method of printing symbols on the side of filters to indicate specific type of servicing instructions. Figure 7.2 shows an example of filter pictograph.

Check the service indicator (2): Verify that the OEM specified service interval has been reached by any of the service indicators shown in the figure.

Turn off system pressure (3): Be sure the system is turned off and that there is no pressure present in the system. At least isolate the filter under service if isolation setup is found.

Remove the used filter and gasket (4): Remove the spin-on filter, properly dispose of the filter in accordance with local regulations or recycle it. Recycling used hydraulic filters is environmentally friendly. Check your local disposal regulations for proper disposal and recycling.

Clean the filter mounting head and bowel (5): Clean the surfaces of the filter head or cover. Flush sediments from filter bowls only with a pre-filtered solvent. Use only lint free wipes or filtered air to dry the bowl.

Lubricate the filter gasket (6): Lubricate threads and spin on seal with clean system oil.

Inspect new filter (7): Check the new filter you will be installing for any shipping or handling damage. DO NOT install any filter or filter element that shows any signs of damage. The exterior of the filter housing should be cleaned, preferably with a compatible solvent wash. Avoid touching or handling a new element if possible. If the element is supplied in a plastic bag, remove the bag after the filter is in place. After replacing the new filter element, secure the bowl immediately.

Install new filter for instructions (8): Install the spin-on filter until the top of the gasket first contacts the sealing surface. Then for final tightening in accordance with the given instructions. Do not overtighten. Once the filter housing is secured, attempt to bleed air from the housing on initial system start-up, to prevent the air from entering the system, particularly cylinders.

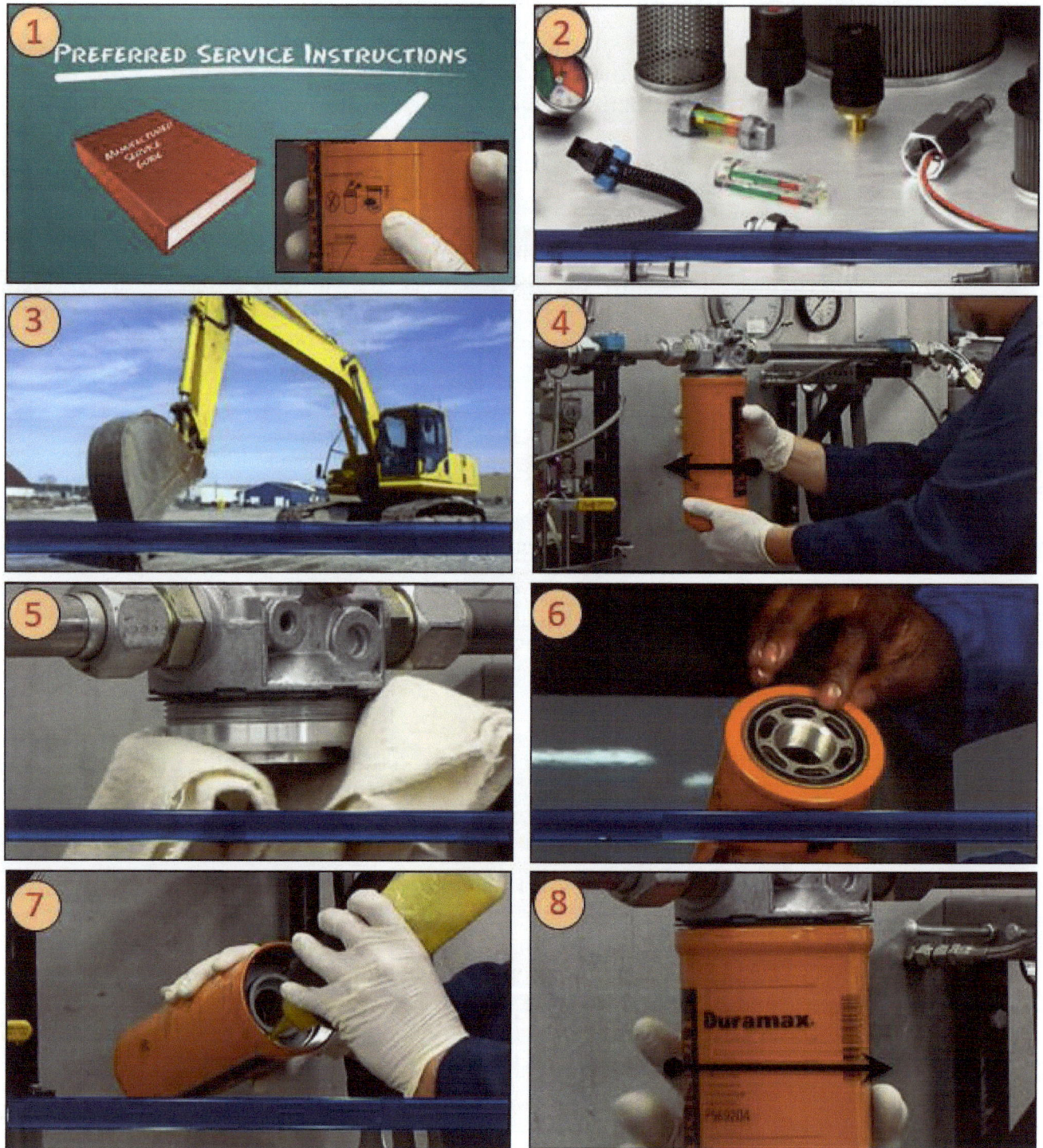

Fig. 7.1- Hydraulic Spin-On Filter Replacement Steps (Courtesy of Donaldson)

Fig. 7.2- Spin-On Hydraulic Filters Service Pictograms (Courtesy from Donaldson)

7.4-BP-Filters-04-Standard Tests and Calibration

Performance characteristics of hydraulic filters are evaluated by several test. The following is a list of standard test methods:

- **ISO 2942:** Filter Element Structural Integrity (Bubble Point) Test.
- **ISO 2943:** Hydraulic Fluid Compatibility Test
- **ISO 16889:** Efficiency and Capacity (Multipaas) Test
- **ISO 3723-2015:** End Load Test
- **ISO 3968:** Differential Pressure Test
- **NFPA (T-2.6.1):** Rated Burst Pressure (RBP) of a Filter Housing
- **ISO 1077-1:** Rated Fatigue Pressure (RFP) of a Filter Housing
- **Cyclic Test Pressure** (CTP) of a Filter Housing
- **ISO 2941:** Collapse Pressure of a Filter Element
- **ISO 3724 OR ISO 23181:** Flow Fatigue Test for Filter Element

Tests shall be run in accordance with the sequence given in Fig 7.3. The table shows the logic sequence of conducting these tests. For example, there is no meaning of conducting burst pressure test if the filter isn't compatible with the fluid.

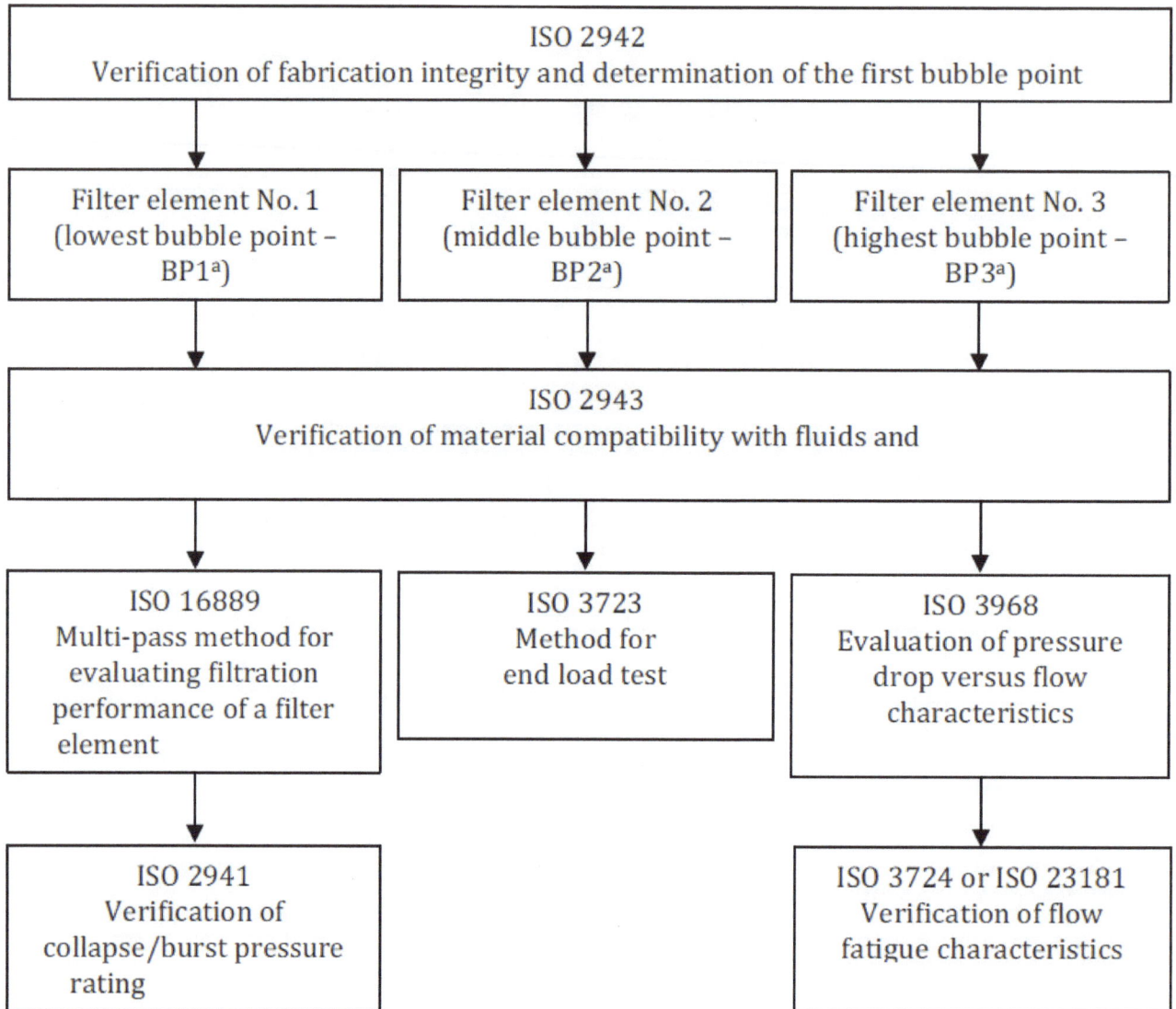

Fig. 7.3- Sequence of Conducting Standard Tests for Hydraulic Filters (Courtesy from Donaldson)

The following sections present the tests based on the sequence they shall be run.

7.4.1- ISO 2942: Filter Element Structural Integrity (Bubble Point) Test.

Verification of fabrication integrity is used to define the acceptability of filter elements for further use or testing. That is why this test is conducted first.

Bubble Point Resistance Test is used to verify the fabrication integrity of a filter element (by checking the absence of bubbles). The fabrication integrity test determines whether a filter element meets the manufacturer's prescribed maximum allowable pore size. Damage created during shipping or manufacturing is identified using the bubble point test.

As shown in Fig. 7.4, the bubble point test is performed by submerging the test element in isopropyl alcohol, or other suitable fluid, while applying air pressure to the inside of the element through a special adapter fitted into the open end of the element. No evidence of a steady stream of bubbles should be detected at the minimum bubble point level, designated by the manufacturer.

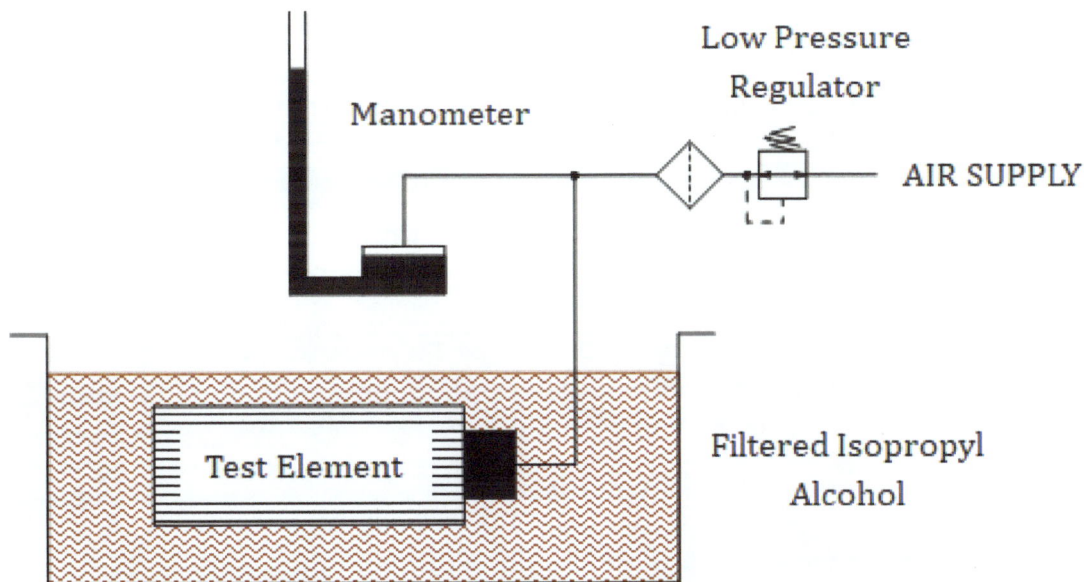

Fig. 7.4- Typical Bubble Point Test Setup

7.4.2- ISO 2943: Hydraulic Fluid Compatibility Test

After passing the structural integrity test, the element compatibility with the hydraulic fluid must be verified. *Fluid Compatibility Test* is conducted according to ISO 2943 standard to verify the compatibility of a specific hydraulic fluid with the component materials of a filter element at maximum expected fluid temperature. Filter elements are submerged in the fluid of interest and subjected to a temperature 15° C (59° F) above the recommended maximum operating temperature of the fluid for a 72-hour period. The element passes the test when no visual evidence of structural failure or material degradation should be present.

7.4.3- ISO 16889: Efficiency and Capacity (Multipaas) Test

7.4.3.1- Multipass Test Purpose and Procedure

The *Multipass Test* is the worldwide recognized method of characterizing hydraulic filter element filtration performance including efficiencies and dirt capacity. To obtain the beta ratio, particulate contaminants of specific sizes of interest must be counted at the upstream (N_u) and downstream (N_d) sizes of the filter. In this test, as shown in Fig. 7.5, hydraulic fluid (Mil-H-5606) is injected with a uniform amount of contaminant (such as ISO 12103-A3 MTD, ISO Medium Test Dust). As shown in the figure, the contaminated fluid is pumped through the filter unit being tested. An automatic particle counter is used to count the particles of certain sizes in both upstream and downstream sides of the filter to determine the *Beta Ratio*.

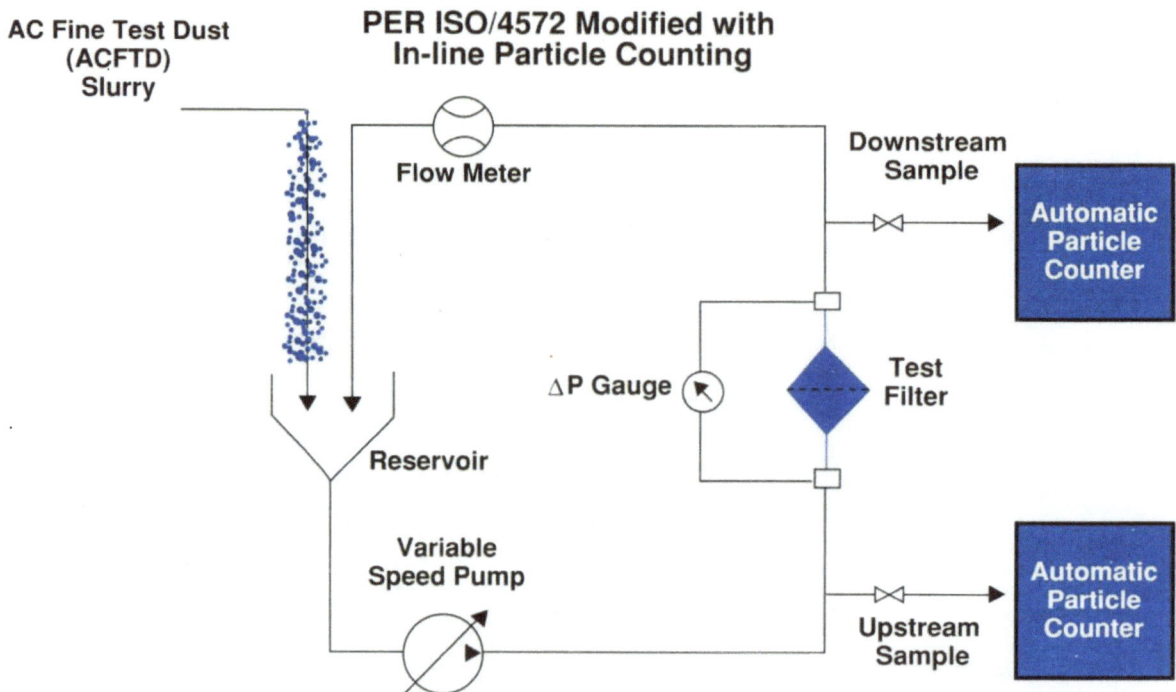

Fig. 7.5- Typical Multipass Performance Test Setup (Courtesy from Pall)

7.4.3.2- Calculation of Beta Ratio

As shown in Fig. 7.6 and Eq. 7.1, *Beta Ratio* is calculated by dividing the number of particles greater than a given size (x) that enter the filter (Nu) by the number of the particles of that same size that leave the filter (Nd). Figure 7.7 shows an example of calculating the beta ratio.

$$\beta_x = \frac{N_U}{N_D} \qquad\qquad 7.1$$

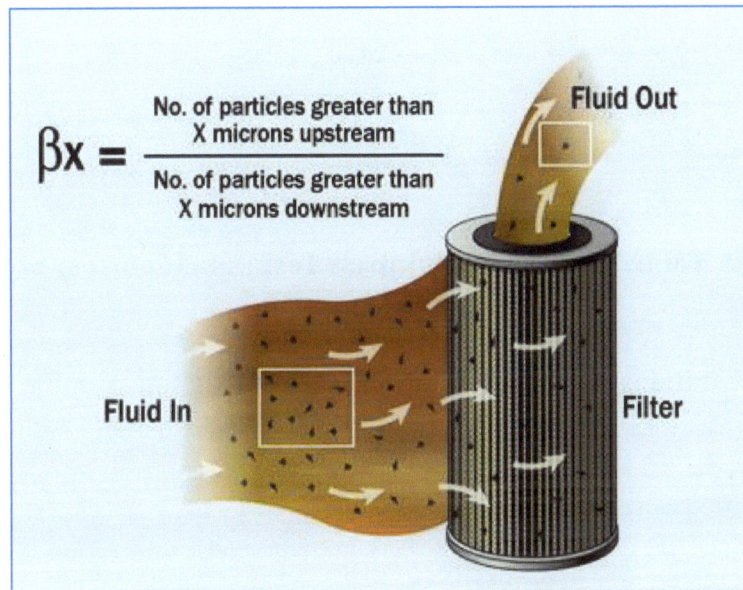

Fig. 12.6- Calculation of Beta Ratio (www.magneticfiltration.com)

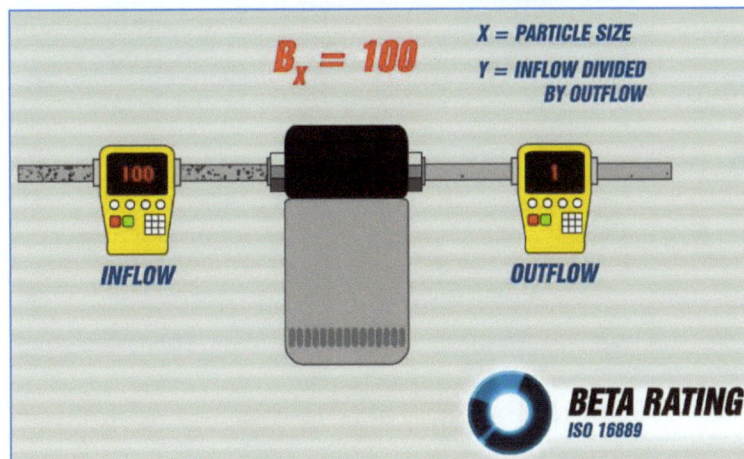

Fig. 7.7- Example of Beta Ratio Calculation (Courtesy of Noria Corporation)

Table 7.3 shows some typical data from a Multipass test. The data is interpreted as follows, the filter has a beta ratio equal to 12 for particle size > 2 μm, a beta ratio equal 100 for particle size > 5 μm, and a beta ratio equal 3000 for particle size > 10 μm.

Particle Size (μm)	Particle Counts (#/ml)		Beta Ratio
2	upstream downstream	15,200 1,267	$\beta_2 = 12$
5	upstream downstream	8,000 80	$\beta_5 = 100$
10	upstream downstream	3,000 1	$\beta_{10} = 3000$

Table 7.3- Typical Multipass Test Data (Courtesy of Pall)

Obviously, as shown in Fig. 7.8, improving the filter rating may double the bearing service life.

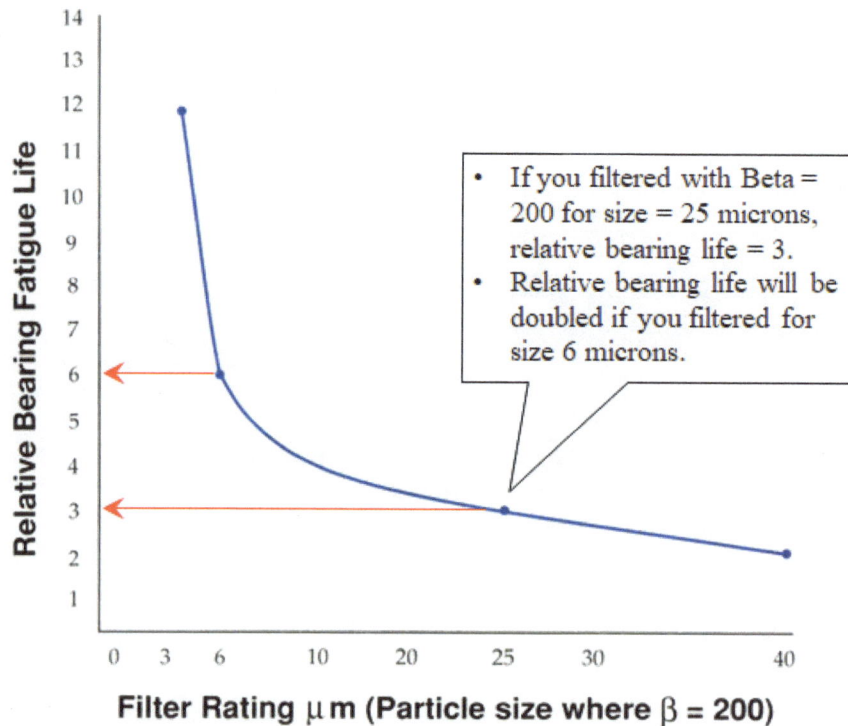

- If you filtered with Beta = 200 for size = 25 microns, relative bearing life = 3.
- Relative bearing life will be doubled if you filtered for size 6 microns.

Ref: Macpherson, P.B., Bhachu, R., Sayles, R., "The Influence of Filtration on Rolling Element Bearing Life"

Fig. 7.8- Effect of Beta Ratio on Bearing Life

7.4.3.3- Beta Ratio Stability

The Multipass test is performed under controlled laboratory conditions and does not take into account some of the challenges an inline pressure filter will experience in most hydraulic systems, such as air bubbles, vibrations, pressure and flow surges. Surge pressure and flow can occur during normal operation, e.g. during start-stop, and when pressure compensated pumps are used.

However, Beta Ratio stability is important because it relates to how well a filter element will perform in service over time. Therefore, beta ratio of a filter should be defined within range of working temperature (such as in cold start) and differential pressure across the filter element.

As shown in Fig. 7.9, cyclic or *Surge Flow* affects the Beta ratio and degrades filter performance dramatically unless the filter is properly designed to resist this action. Such design involves filter medium support and resin bonding, as well as smaller pores.

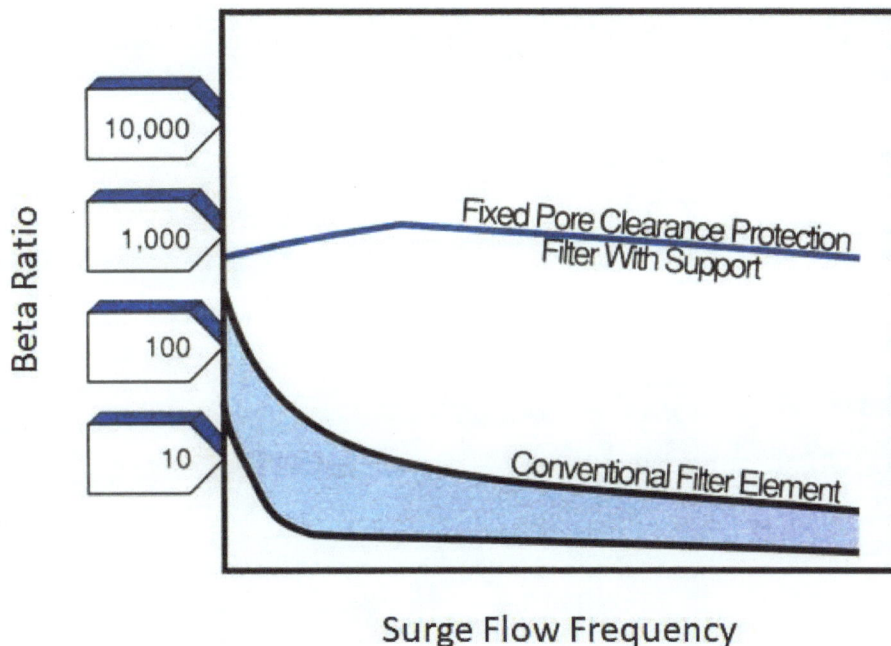

Fig. 7.9- Effect of Surge Flow on Beta Ratio (Courtesy of Pall)

7.4.3.4- Filter Efficiency

Considering of *Filter Efficiency* as being straight forward and easier than beta ratio, it can be calculated using Eq. 7.2. For a beta ratio equal 2, this means that the filter holds 50% of the number of particles introduced to the filter. If Beta Ratio equals 3, then the Filter Efficiency is 67%.

$$E_x = \left[1 - \frac{N_D}{N_U}\right] \times 100 = \left[1 - \frac{1}{\beta_x}\right] \times 100 = \left[\frac{\beta_x - 1}{\beta_x}\right] \times 100 \qquad 7.2$$

Figure 7.10 shows the results of applying the previous equation for a filter element. The figure shows that, in spite of the large change in the beta value from 200 to 1000, the corresponding change in the efficiency is very small (0.4%). Therefore, differentiating between two filters based on beta ratio above 200 is somewhat deceiving.

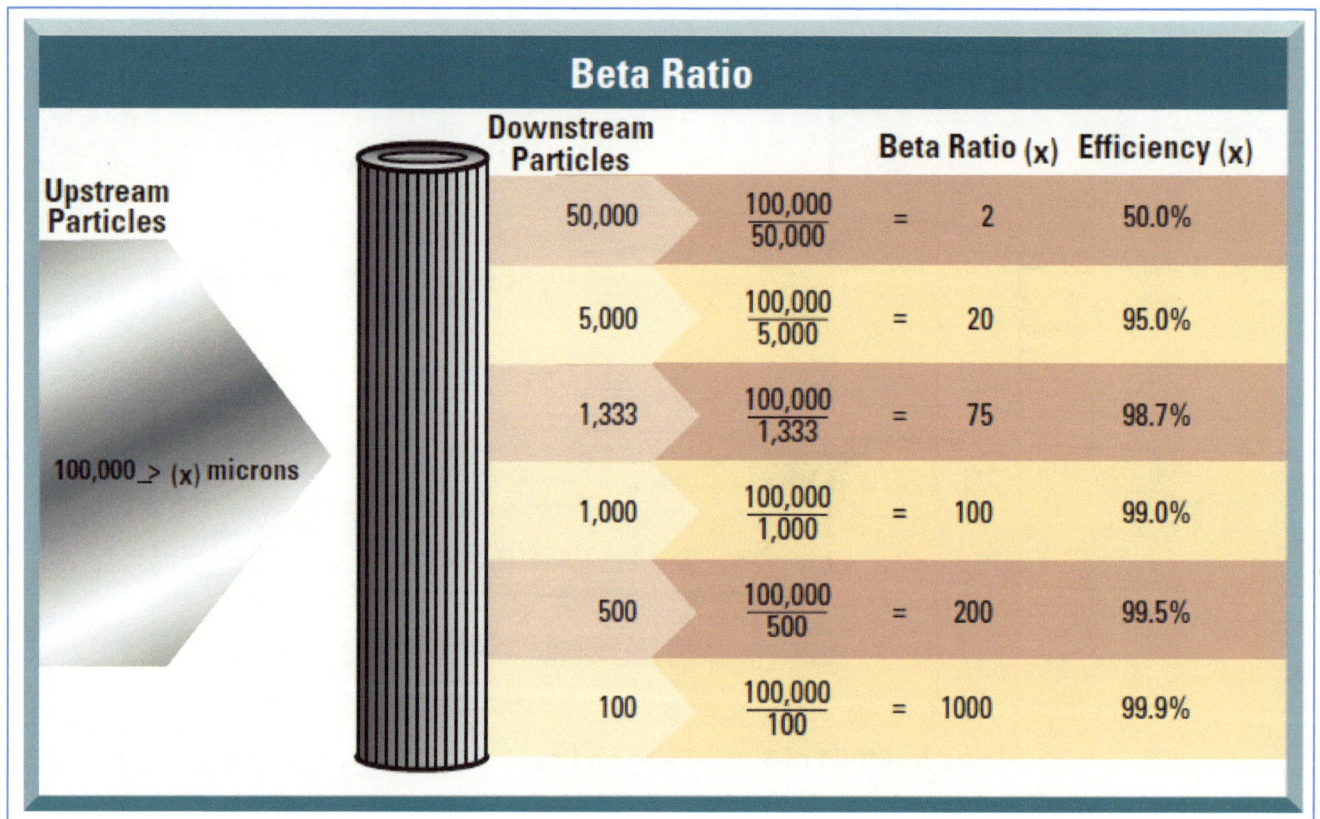

Fig. 7.10- Filter Efficiency vs. Beta Ratio (Courtesy of Parker)

Filter efficiency versus fitter beta ratio can be graphically represented as shown in Fig. 7.11.

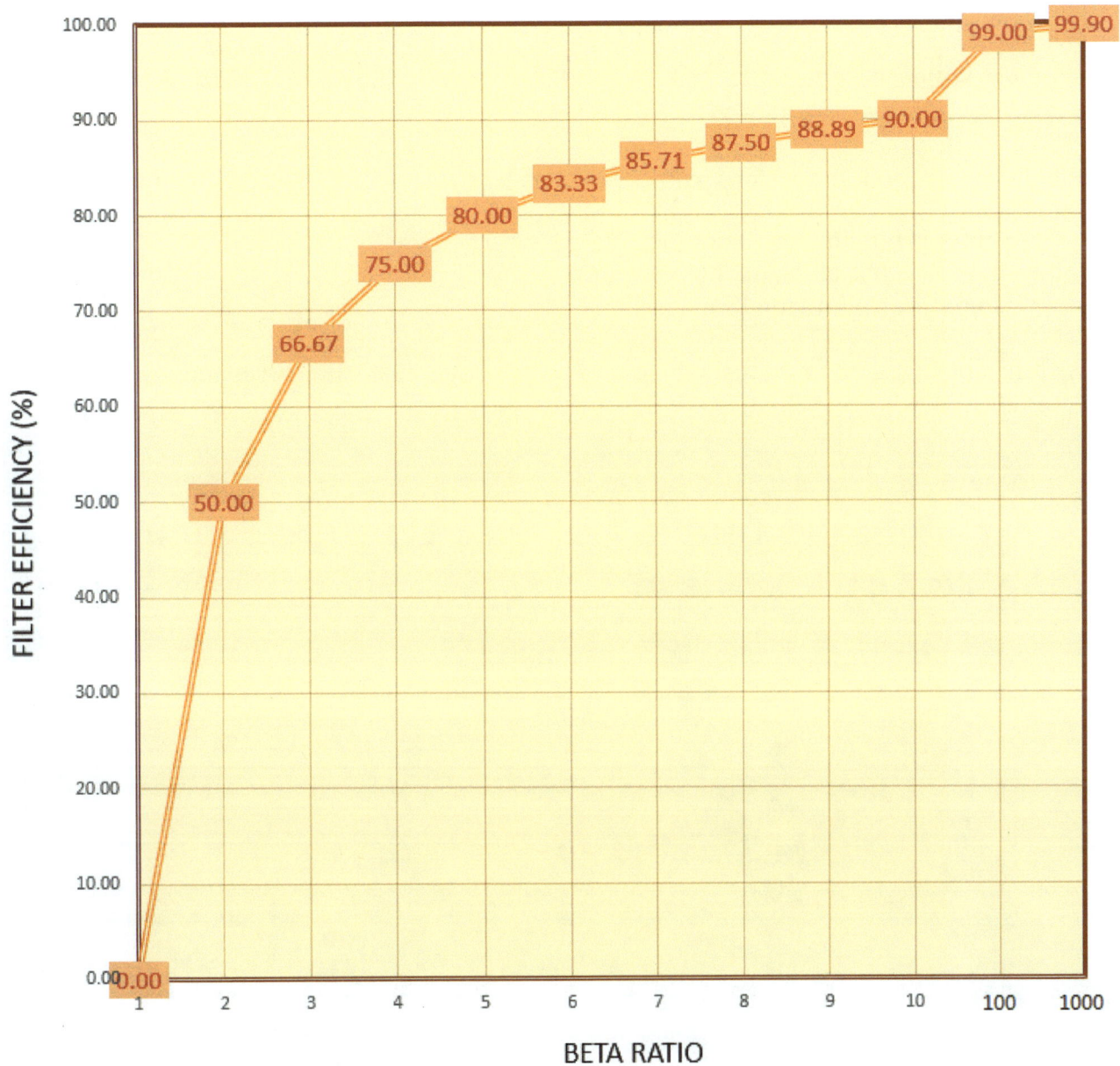

Fig. 7.11- Filter Efficiency versus Beta Ratio

7.4.3.5- Nominal and Absolute Ratings

As shown in Table. 7.4, *Nominal Rating* is the particle size (x) where the filter has 50% efficiency, i.e. $\beta_x = 2$ ($E_x = 50\%$).

Absolute Rating, based upon a historical military standard, is particle size (x) where the filter has 98.7% efficiency, i.e. $\beta_x = 75$ ($E_x = 98.7\%$).

Filters can be rated for various particle sizes as B3/6/15 = 2, 10, 75. This means:

- Filter is nominal at 3 microns.
- Filter is 90% efficient at 6 microns.
- Filter is absolute for 15 microns.

As it has been stated previously for Beta Ratios above 75%, the corresponding increase in the filter efficiency is very slight.

Filtration Ratio (at a given particle size)	Capture Efficiency (at the same particle size)
2	50 %
5	80%
10	90%
20	95%
75	98.7 %
100	99%
200	99.5%
1000	99.9%

Table 7.4- Nominal and Absolute Ratings

7.4.3.6- Filter Dirt Holding Capacity

Another important characteristic on which a filter is evaluated is the *Dirt Holding Capacity (DHC)*. It is defined as the weight of dirt that a filter element can hold before the pressure drop (*Terminal Pressure*) across the filter element reaches a predetermined (saturation) limit. As shown in Fig. 7.12, to measure the DHC of a filter element, ISO MTD Test Dust is added to the system to bring the test filter element to a specified maximum differential pressure drop. The total grams of dirt that a filter held is measured. This is part of ISO 16889.

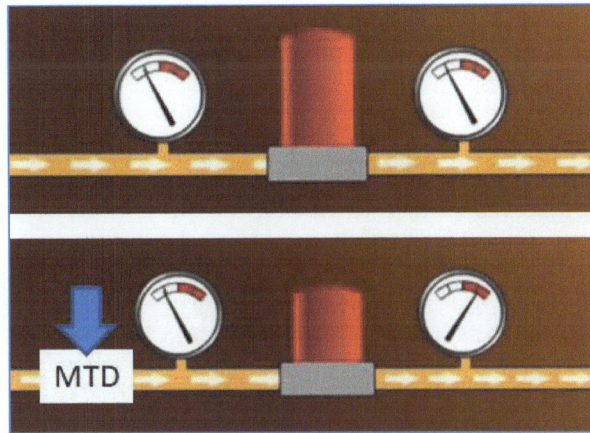

Fig. 7.12- Dirt Holding Capacity Test (Courtesy of Parker)

Since elements with higher DHC need to be changed less frequently, DHC has a direct impact on the overall cost of operation. Equation 7.3 shows the calculation of the cost of removing 1 kg of dirt.

$$\textbf{Cost of Removing 1 kg or lb of Dirt} = \frac{\textbf{Cost of Filter Element (Installation \& Disposal)}}{\textbf{Dirt Holding Capacity in kg or lb}} \qquad \textbf{7.3}$$

Table 7.5 shows the results of applying the previous equation on two different filters as follows:

- While most conventional pressure line filter elements can retain less than hundred grams of dirt (<0.2 lbs), they may be fairly inexpensive to replace. However, if the cost of removing 1 kg or pound of oil contamination is calculated, these conventional pressure filter elements will suddenly appear quite expensive.

- A good quality cellulose based, microfiber, or synthetic offline filter elements can retain up to several kgs/lbs of dirt, so even though the purchase price is higher, the calculated cost for removing one kg or pound of contamination will be considerably lower than that of a pleated pressure filter element, giving lower lifetime costs.

	Example 1	Example 2
Filter type	Glass fiber based pressure filter insert	Cellulose based offline filter insert
Cost of element/insert	€ 35 / $ 50	€ 200 / $ 300
Dirt holding capacity	0.085 kg / 0.18 lbs	4 kg / 8 lbs
Cost per kg/lb removed dirt	€ 412 / $ 278	€ 50 / $ 40

Table 7.5- Cost of Removing Dirt (Courtesy of C.C. Jensen Inc.)

Figure 7.13 shows a stacked disc filter element that is 3 μm nominal and 8 μm absolute. This means that 50% of all particles larger than 3 μm and 98.7% of all solid particles larger than 8 μm are retained. The filter can hold anywhere from 1.5-8 kg of dirt depends on the filter size. Such types of filter elements have high efficiency and DHC but their flow is very low. That is why they commonly used for offline filtration in parallel with the main filter in the system.

Before   After

**Fig. 7.13- Example of Stack Disc Elements
(Courtesy of C.C. Jensen Inc.)**

7.4.3.7- Filter Flow Rate

The fluid flow in the system is a major factor in determining the appropriate filter to use. It is important to have the right size filter to meet the system's requirements. Because fluid can only travel through the filter media so fast, a system with a higher flow rate will need physically larger filters compared to a system with a lower flow rate. If the filter is too small, it will not be able to handle the system flow rate and will create excessive pressure drop, possibly even opening the bypass valve allowing unfiltered fluid through.

Therefore, the filter shall be selected such that the initial differential pressure recommended by the filter manufacturer is not exceeded at the intended flow rate and maximum fluid viscosity.

It is to be noted that, in some hydraulic systems, the maximum flow rate in a return line filter can be greater than the maximum pump flow rate. Examples of these systems are when using differential area cylinders, large single acting cylinders that retract faster than extending, and rapid accumulator discharges.

7.4.3.8- Filter Capacity versus Efficiency

Generally speaking, normal size filters in the system are not designed to adequately deal with large quantities of dirt that occur in connection with component machining, system assembly, system filling, system commissioning, or repair work. Such a large amount of dirt is handled by special large filters during system flushing or offline filtration process.

As shown in Fig. 7.14, a highly restrictive media has better efficiency, but it will be plugged by a small amount of dirt. So, it has low DHC. On the other side, a less restrictive media has lower efficiency, but it can retain more dirt before it gets blocked. Therefore, when selecting a filter, a balance between DHC and efficiency of a filter must be considered.

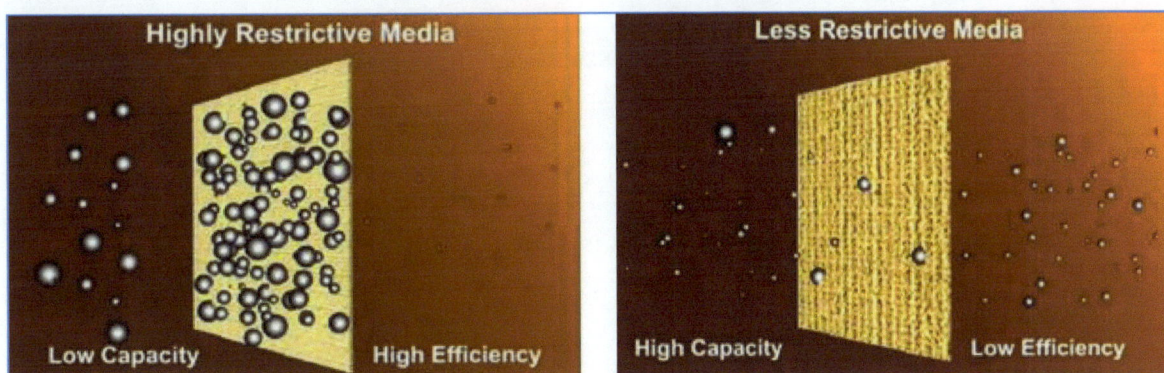

Fig. 7.14- Filter Efficiency vs. DHC (Courtesy of Parker)

7.4.4- ISO 3968: Differential Pressure Test

Differential Pressure, as shown in Fig. 7.15, is the difference between the pressure at the upstream and the downstream sides of the filter. Filter differential pressure depends on:
- Construction of the filter housing.
- Construction and type of filter element.
- Filter size and flow rate through the filter.
- Viscosity and specific gravity (SG) of the fluid flowing through the filter.

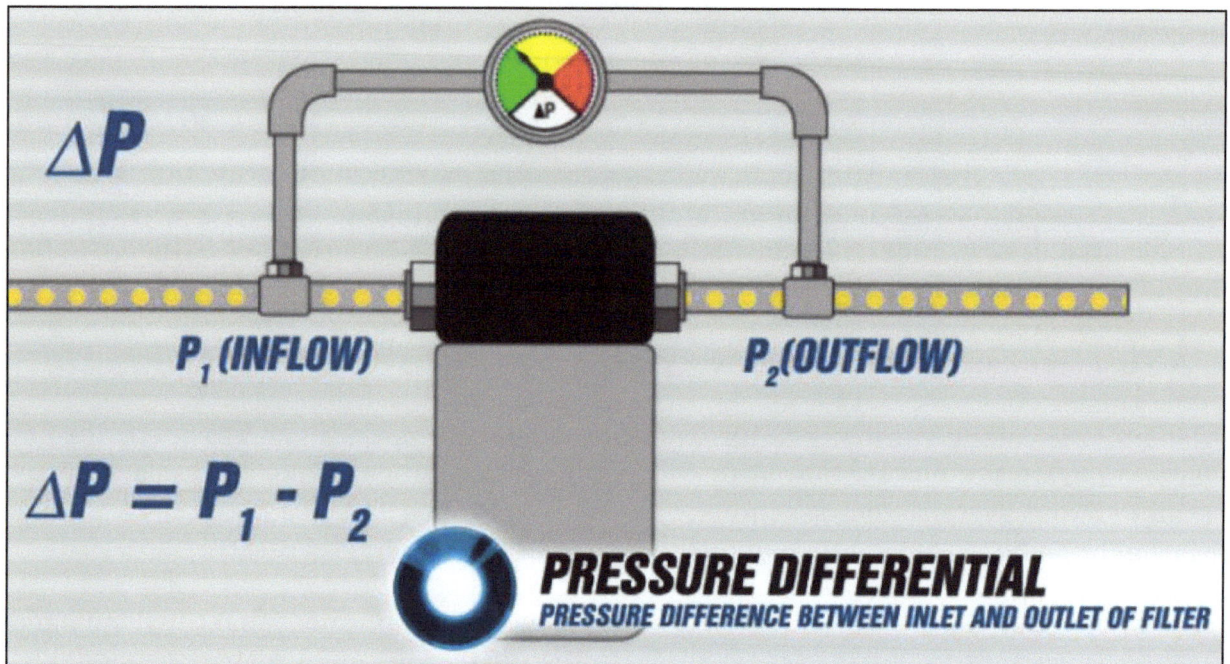

Fig. 7.15- Typical Filter Differential Pressure Test Setup (Courtesy of Noria Corporation)

As shown in Eq. 7.4, pressure drop across a filter is due to both the filter housing and the element.

$$\Delta p_{total} = (\Delta p_H + \Delta p_E) \hspace{4cm} 7.4$$

Where, for a specific filter size, fluid flow, viscosity, and specific gravity:
- Δp_{total} is the total differential pressure across the filter assembly.
- Δp_H is the differential pressure across the filter housing (corrected based on SG).
- Δp_E is the differential pressure across the filter element (corrected based on SG & viscosity).

Figure 7.16 shows a typical flow-pressure drop curve for a specific filter size, a specific clean filter media, and a specific fluid.

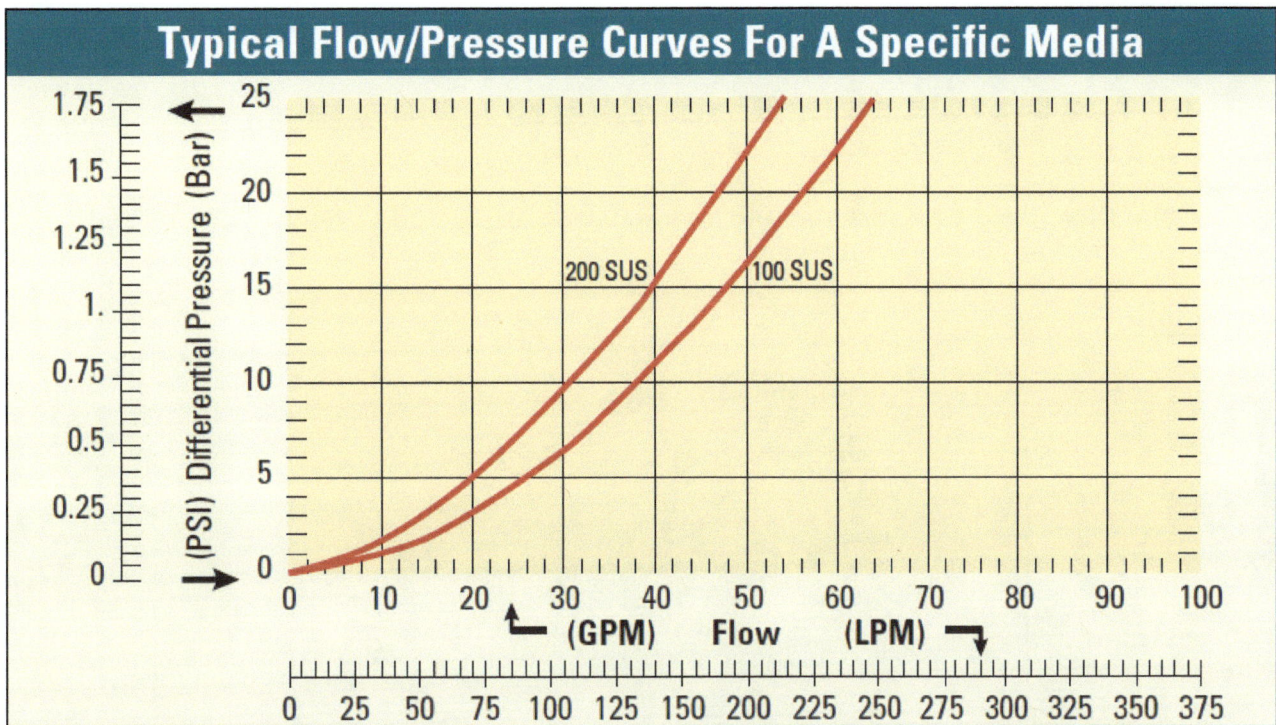

Fig. 7.16- Typical Flow-Pressure Curve for a Specific Filter (Courtesy of Parker)

Catalog data is generally provided for "clean" filter elements for specified media and at a given viscosity. Pressure drop is highly dependent on viscosity, so corrections should be made to the actual fluid being used. In addition, the worst-case viscosity condition is at the coldest anticipated operating temperature, which will need to be considered.

There is no one equation that is applicable for all brands of filters. However, filter manufacturers provide instructions on how to calculate the pressure drop and correct it based on the actual operating conditions. The following examples show different ways to calculate the differential pressure for various filter brands.

Example 1 (Ref. Donaldson):

Given Data:
- Filter Data Sheet for a spin on filter (5 μm) shown in Fig. 7.17.
- Test fluid viscosity = 32cSt [150 SSU] at 100°F (37.7°C).
- Test fluid specific gravity = 0.9 at 100°F (37.7°C).

Exercise:
Find the filter head pressure drop for an actual hydraulic oil of 64 cSt viscosity and 1.1 specific gravity. Estimated flow rate is 150 gpm.

Solution:

$$\Delta p_{\text{Fiter Head}} = 3 \ \times \frac{64}{32} \ \times \frac{1.1}{0.9} \ = \ 7.33 \ \text{psid}$$

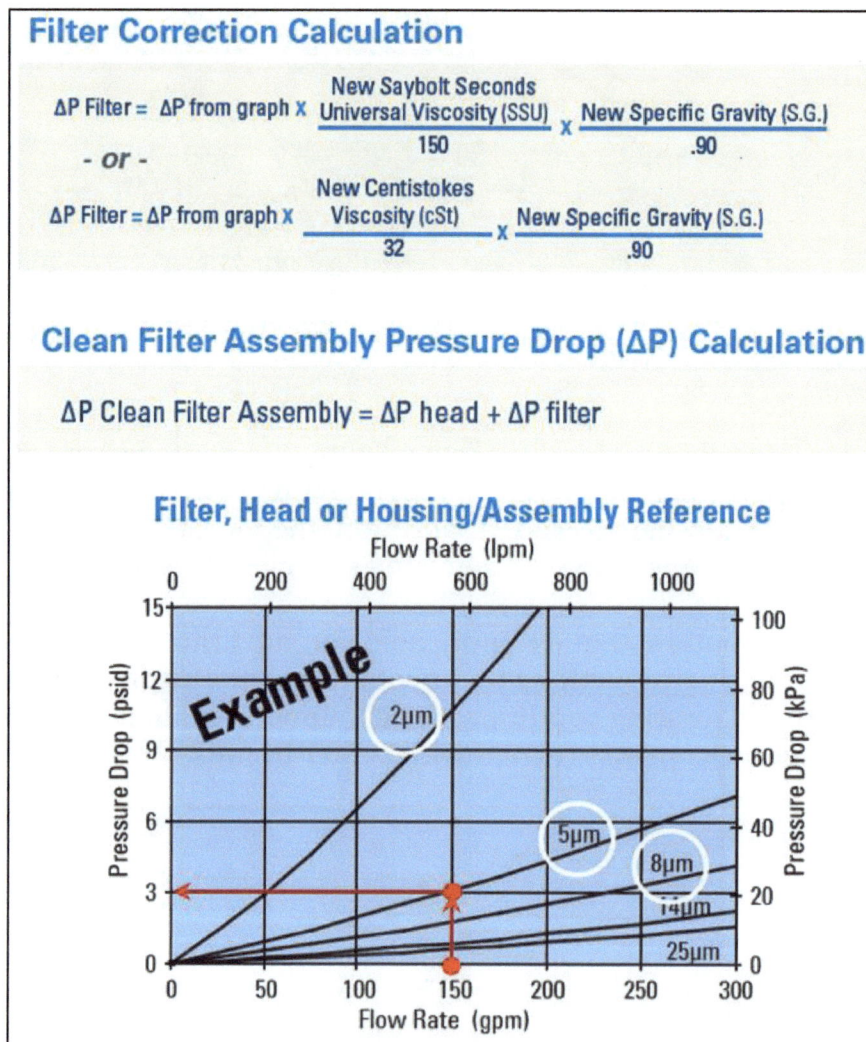

Fig. 7.17- Example of Pressure Drop Calculation (Courtesy of Donaldson)

Example 2 (Ref. Schroeder):

Given Data:
- Filter Data Sheet shown in Fig. 7.18.
- Test fluid viscosity = 32cSt [150 SSU] at 100°F (37.7°C).

Exercise: For a filter NZ25-1N series, find the filter assembly total pressure drop for an actual hydraulic oil of 44 cSt (200 SUS) and 0.86 specific gravity. Estimated flow rate is 15 gpm.

Solution: See the figure below.

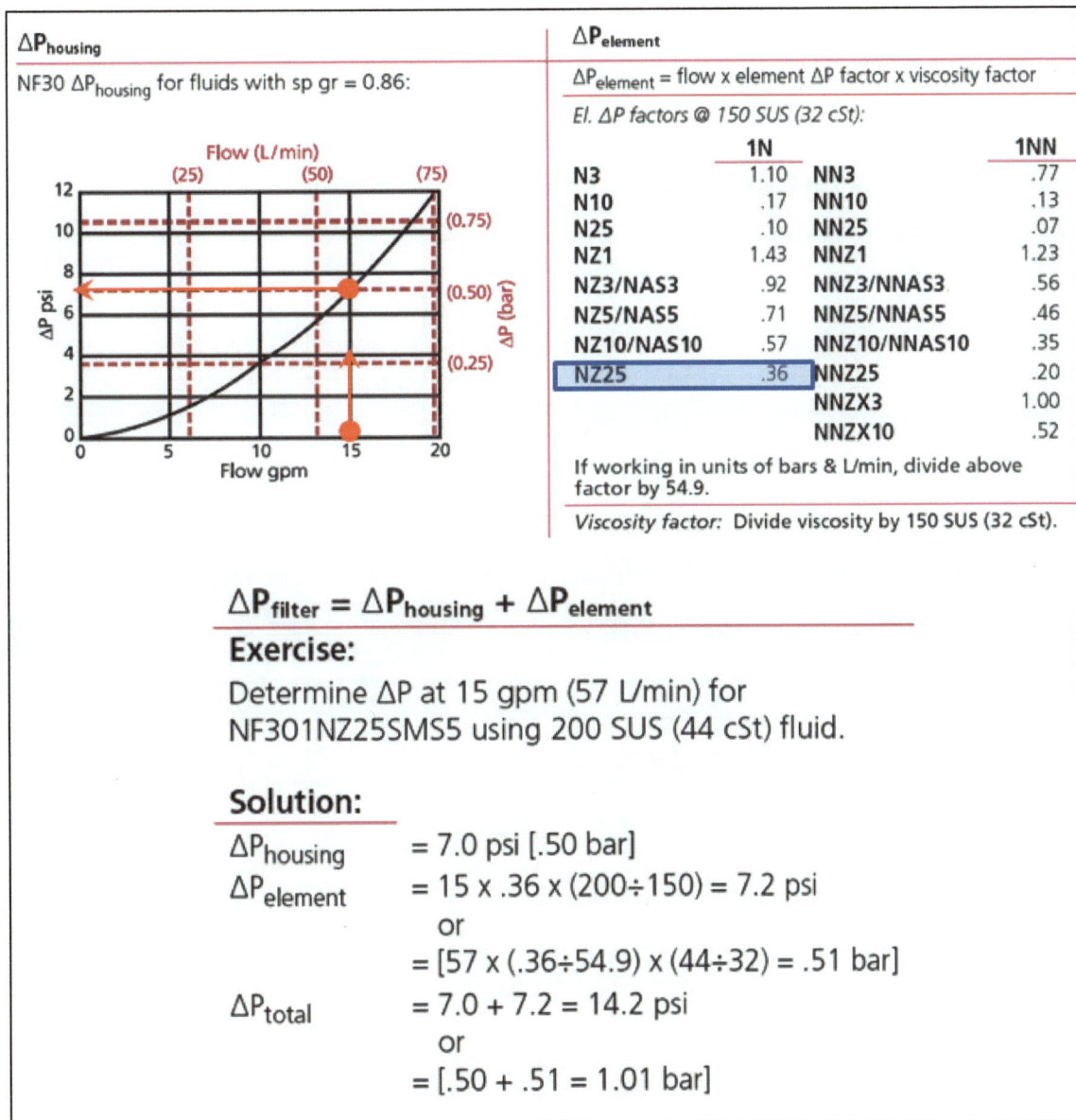

$\Delta P_{housing}$

NF30 $\Delta P_{housing}$ for fluids with sp gr = 0.86:

$\Delta P_{element}$

$\Delta P_{element}$ = flow x element ΔP factor x viscosity factor

El. ΔP factors @ 150 SUS (32 cSt):

	1N		1NN
N3	1.10	NN3	.77
N10	.17	NN10	.13
N25	.10	NN25	.07
NZ1	1.43	NNZ1	1.23
NZ3/NAS3	.92	NNZ3/NNAS3	.56
NZ5/NAS5	.71	NNZ5/NNAS5	.46
NZ10/NAS10	.57	NNZ10/NNAS10	.35
NZ25	.36	NNZ25	.20
		NNZX3	1.00
		NNZX10	.52

If working in units of bars & L/min, divide above factor by 54.9.

Viscosity factor: Divide viscosity by 150 SUS (32 cSt).

$\Delta P_{filter} = \Delta P_{housing} + \Delta P_{element}$

Exercise:

Determine ΔP at 15 gpm (57 L/min) for NF301NZ25SMS5 using 200 SUS (44 cSt) fluid.

Solution:

$\Delta P_{housing}$	= 7.0 psi [.50 bar]
$\Delta P_{element}$	= 15 x .36 x (200÷150) = 7.2 psi
	or
	= [57 x (.36÷54.9) x (44÷32) = .51 bar]
ΔP_{total}	= 7.0 + 7.2 = 14.2 psi
	or
	= [.50 + .51 = 1.01 bar]

Fig. 7.18- Example of Pressure Drop Calculation (Courtesy of Schroeder)

Example 3 (Ref. Hydac):

Figure 7.19 shows another example for calculating the total pressure drop of a filter including the given catalog data, the application data, and the solution.

EXAMPLE - an application with the following criteria would be sized as shown.

Conditions: **Fluid** – Hydraulic Oil (ISO-32)

Specific Gravity – 0.86

Viscosity – 141 SSU

Flow Rate – 30 GPM

Fluid Temperature - 104°F normal

Filter Type Selected - Pressure Filter

HYDAC Model No. DF ON 240 TE 10 D 1.0 / 12 V -B6

HOUSING

$$\Delta P \text{ Housing} = \Delta P \text{ Calculation } \textit{(From Curve in catalog)} \times \frac{\text{Actual Specific Gravity}}{0.86}$$

$$\Delta P \text{ Housing} = 1.5 \text{ psid} \times \frac{0.86}{0.86} = 1.5 \text{ psid}$$

ELEMENT

$$\Delta P \text{ Clean Element} = \Delta P \text{ Calculation } \times \frac{\text{Actual Specific Gravity}}{0.86} \times \frac{\text{Actual Viscosity}}{141 \text{ SSU}}$$

$$\Delta P \text{ Clean Element} = 30 \text{ GPM} \times 0.175 \times \frac{0.86}{0.86} \times \frac{141 \text{ SSU}}{141 \text{ SSU}}$$

$$\Delta P \text{ Clean Element} = 5.25 \times 1 \times 1 = 5.25 \text{ psid}$$

FILTER ASSEMBLY

$$\Delta P \text{ Filter Assembly} = \Delta P \text{ Housing} + \Delta P \text{ Clean Element}$$
$$1.5 \text{ psid} + 5.25 \text{ psid} = 6.75 \text{ psid}$$

Fig. 7.19- Example of Pressure Drop Calculation (Courtesy of Hydac)

Example 4 (Ref. Pall):

Given Data:
- Filter Data Sheet shown in Fig. 7.20.
- Test fluid viscosity = 32cSt [150 SSU] at 100°F (37.7°C),
- Test fluid specific gravity = 0.9 at 100°F (37.7°C).
- Fluid flow = 100 l/min.

Exercise: Find the filter assembly pressure drop for a Series UH210 housing with -20 port sizes housing and an AN grade element of 13" length. Actual hydraulic fluid used has 50 cSt and specific gravity of 1.2. Estimated flow rate is 100 l/min.

Solution: see the figure below.

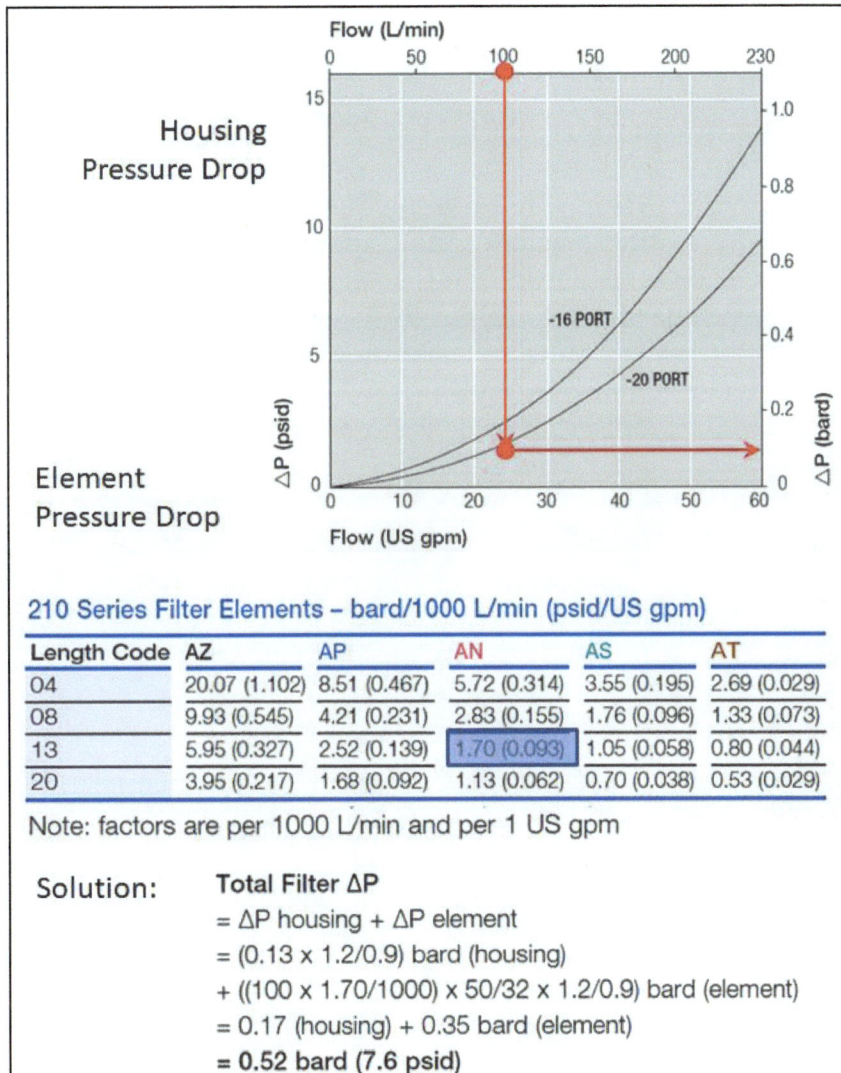

210 Series Filter Elements – bard/1000 L/min (psid/US gpm)

Length Code	AZ	AP	AN	AS	AT
04	20.07 (1.102)	8.51 (0.467)	5.72 (0.314)	3.55 (0.195)	2.69 (0.029)
08	9.93 (0.545)	4.21 (0.231)	2.83 (0.155)	1.76 (0.096)	1.33 (0.073)
13	5.95 (0.327)	2.52 (0.139)	1.70 (0.093)	1.05 (0.058)	0.80 (0.044)
20	3.95 (0.217)	1.68 (0.092)	1.13 (0.062)	0.70 (0.038)	0.53 (0.029)

Note: factors are per 1000 L/min and per 1 US gpm

Solution: Total Filter ΔP
= ΔP housing + ΔP element
= (0.13 x 1.2/0.9) bard (housing)
+ ((100 x 1.70/1000) x 50/32 x 1.2/0.9) bard (element)
= 0.17 (housing) + 0.35 bard (element)
= 0.52 bard (7.6 psid)

Fig. 7.20- Example of Pressure Drop Calculation (Courtesy of Pall)

7.4.5- NFPA (T-2.6.1): Rated Burst Pressure (RBP) of a Filter Housing

Rated Burst Pressure (RBP) of a filter housing is the static pressure at which <u>filter housing</u> structural failure occurs. Burst Pressure is determined by a test according to NFPA (T-2.6.1) standards.

7.4.6- ISO 1077-1: Rated Fatigue Pressure (RFP) of a Filter Housing

Rated Fatigue Pressure (RFP), as shown in Eq. 7.5, is the maximum allowable pressure for a <u>filter housing</u> according to <u>(ISO 10771-1)</u>. Safety factor is typically 4 – 6.

$$\mathbf{RFP} = \frac{\mathbf{RBP}}{\mathbf{Safety\ Fator}}$$

7.5

7.4.7- Cyclic Test Pressure (CTP) of a Filter Housing

There are many hydraulic systems that use highly repetitive functions such as plastic injection molding machines, die-casting machines, and hydraulic presses. In such systems, *Cyclic Test Pressure (CTP)* should be considered when selecting a filter. CTP is the maximum pressure applied for certain number of cycles (typically 1 million cycles) before housing failure occurs. CRP is experimentally tested. However, as shown in Eq. 7.6, CTP can be mathematically calculated. CTP equals RFP multiplied by a factor K that is obtained from tables associated with the above-mentioned standard based on confidence, assurance levels, materials of construction, and number of units tested.

$$\mathbf{CTP} = \mathbf{RFP} \times \mathbf{K}$$

7.6

Example:
- RBP = 20,000 psi.
- Safety Factor = 4 →
- RFP = 20,000/4= 5,000 psi.
- K = 1.1 – 1.4 →
- CTP = 5,000 X 1.5 = 7,500 PSI

12.4.8- ISO 2941: Collapse Pressure of a Filter Element

As shown in Fig. 7.21, differential pressure is used as indicator for the state of the filter. It indicates whether the filter is ok to continue to operate or if it should be replaced.

When a filter reaches a level of plugging or a cold start occurs or a combination of both, an increase in pressure is seen between the inlet (dirty side) and the outlet (clean side). If this differential pressure is high enough, the filter element and/or center tube can rupture or collapse. This is serious because unfiltered fluid and damaged filter components can then be routed back into the system.

A filter assembly whose element cannot withstand, without damage, the maximum differential pressure in its part of the system shall be equipped with a filter bypass valve. Ideally, a filter element should be sized so that the initial differential pressure across the clean element (plus the filter housing drop) is less than half the bypass valve setting in the filter housing.

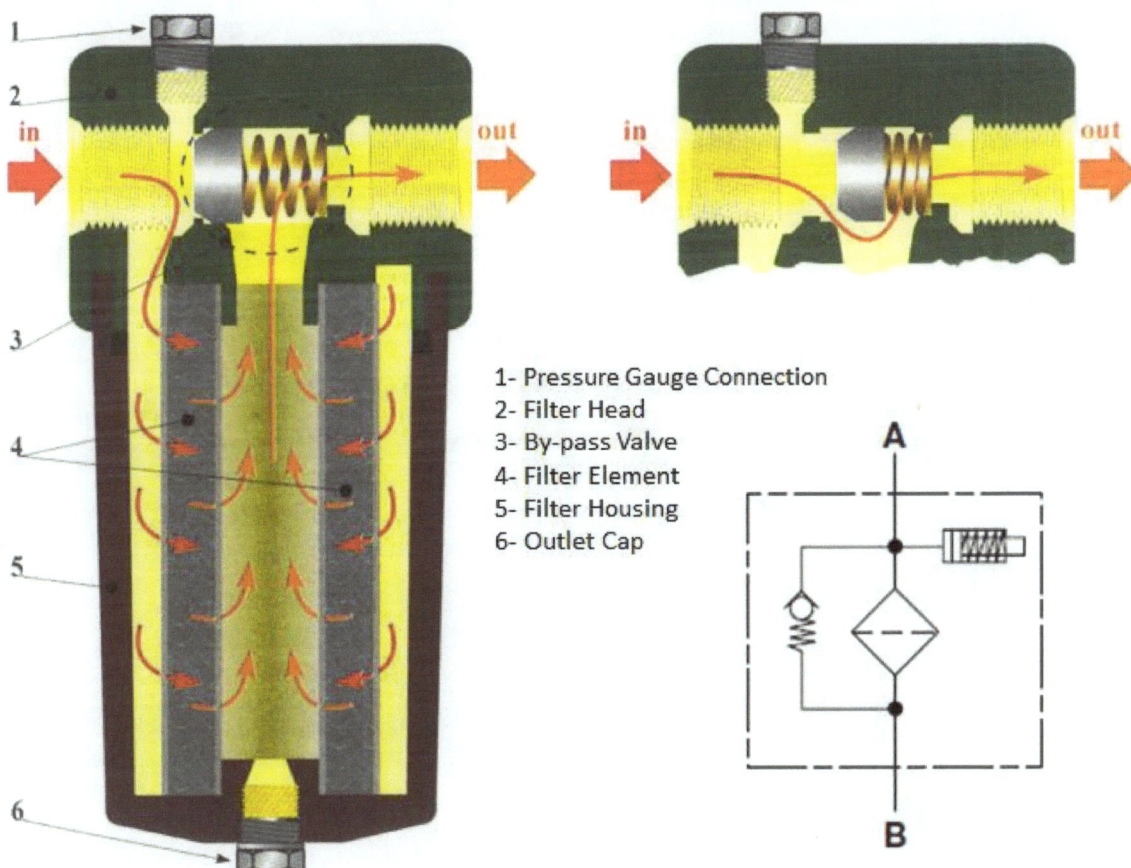

1- Pressure Gauge Connection
2- Filter Head
3- By-pass Valve
4- Filter Element
5- Filter Housing
6- Outlet Cap

**Fig. 7.21- Filter Housing Equipment with Bypass Valve and Clogging Indicator
(Courtesy of Assofluid)**

Collapse Pressure of a <u>filter element</u> is the differential pressure at which a structural failure of the filter element and/or center tube occurs. Collapse pressure is determined by ISO 2941 standards.

As shown in Fig. 11.22, the collapse pressure rating of a filter element installed in a filter housing, with a bypass valve, should be at least two times greater than the full flow bypass valve pressure drop.

Pressure filters with no bypass are recommended with the use of servo valves. The collapse pressure rating for filter elements used in filter housings with no bypass valves must be at least the same as the setting of the system relief valve upstream of the filter high-crush element. When a high-pressure collapse element becomes clogged with contamination all functions downstream of the filter will become inoperative.

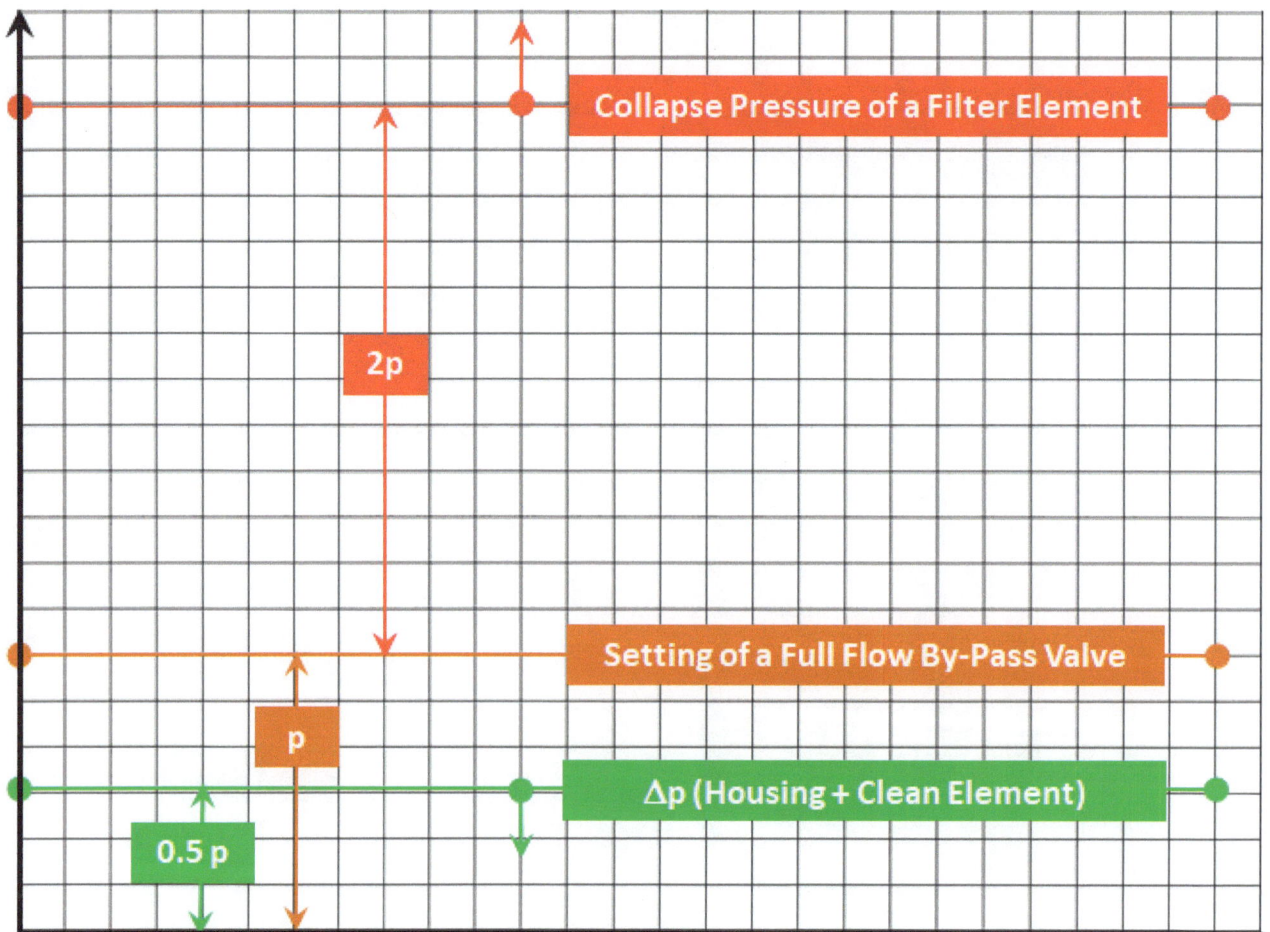

Fig. 7.22- Collapse Pressure of a Filter Element versus By-Pass Setting

Contamination Loading Curve: As shown in Fig. 7.23, as dirt is trapped by the filter, differential pressure (ΔP) increases. As shown in the figure, an alarming pressure must be activated before the bypass valve open, the bypass valve must open before the collapse pressure of the filter media is achieved, and the collapse pressure (1) of the filter media must be lower than the core collapse pressure (2). The *core collapse pressure* is the pressure at which the supporting center tube is crashed or ruptured.

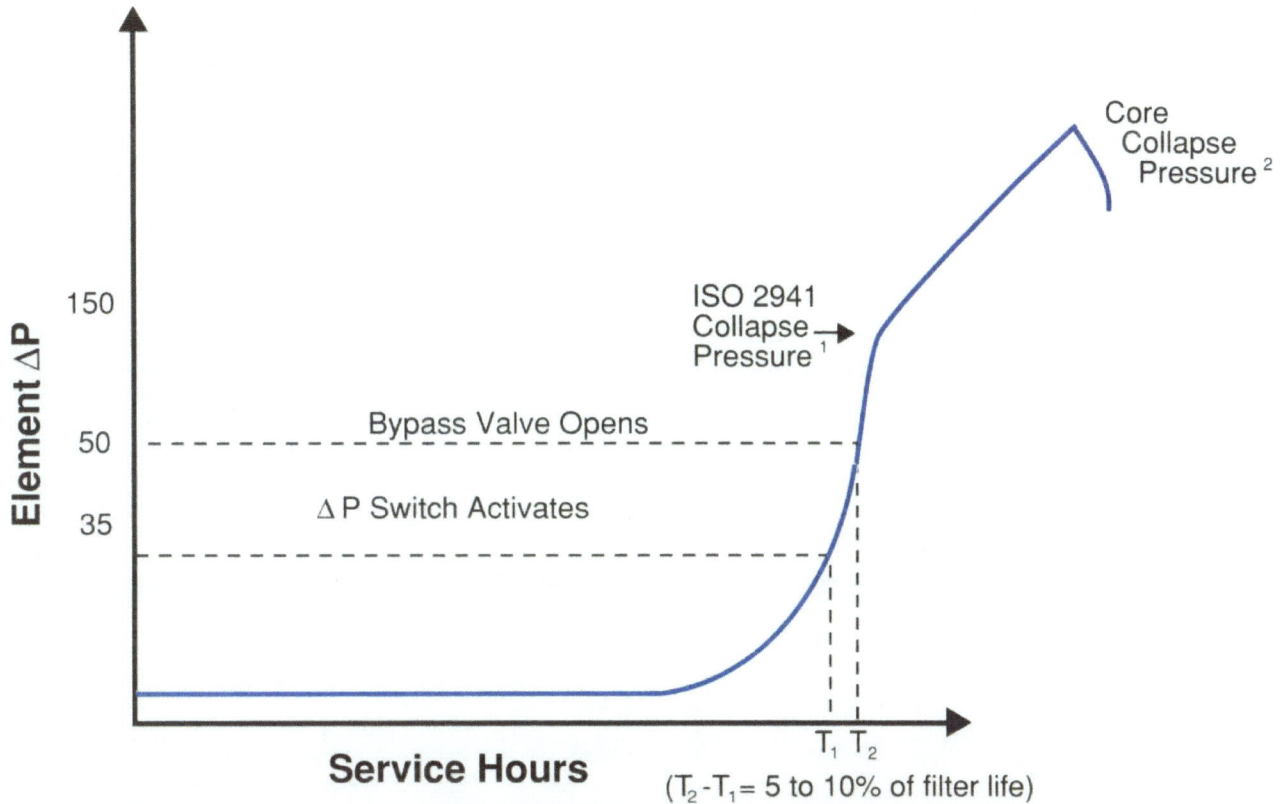

Fig. 7.23- Filter Service Life versus Pressure Drop

7.4.9- ISO 3723-2015: End Load Test

The *End Load Test* is conducted to heck the ability of a filter element to resist axial deformation caused by differential pressure across the element. Weights or other load generating means are used to simulate an axial force on the same surfaces receiving the load when installed in the proper housing. The element is considered pass the test when it is subjected to the maximum axial load specified by the manufacturer without permanent deformation, structural damage, or seal failure.

7.4.10- ISO 3724 OR ISO 23181: Flow Fatigue Test for Filter Element

Due to pulsations of flow, <u>filter media</u> may fail prior to replacement time. High fatigue stability is achieved by better filter element design including supporting both sides of the element and high inherent stability of the filter materials.

Flow Fatigue Test is used to predict the ability of a filter element to withstand structural failure due to flexing of the pleats caused by cyclic flow. Flow Fatigue Test ran in accordance with (ISO 3724 OR ISO 23181).

As show in Fig. 7.24, This test requires that an element be contaminated to its terminal differential pressure. The element is then subjected to a cyclic change in flow from zero to its maximum rated flow and then back to zero for the number of cycles prescribed by the element manufacturer, usually based on (10-200) thousand cycles. An element is considered to have passed the test if there is no visual evidence of structural, seal or filter medium failure.

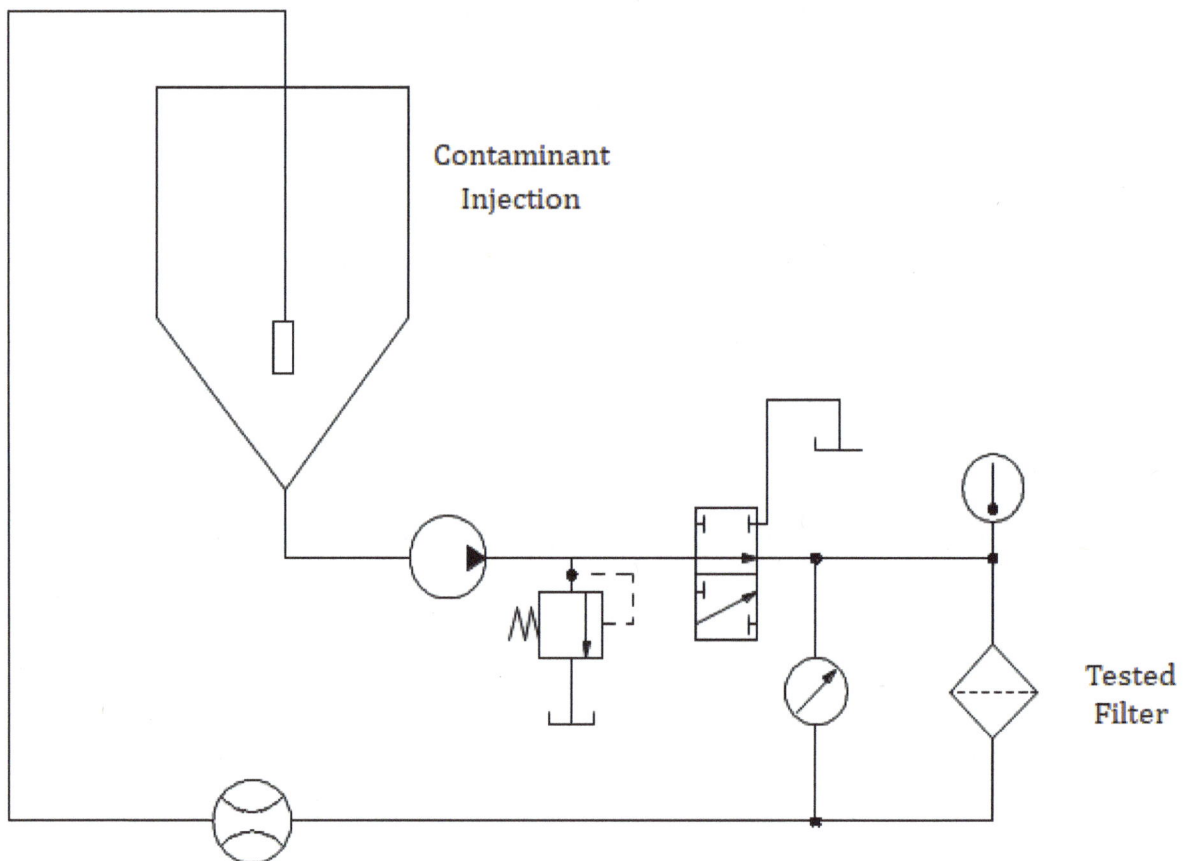

Fig. 7.24- Typical Flow Fatigue Test Setup

7.4.11- Filter Tests Pictogram

As shown in Fig. 7.25, filter tests may be presented in form of pictograms in the filter datasheet or service manual. The shown pictogram shows the standards and the measure of the dirt holding capacity and efficiency via multipass test (1), filter element collapse pressure (2), filter element differential pressure (3), and filter element structural integrity test via bubble point test (4).

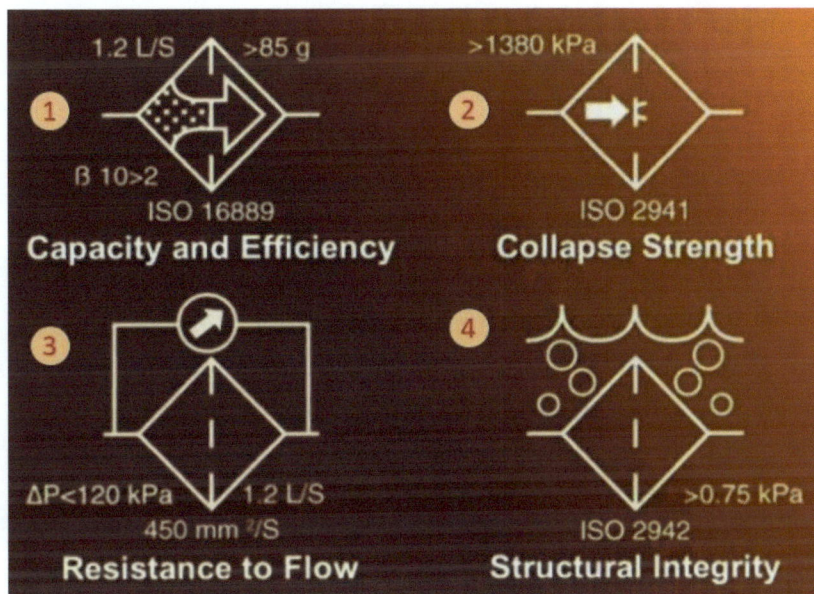

Fig. 7.25- Hydraulic Filters Tests Pictograms (Courtesy from Parker)

7.5-BP-Filters-05-Transportation and Storage

It's important to practice good storage and handling techniques when it comes to filters. As reported by and excerpted from Donaldson service manuals, the following tips are considered best practices for storage and transportation of filters for the sake of contamination control:

Storage:
- Never store a filter on a shelf without it being in a box or totally sealed from outside contaminant.
- When you see an open box of filters on the shelf, tape it shut–unless the filters inside the box are individually sealed.
- Make sure labels with product information and manufacturing dates are visible to personnel selecting from the shelves.
- Metal storage shelves may cause condensation to form on filters if sitting directly on metal. So, it is recommended to use wooden or plastic shelves.
- Over time the filter may get rusty. This is another good reason to store filters in boxes.

Shipping:
- If transporting filters from one job site to another, don't let them roll around on the floorboard or in the back of a truck as it may damage the filter.

Handling:
- Practice "first-in, first-out" with your inventory. When possible, always use the oldest inventory first.
- If a product box has layers of contaminant, take care that the contaminant doesn't get on the new filter as you remove it from the box.
- Handle filters with care to prevent filter damage. For example, don't throw filters into the back of a truck.

Chapter 8

Filter Selection Criteria

Objectives

This chapter presents a selection checklist as a guide for selecting proper filters. The chapter also discusses briefly the concepts for cost-effective filtration and selecting a filter cleanliness level based on system requirements. This chapter presents several examples of filtration solution for hydraulic systems.

Brief Contents

8.1- Filter Selection Checklist

8.2- Cost-Effective Filtration

8.3- Filter Selection Based on Cleanliness Requirements

8.4- Examples of Filtration Solutions

Chapter 8 – Filter Selection Criteria

8.1- Filter Selection Checklist

The proper choice of a filter is essential as early as the design stage of a system. When selecting a filter, the following questions in *Filter Selection Checklist* must be answered in order determine the proper filter

- Filter Purpose:
  - For regular operation, for flushing, for water or varnish removal?
  - For mobile or industrial application?

- Filter Location:
  - Inline filter (suction -pressure – return), or offline filter?

- System Operating Conditions:
  - What is the maximum system pressure, temperature, and flow?
  - Are there possible pressure fluctuation and/or pressure spikes?
  - Are there possible flow fluctuation or flow surges?
  - Are there sensitive components in the system that are intolerant to contamination?
  - What type of the hydraulic fluid is used and its viscosity?

- System Cleanliness Requirements:
  - What is the required absolute/nominal beta ratio and filter efficiency?
  - What is the mesh size (for screens)?
  - Requirements for bypass?
  - Requirements for clogging indicators/alarms?
  - What is the anticipated dirt holding capacity?

- Filter Media:
  - Collapse pressure and flow fatigue resistance?
  - Moisture absorbance characteristics and Anti-Static characteristics?

- Filter Housing:
  - Burst pressure and fatigue pressure?
  - Method of mounting and body style (inside tank, line mounted, on-tank top)
  - Port size?

8.2- Cost-Effective Filtration

The initial and running cost of efficient filtration in hydraulic systems are paid back by increasing the service life of important and expensive components and improving system reliability. The following bullets discuss solutions for *cost-effective* filtration.

Service Life Versus Filter Area: Dirt holding capacity and service life will vary greatly between filter types, media and manufacturers. The service life of a filter is the length of time that a filter element will last in actual system service before the allowable differential pressure is reached. Dirt holding capacity alone may not indicate which filter would give the best service life. In order to achieve the highest service life, element area should be increased. As shown in Fig. 8.1, a filter element with three times the area will yield 4 to 6 times the service life. Then, price of element with large area must be compromised with the cost paid back by extended filter lifetime.

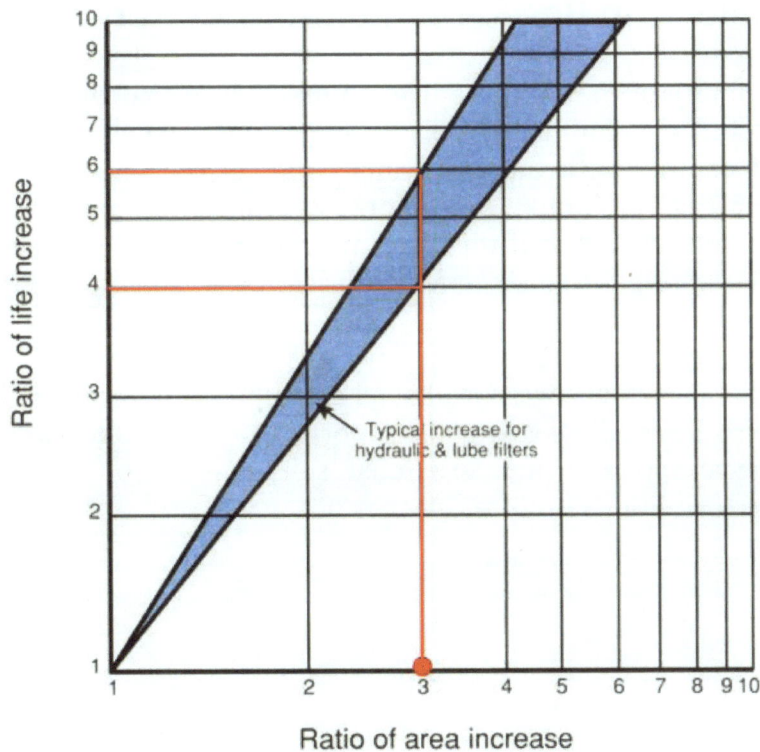

Fig. 8.1- Filter Element Service Life versus Filter Area

Staged Filtration: Some system designers prefer to use "*staged*" filtration rather than rely on a single filter element. Staged filtration utilizes two or more filters in series with the lower or finer filtration rating downstream of the first filter. The first filter acts as a pre-filter for the fine filter, removing the majority of coarse particulates. The initial cost of staged filtration is greater due to multiple filter housings. However, this initial cost is compensated with the extended filter element service life and clean operating system.

Filter Location: By selecting the proper location of filters in the system, filter costs can be minimized. Most cost-effective filtration solutions are either return filters or offline filters. However, a pressure filter might be mandatory in cases where sensitive components are used such as servo valves. If pressure filters are used, the filter housing must be selected for the maximum system operating pressure. Pressure filters are the most expensive.

Bypass OR Non-Bypass: Having a bypass valve in the filter construction limits the maximum differential pressure across the filter element. That helps reducing cost by selecting a filter element with relatively low collapse pressure. However, collapse pressure of the filter element must be higher than the bypass valve cracking pressure. However, non-bypass filters are required for cases where sensitive components are located downstream of the filter. An alternative solution is to use a bypass-to-tank filter.

8.3- Filter Selection Based on Cleanliness Requirements

It is the responsibility of the machine user to make sure that the oil continues to be clean as per the manufacturer' recommendation during machine operation. Filtration system in the machine should comply with the component that is the most sensitive (least tolerant) to contamination. In case of component failure, warranty may be voided if fluid cleanliness is greater than recommendations. If there are no recommendations given by the manufacturer, the following section provide some general guidelines.

Table 8.1 shows system-based and component-based cleanliness level recommendations, respectively. Table 8.2 shows pressure-based cleanliness requirements. Presented information shows that electro-hydraulic servo and proportional valves and variable pumps, particularly piston type, are components that are most sensitive to contamination. Additionally, high working pressure conditions requires cleaner oil.

Application	Oil cleanliness required in accordance with ISO 4406
Systems with extremely high dirt sensitivity and very high availability requirements	≤ 16/12/9
Systems with high dirt sensitivity and high availability requirements, such as servo valve technology	≤ 18/13/10
Systems with proportional valves and pressures > 160 bar	≤ 18/14/11
Vane pumps, piston pumps, piston engines	≤ 19/16/13
Modern industrial hydraulic systems, directional valves, pressure valves	≤ 20/16/13
Industrial hydraulic systems with large tolerances and low dirt sensitivity	≤ 21/17/14

Pumps	ISO Ratings
Fixed Gear Pump	19/17/15
Fixed Vane Pump	19/17/14
Fixed Piston Pump	18/16/14
Variable Vane Pump	18/16/14
Variable Piston Pump	17/15/13
Valves	
Directional (solenoid)	20/18/15
Pressure (modulating)	19/17/14
Flow Controls (standard)	19/17/14
Check Valves	20/18/15
Cartridge Valves	20/18/15
Load-sensing Directional Valves	18/16/14
Proportional Pressure Controls	18/16/13
Proportional Cartridge Valves	18/16/13
Servo Valves	16/14/11*
Actuators	
Cylinders	20/18/15
Vane Motors	19/17/14
Axial Piston Motors	18/16/13
Gear Motors	20/18/15
Radial Piston Motors	19/17/15

Table 8.1- System-Based and Component-Based Cleanliness Requirements

	ISO Target Levels		
	Low/Medium Pressure Under 2000 psi (moderate conditions)	High Pressure 2000 to 2999 psi (low/medium with severe conditions)	Very High Pressure 3000 psi and over (high pressure with severe conditions)
Pumps			
Fixed Gear or Fixed Vane	20/18/15	19/17/14	18/16/13
Fixed Piston	19/17/14	18/16/13	17/15/12
Variable Vane	18/16/13	17/15/12	not applicable
Variable Piston	18/16/13	17/15/12	16/14/11
Valves			
Check Valve	20/18/15	20/18/15	19/17/14
Directional (solenoid)	20/18/15	19/17/14	18/16/13
Standard Flow Control	20/18/15	19/17/14	18/16/13
Cartridge Valve	19/17/14	18/16/13	17/15/12
Proportional Valve	18/16/13	17/15/12	16/14/11
Servo Valve	16/14/11	16/14/11	15/13/10
Actuators			
Cylinders, Vane Motors, Gear Motors	20/18/15	19/17/14	18/16/13
Piston Motors, Swash Plate Motors	19/17/14	18/16/13	17/15/12
Hydrostatic Drives	16/15/12	16/14/11	15/13/10
Test Stands	15/13/10	15/13/10	15/13/10
Bearings			
Journal Bearings	17/15/12	not applicable	not applicable
Industrial Gearboxes	17/15/12	not applicable	not applicable
Ball Bearings	15/13/10	not applicable	not applicable
Roller Bearings	16/14/11	not applicable	not applicable

Table 8.2- Pressure-Based Cleanliness Requirements (Courtesy of Hydac)

8.4- Examples of Filtration Solutions

Example 1 (Fig. 8.2): Filtration Solutions for Injection Molding Machines.

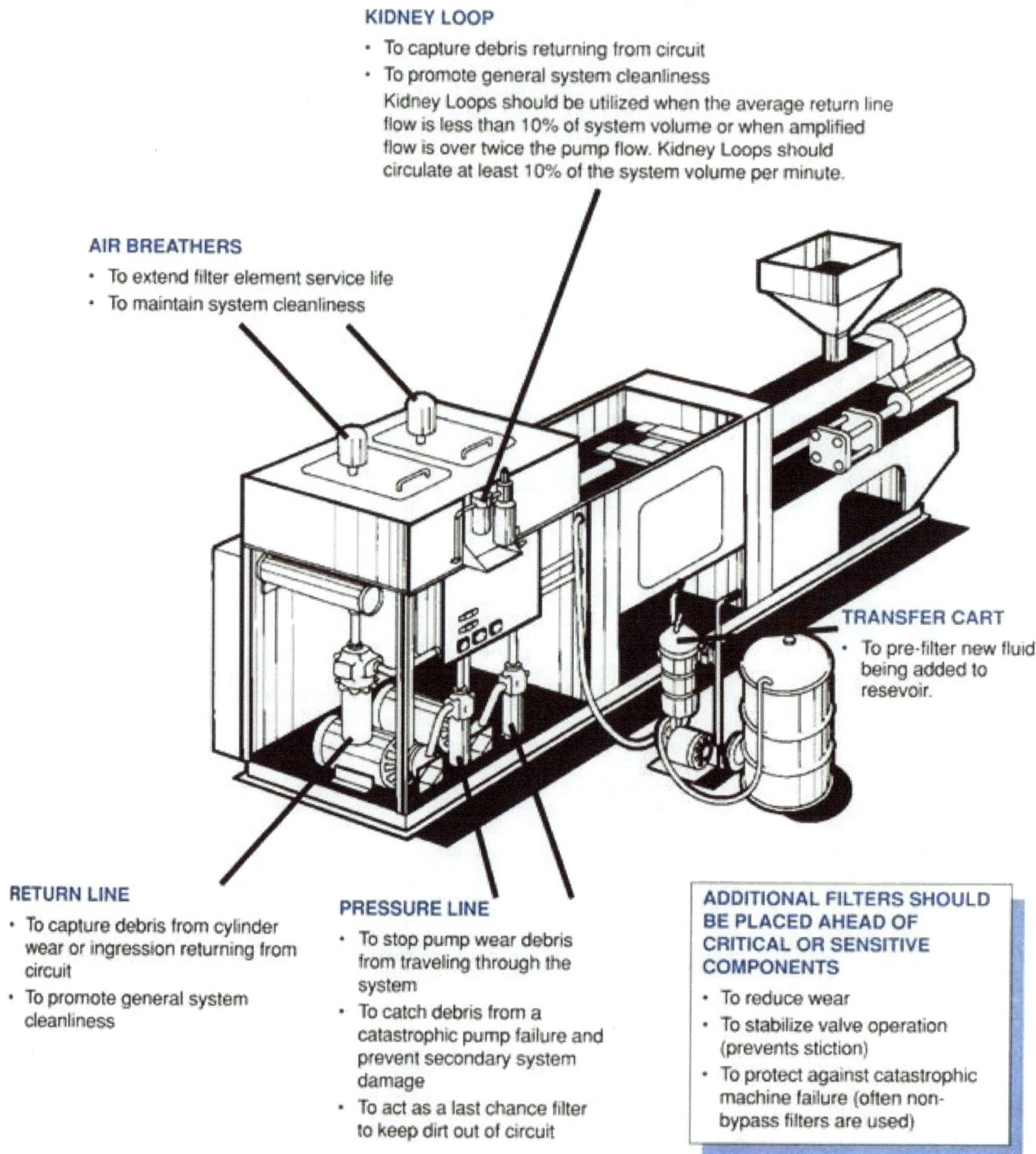

KIDNEY LOOP
- To capture debris returning from circuit
- To promote general system cleanliness
  Kidney Loops should be utilized when the average return line flow is less than 10% of system volume or when amplified flow is over twice the pump flow. Kidney Loops should circulate at least 10% of the system volume per minute.

AIR BREATHERS
- To extend filter element service life
- To maintain system cleanliness

TRANSFER CART
- To pre-filter new fluid being added to resevoir.

RETURN LINE
- To capture debris from cylinder wear or ingression returning from circuit
- To promote general system cleanliness

PRESSURE LINE
- To stop pump wear debris from traveling through the system
- To catch debris from a catastrophic pump failure and prevent secondary system damage
- To act as a last chance filter to keep dirt out of circuit

ADDITIONAL FILTERS SHOULD BE PLACED AHEAD OF CRITICAL OR SENSITIVE COMPONENTS
- To reduce wear
- To stabilize valve operation (prevents stiction)
- To protect against catastrophic machine failure (often non-bypass filters are used)

Fig. 8.2- Filtration Solutions for Injection Molding Machines (Courtesy of Pall)

Example 2 (Fig. 8.3) - Open Circuit Systems with Solenoid Valves:

System Parameters:
- Pump flow = 30 gpm.
- Maximum pressure = 1800 psi.
- Fluid type is petroleum-based fluid.

Selected Filters:
1- Pressure Line Filter (HH9660A20DNTBPT).
2- Return Line Filter (HH8800A2DNTBPL).
3- Air Breather Filter (HC7500S038H-B).

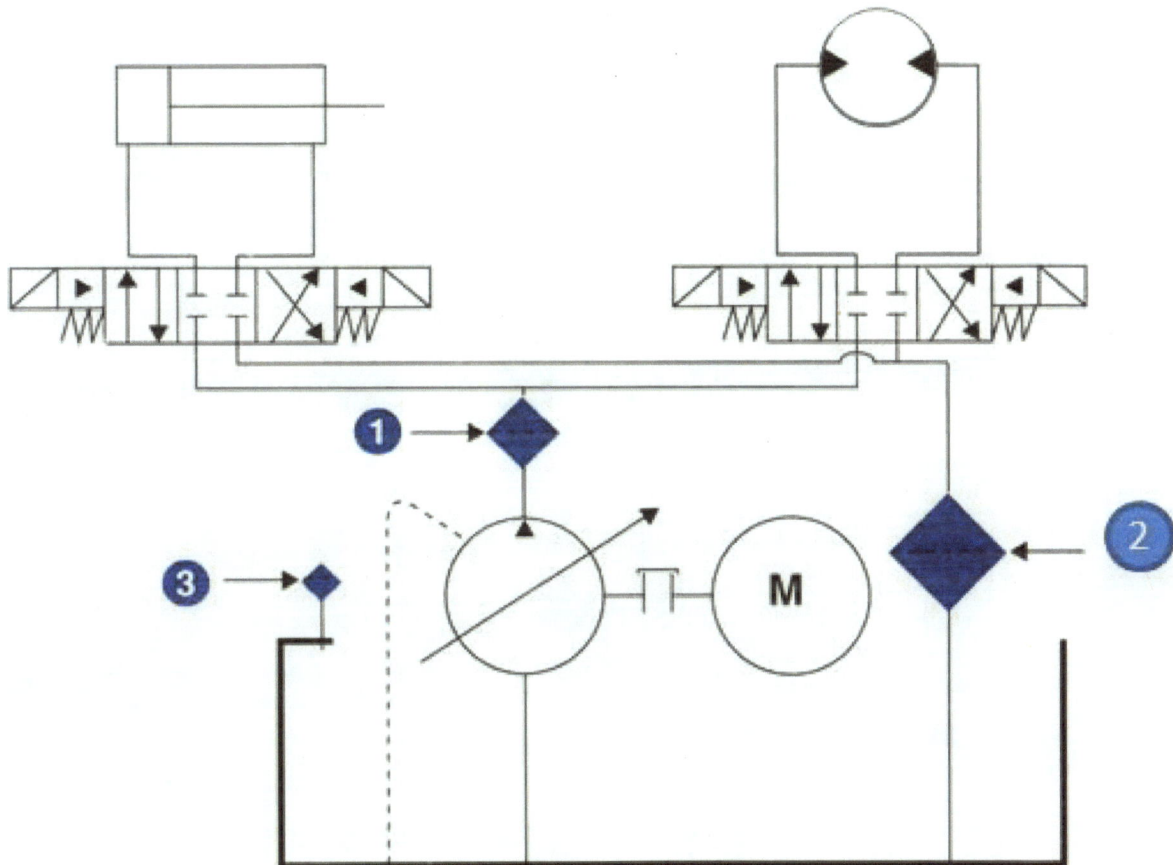

Fig. 8.3- Filtration Solutions for Open Circuit with Solenoid Valves (Courtesy of Pall)

Example 3 (Fig. 8.4) - Open Circuit Systems with Servo Valve:

System Parameters:
- Pump flow = 15 gpm.
- Maximum pressure = 1900 psi.
- Fluid type is petroleum-based fluid.

Selected Filters:
1- Pressure Line Filter (HH9850A 16DPSBPT).
2- Remote Mounted Non-Bypass Filter (HH9021 A 12DPRWPT).
3- Return Line Filter (HH8200A20DPSBPL).
4- Aire Breather Filter (HC7500S038H-B).

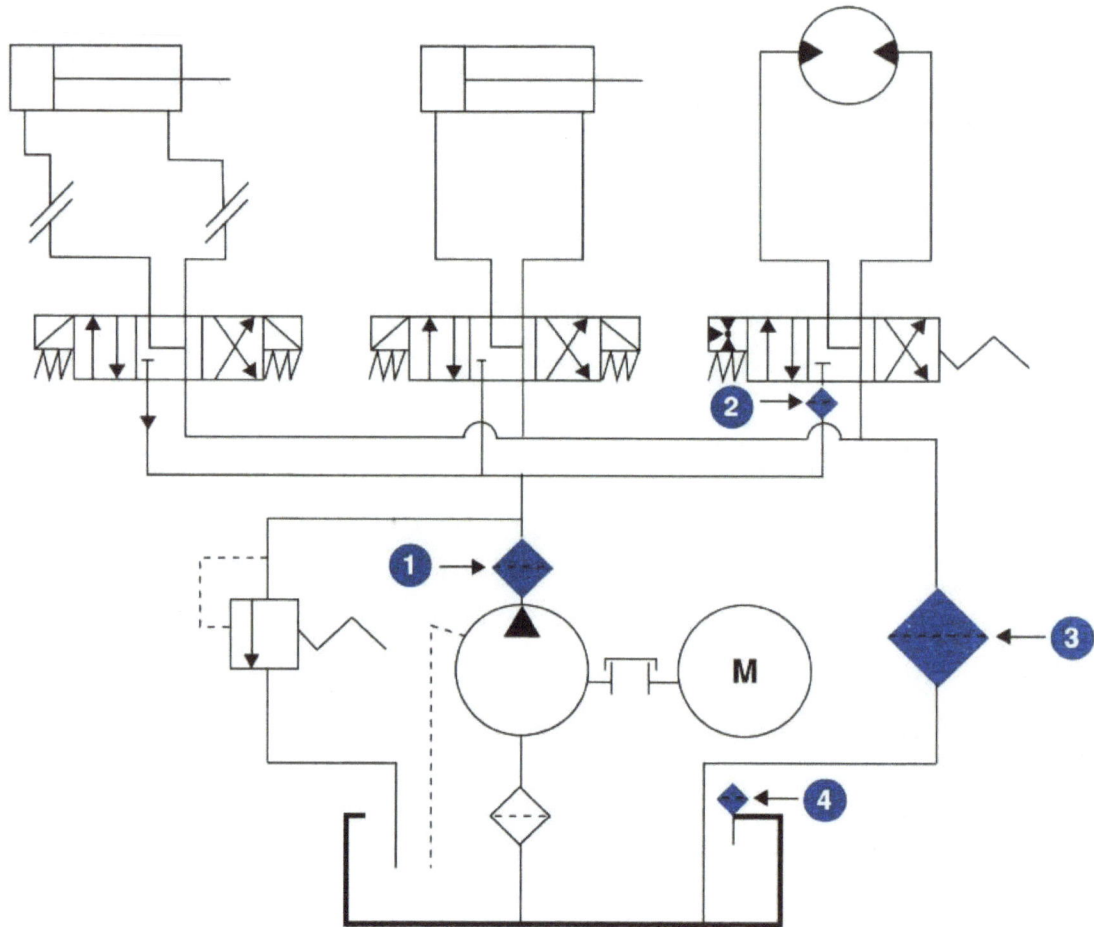

Fig. 8.4- Filtration Solutions for Open Circuit with Servo Valve (Courtesy of Pall)

Example 4 (Fig. 8.5) – Clamp and Hold Hydraulic Circuit:

System Parameters:
- Pump Type: Piston, variable displacement pressure compensated.
- Pump flow = 50 gpm.
- Maximum pressure = 2800 psi.
- Fluid type is Water Glycol.
- Total System Fluid Volume = 300 Gallons.

Selected Filters:
1- Pressure Line Filter (HH971 OA24DP2BDT).
2- Kidney Loop Filter (HH8900D32DPUBDL).
3- Aire Breather Filter (HC7500S038H-B).

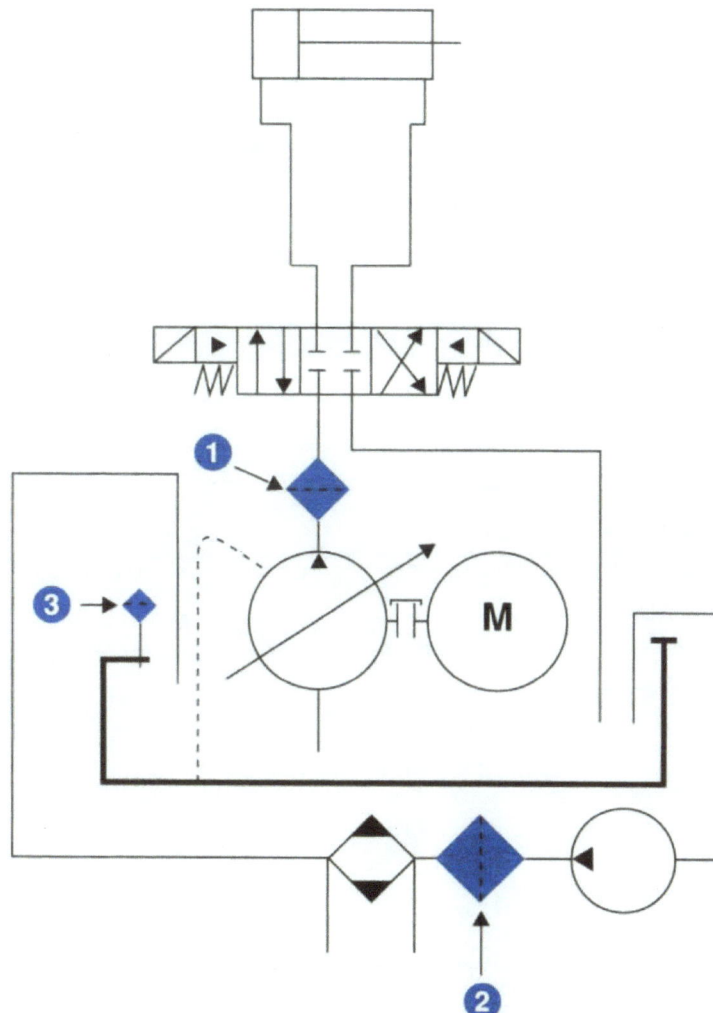

Fig. 8.5- Filtration Solutions for Clamp and Hold Circuit (Courtesy of Pall)

Example 5 (Fig. 8.6) – Closed Circuit Hydrostatic Transmission:

System Parameters:
- Charge pressure = 125 psi.
- Charge flow = 4 gpm.
- System pressure = 3200 psi.
- System flow = 32 gpm.
- System flow direction is bidirectional.
- Fluid type is petroleum-based.

Selected Filters:
1 and 2 in-loop filters with reverse flow valve (HH9660E20DPTCPT).

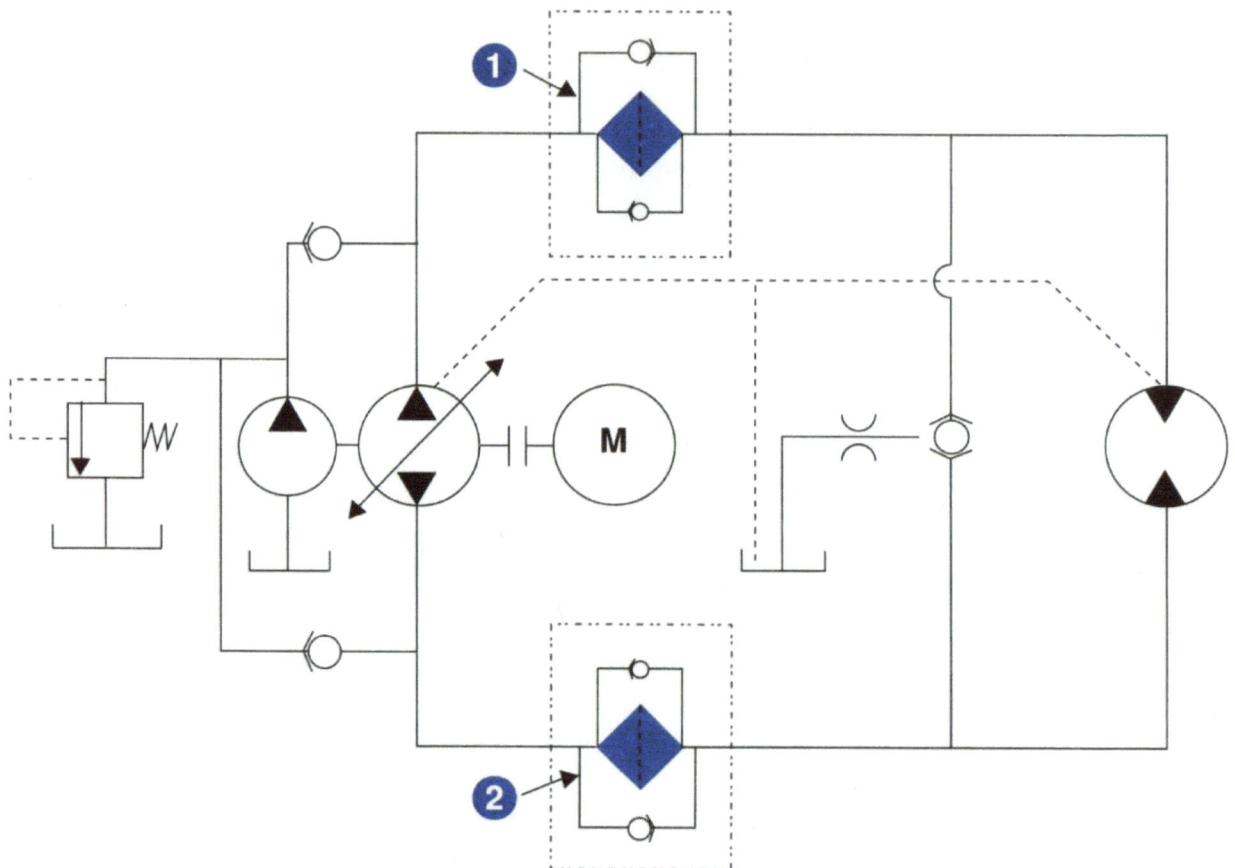

Fig. 8.6- Filtration Solutions for Hydrostatic Transmission (Courtesy of Pall)

Example 6 (Fig. 8.7) – Bearing Lubrication Circuit:

System Parameters:
- Pump flow = 120 gpm.
- System pressure = 45 psi.
- Maximum pressure at the pump = 120 psi
- Fluid type is petroleum-based.

Selected Filters:
1- Duplex filter in Pressure Line (HH8342D64DNXAPT)
2- Air Breathers - (2) HC7500S038H-B

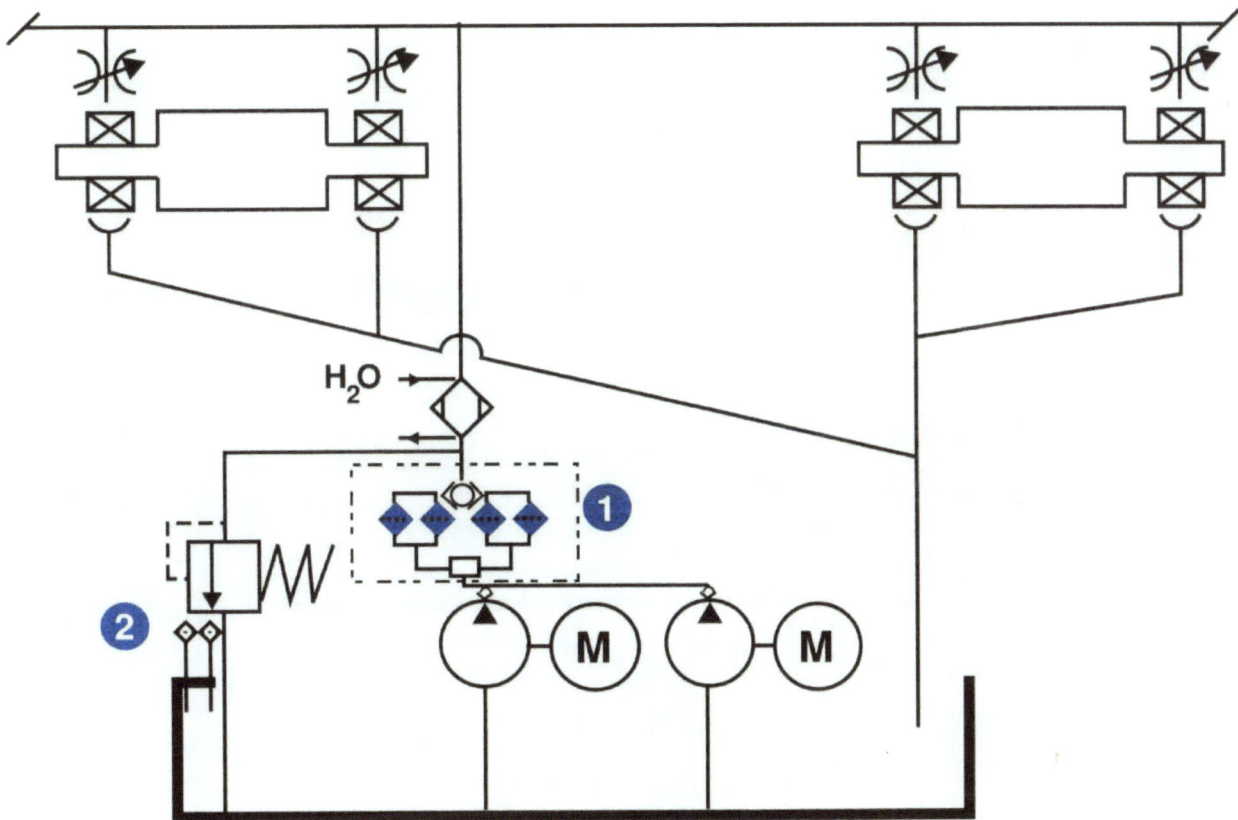

Fig. 8.7- Filtration Solutions for Bearing Lubrication Circuit (Courtesy of Pall)

Example 7 (Fig. 8.8) – Turbine System Lubrication:

System Parameters:
- Main reservoir volume = 6000 Gallons.
- Operating pressure = 15 psi.
- Kidney loop flow (added) = 100 Gallons.
- Fluid type is petroleum-based.

Selected Filters:
1 and 2 kidney loop filters (HH8300D40DPXBPT).

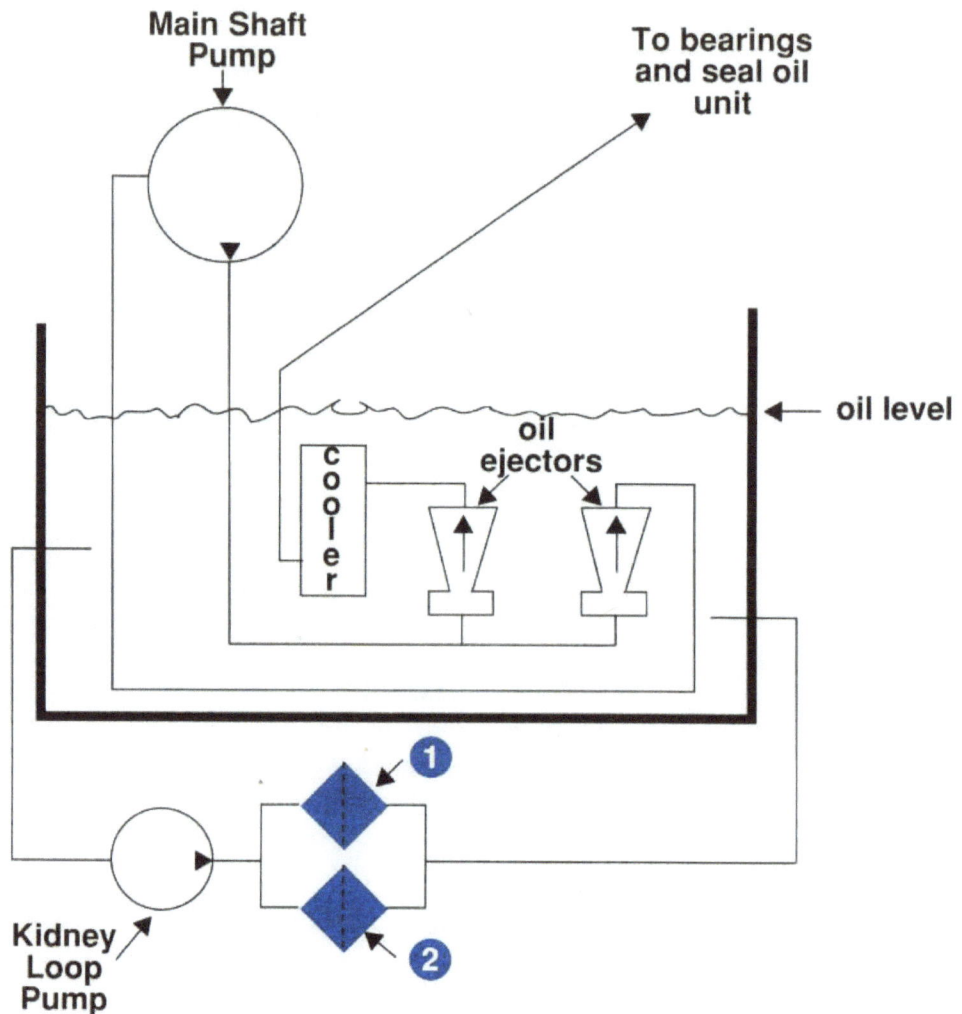

Fig. 8.8- Filtration Solutions for Turbine System Lubrication (Courtesy of Pall)

Chapter 9

Troubleshooting and Failure Analysis of Filters

Objectives

This chapter discusses hydraulic *filters* inspection, troubleshooting, and failure analysis. In this chapter, a troubleshooting chart for filter faults is presented. This chapter also presents examples of defective filter.

Brief Contents

9.1- Filters Inspection

9.2- Filters Troubleshooting

9.3- Filters Failure Analysis

Chapter 9: Troubleshooting and Failure Analysis of Filters

9.1- Filters Inspection

Hydraulic *filters* can be suction, pressure, or return. Filters are available in various sizes, efficiency, and dirt holding capacity. Volume 5 of this series of textbooks provides guidelines for filter maintenance and safety. Table 9.1 shows typical inspection sheet for filters.

Filters Inspection Sheet	
Manufacturer	
Model #	
Serial #	
Location	
Filter Type	☐ Suction ☐Pressure ☐Return ☐Breather
Filter Body Type	
Filter Rated Flow Rate	
Filter Beta Ratio/Efficiency	
Conditions of the Filter	

Table 9.1 – Filters Inspection Sheet

9.2- Filters Troubleshooting

Table 9.2 shows troubleshooting guidelines for filters.

T-Filters-01: Filters Troubleshooting	
▪ Pressure drop across the filter is larger than the rated value. ▪ OR Clogging Indicator is activated.	▪ Check if the filter cartridge is clogged due to particulate contaminates, sludge, or varnish. ▪ Check if the flow rate is above the rated value. ▪ Check if the check valve is stuck closed.
▪ Media cracks. ▪ OR Media migrates downstream the filter.	▪ Check if filter element is subjected to fatigue due to cyclic flow, such as when a pressure compensated pump is stroked/de-stroked very frequently.
▪ Improper filtration process.	▪ Small dirt holding capacity of the cartridge. ▪ Bypass check valve stuck open.
▪ Broken filter housing.	▪ Too high pressure. ▪ Shock pressure.

Table 9.2– Filters Troubleshooting Chart

9.3- Filters Failure Analysis

Filters can tell us too much about what is going on inside a machine as well as in the oil.

Filter Clogging due to Particulate Contamination: Figure 9.1 shows an example of a filter that has been clogged by dirt. The filter appears normal but the particles clogging it are smaller than the limit of vision. In operation, this filter will by-pass due to high differential pressure, thus a pressure indicator is needed to detect when the filter has reached maximum dirt holding capacity. It is advisable to unroll the pleated filter media and check what kind of material it catches. If metal flacks were found, this is an indication of machine wear. Rubber flacks indicates seal deterioration.

**Fig. 9.1- Example of Filter Blockage due to Particulate Contamination
(Courtesy of Noria Corporation)**

Filter Clogging due to Sludge: If the hydraulic fluid is exposed to high temperatures, many fluids will break-down and release resinous materials. When combined with other contaminates, sludge is formed. Sludge tends to plug small openings and orifices and interfere with heat transfer. As shown in Fig. 9.2, *Sludge* is thick polymerized compounds dissolved in warm oil. Sludge is a major source of clogging filters, strainers, and control orifices causing sudden system failure.

Fig. 9.2- Example of Filter Blockage due to Sludge

Filter Clogging due to Varnish: *Varnish* is a product of chemical degradation of hydraulic fluids. As shown in Fig. 9.3, varnish clogs filters. Furthermore, varnish acts as an insulator reducing the effect of heat exchangers.

Fig. 9.3- Example of Filter Blockage due to Varnish

Filter Media Collapse due to Cyclic Flow: As shown in Fig. 9.4, *cyclic flow* can cause fatigue of filter element structure and result in cracking of the pleats unless proper filter medium support is included within the element. Surge or cyclic flow occurs in cases such as when a pressure compensated pump is stroked/de-stroked very frequently.

Fig. 9.4- Example of Filter Media Collapse due to Cyclic or Surge Flow

APPENDIXES

APPENDIX A: LIST OF FIGURES

Chapter 6: Particulate Contamination

Chapter 7: Maintenance of Filters

APPENDIX B: LIST OF TABLES

APPENDIX C: LIST OF REFERENCES

Hydraulic Systems Volume 1- Introduction to Hydraulics for Industry Professionals
Author: Dr. Medhat Kamel Bahr Khalil, 2016.
Publisher: Compudraulic, USA.
ISBN 978-0-692-62236-0

Hydraulic Systems Volume 2- Electro-Hydraulic Components and Systems
Author: Dr. Medhat Kamel Bahr Khalil, 2016.
Publisher: Compudraulic, USA.
ISBN: 978-0-9977634-2-3

Hydraulic Systems Volume 3- Hydraulic Fluids and Contamination Control
Author: Dr. Medhat Kamel Bahr Khalil, 2016.
Publisher: Compudraulic, USA.
ISBN: 978-0-9977816-3-2

Hydraulic Systems Volume 4- Hydraulic Fluids Conditioning
Author: Dr. Medhat Kamel Bahr Khalil, 2022.
Publisher: Compudraulic, USA.
ISBN: 978-0-9977634-8-5

Hydraulic Systems Volume 5- Safety and Maintenance
Author: Dr. Medhat Kamel Bahr Khalil, 2022.
Publisher: Compudraulic, USA.
ISBN: 978-0-9977816-5-6

Hydraulic Systems Volume 6- Troubleshooting and Failure Analysis
Author: Dr. Medhat Kamel Bahr Khalil, 2022.
Publisher: Compudraulic, USA.
ISBN: 978-0-9977634-6-1

Hydraulic Systems Volume 7- Modeling and Simulation for Application Engineers
Author: Dr. Medhat Kamel Bahr Khalil, 2016.
Publisher: Compudraulic, USA.
ISBN: 978-0-9977816-3-2

R01- Basic Electronics for Hydraulic Motion Control
Author: Jack L. Johnson, PE 1992.
Publisher: Penton Publishing Inc. 1100 Superior Avenue. Cleveland, OH 44114.
ISBN No. 0-932905-07-2.

R02- Closed Loop Electro-hydraulics Systems Manual
Author: Vickers/Eaton.
Publisher: Vickers Inc. 1992.
Training Center, 2730 Research Drive, Rochester Hills, MI 48309-3570.
ISBN 0-9634162-1-9

R03- Bosch Automation Technology
Author: Werner Gotz, Steffen Haack, Ralph Mertlick.
Publisher: Bosch.
ISBN 3-933698-05-7.

R04- Electrohydraulic Proportional and Control Systems
Publisher: Bosch Automation 1999.
ISBN 0-7680-0538-8.

R05- Proportional and Servo Valve Technology – The Hydraulic Trainer Volume 2
Author: R. Edwards, J. Hunter, D. Kretz, F. Liedhegener, W. Schenkel, A. Schmitt.
Publisher: Mannesman Rexroth AG 1988. D-8770 Lohr a. Main.
ISBN 3-8023-0266-4.

R06- Proportional Hydraulics
Author: D. Scholz.
Publisher: Festo Didactic KG, Esslingen, Germany.
R07- Electricity, Fluid Power, and Mechanical Systems for Industrial Maintenance
Author: Thomas Kissell.
Publisher: Prentice Hall, Inc. 1999, Upper Saddle River, NJ 07458.
ISBN 0-13-896473-4.

R08- Fluid Power in Plant and Field – First Edition
Author: Charles S. Hedges, R.C. Womack.
Publisher: Womack Machine Supply Co. 1968.
Womack Educational Publication, 2010 Shea Road, Dallas, TX 75235.
ISBN   68-22573 (Library of Congress Card Catalog No.).

R09- Hydraulics, Fundamentals of Service
Author: Deere and Company.
Publisher: John Deere Publishing 1999.
Almon TIAC Bldg. Suite 104, 1300-19th Street, East Moline, IL 61244.
ISBN 0-86691-265-7.

R10- Industrial Hydraulics Troubleshooting
Author: James E. Anders, Sr.
Publisher: McGraw-Hill, Inc.
ISBN 0-07-001592-9.
R11- Power Hydraulics
Author: John Ashby.
Publisher: Prentice Hall 1989. Prentice Hall International, (UK) Ltd.
66 Wood Lane End, Hemel Hempstead, Hertfordshire, HP2 4RG.
ISBN 0-13-687443-6.

R12- Fluid Power with Application
Author: Anthony Esposito.
Publisher: Prentice Hall.
ISBN 0-13-060899-8.

R13- Hydraulic Component Design and Selection
Author: E.C. Fitch.
Publisher: BarDyne Inc. 5111 North Perkins Rd. Stillwater, OK 74075.
ISBN 0-9705922-3-X.

R14- Planning and Design of Hydraulic Power Systems – The Hydraulic Trainer, Vol. 3
Author: Mannesmann Rexroth GmbH.
Publisher: Mannesman Rexroth AG 1988.
D-97813 Lhr a. Main, Jahnsrtrabe 3-5 D-97816 Lohr a. Main.

ISBN 3-8023-0266-4.

R15- Logic Element Technology: Hydraulic Trainer, Volume 4
Author: Mannesmann Rexroth GmbH.
Publisher: Mannesmann Rexroth GmbH 1989.
.Postfach 340, D 8770 Lohr am Main, Telefon (09352) 180.
ISBN 3-8023-0291-5.

R16- Hydrostatic Drives with Control of the Secondary Unit. The Hydraulic Trainer, Volume 6
Author: Dr. Alfred Feuser, Rolf Kordak, Gerold Liebler.
Publisher: Mannesmann Rexroth GmbH 1989.
Postfach 340, D 8770 Lohr am Main.

R17- Control Strategies for Dynamic Systems: Design and Implementation
Author: John H. Lumkes, Jr.
Publisher: Marcel Dekker, Inc. 2002.
Marcel Dekker, Inc. 270 Madison Avenue, New York, NY 10016.
ISBN 0-8247-0661-7.

R18- Feedback Control Of Dynamic Systems
Author: Gene F. Franklin, J. David Powell, Abbas Emami-Naeini.
Publisher: Prentice-Hall, Inc.
Upper Saddle River, New Jersey.
ISBN 0-13-032393-4.

R19- Modeling and Analysis of Dynamic Systems
Author: Charles M. Close, Dean. Frederick
Rensselaer Polytechnic Institute
Publisher: John Wiley & Sons, Inc.
ISBN 0-471-12517-2.

R20- Design of Electrohydraulic Systems For Industrial Motion Control
Author: Jack L. Johnson, PE.
Milwaukee School of Engineering.
Publisher: Parker.
Copyright © Jack L. Johnson, PE 1991.

R21- Basic Pneumatics
Author: Kjell Evensen & Jul Ruud.

Publisher: AB Mecmann Stockholm 1991.
S-125 81 Stockholm, Sweden.
ISBN 91-85800*21-X.

R22- Basic Pneumatics: The Pneumatic Trainer, Volume 1
Author: Ing. –Buro J.P. Hasebrink.
D7761 Moos.
Editor: Mannesmann Rexroth Pneumatik GmbH.
Bartweg 13, W 3000 Hannover 91.

R23- Electro-Pneumatics: The Pneumatic Trainer, Volume 2
Author: Rolf Balla.
Publisher: Mannesmann Rexroth 1990, Pneumatik GmbH.
Publication No: RE 00 262/01.92.

R24- Pneumatics Theory and Applications
Author: Bosch Automation.
Publisher: Robert Bosch GmbH 1998.
Automation Technology Division, Training (AT/VSZ)
ISBN 1-85226-135-8.

R25- Fluid Power Engineering
Author: M. Galal Rabie.
Publisher: McGraw-Hill.
ISBN 978-0-07-162246-2.

R26- Air Motors Ideas with Air
Author: GAST Mfg. Co.
Publisher: GAST Mfg. Co. 1978.
P.O. Box 97, Benton Harbor, MI 49022.
Book No: Booklet #100.

R27- Air Motor Handbook
Author: GAST Mfg. Co.
Publisher: GAST Mfg. Co. 1978.
P.O. Box 117, Benton Harbor, MI 49022.

R28- Troubleshooting Hydraulic Components: Using Leakage Path Analysis Methods
Author: Rory S. McLaren.
Publisher: Rory McLaren Fluid Power Training 1993.
562 East 7200 South, Salt Lake City, UT 84171.

ISBN No. 0-9639619-1-8.

R29- Hydraulics Theory and Application From Bosch
Author: Werner Gotz.
Publisher: Robert Bosch GmbH.
Hydraulics Division K6, Postfach 30 02 40, D-7000 Stuttgart 30.
Federal Republic of Germany, Technical Publications Department, K6/VKD2.

R30- A Complete Guide to ISO and ANSI Fluid Power Symbols
Author: Fluid Power Training Institute.
Publisher: Fluid Power Training Institute 200.
562 East Fort Union Boulevard, Midvale, Utah 84047.

R31- How to Work Safely with Hydraulics
Author: Fluid Power Training Institute.
Publisher: Fluid Power Training Institute 2004.
562 East7200 South, Midvale, Utah 84047.

R32- How to Interpret Fluid Power Symbols
Author: Rory S. McLaren.
Publisher: Fluid Power Training Institute.
Rory S. McLaren 1995.
ISBN 0-9639619-2-6.

R33- Safe Hydraulics
Editor: Gates Rubber Company.
Copyright 1995.
Denver, CO 80217.

R34- Electronically Controlled Proportional Valves. Selection and Application
Author: Michael J. Tonyan.
Publisher: Marcel Dekker, Inc. 1985.
Marcel Dekker, Inc., 270 Madison Avenue, New York, NY 10016.
ISBN 0-8247-7431-0.

R35- Introduction to Closed-Loop Oil Systems
Author: Rory S. McLaren.
Publisher: Rory McLaren Fluid Power Training Institute.
7050 Cherry Tree Lane, P.O. Box 711201, Salt Lake City, UT 84171.

R36- Industrial Hydraulic Technology, Second Edition
Author: Parker Hannifin Corporation.
Publisher: Parker Hannifin Corporation 1997.
6035 Parkland Blvd, Cleveland, OH 44124-4141.

Publication No: Bulletin 0231-B1.

R37- Basic Principle and Components of Fluid Technology – The Hydraulic Trainer, Volume 1
Author: Mannesman Rexroth.
Publisher: Mannesman Rexroth AG 1988.
D-97813 Lhr a. Main, Jahnsrtrabe 3-5 D-97816 Lohr a. Main.
ISBN 3-8023-0266-4.

R38- Safe-T-Bleed Corporation Catalog
Publisher: Safe-T-Bleed Corporation 2001.
Catalog No. STB-PC-1201-1
R39- Industrial Hydraulics Manual – EATON
Publisher: Eaton Fluid Power Training.
ISBN: 0-9788022-0-9.

R40- Vickers-Mobile Hydraulic Manual – Fourth Edition 1998
Author: Vickers.
Publisher: Vickers Inc. 1999.
Training Center, 2730 Research Drive, Rochester Hills, MI 48309-3570.
ISBN No. 0-9634162-5-1.

R41- Industrial Fluid Power Text, Volume 2
Author: Charles S. Hedges, R.C. Womack.
Publisher: Womack Machine Supply Company 1972.
Womack Educational Publications, 2010 Shea Road, Dallas, TX 75235.
ISBN 66-28254 (Library of Congress Card Catalog No.).

R42- Fluid Power Hydraulics and Pneumatics
Author: R. Daines.
Publisher: The Good-heart Willcox Company, Inc.

R43- Hydraulics in Industrial and Mobile Applications
Publisher: ASSOFLUID, Italian Association of Manufacturing and Trading Companies in Fluid Power Equipment and Components

R44- Fluid Power in Plant and Field – Second Edition
Author: Charles S. Hedges, R.C. Womack.
Publisher: Womack Machine Supply Co. 1968.
Womack Educational Publication, 2010 Shea Road, Dallas, TX 75235.
ISBN 68-22573.

R45- Mobile Hydraulics Manual
Author: Eaton.
Publisher: Eaton Corporation Training.
Eden Prairie, Minnesota.
ISBN 0-9634162-5-1.

R46- EH Control Systems
Author: F.D. Norvelle.

R47- Fluid Power Journal
Publisher: International Fluid Power Society.

R48- Fundamentals of Industrial Controls and Automation
Author: Lonnie L. Smith and Mike J. Rowlett.
Publisher: Womack Educational Publications.
Dallas, Texas.
ISBN: 0-943719-04-6.

R49- Lightning Reference Handbook
Publisher: Berendsen Fluid Power.

R50- Pneumatics Basic Level
Author: P. Croser, F. Ebel.
Publisher: Festo Didactic GmbH & Co.

R51- Electro-pneumatics Basic Level
Author: F. Ebel, G. Prede, D. Scholz.
Publisher: Festo Didactic GmbH & Co.

R52- Mechanical System Components
Author: James F. Thorpe.
Publisher: Allyn and Bacon.
Needham Heights, Massachusetts.
ISBN: 0-205-11713-9.

R53- Electrical Motor Controls for Integrated Systems, Third Edition
Author: Gary J. Rockis, Glen A. Mazur.
Publisher: American Technical Publishers, Inc.
ISBN: 0-8269-1207-9.

R54- Instrumentation, Fourth Edition
Author Franklyn W. Kirk, Thomas A. Weedon, Philip Kirk.
Publisher American Technical Publishers, Inc.
ISBN: 0-8269-3423-4.

R55- Introduction to Mechatronics and Measurement Systems, Second Edition
Author David G. Alciatore, Michael B. Histand.
Publisher McGraw-Hill, Inc.
ISBN: 0-07-240241-5.

R56- Study Guides for IFPS Certification

R57- Work Books from Coastal Training Technologies

R58- Industrial Hydraulic Manual – Fourth Edition 1999
Author: Vickers.
Publisher: Vickers Inc. 1999.
 Training center, 2730 Research Drive, Rochester hills, Michigan 48309-3570.
ISBN 0-9634162-0-0.

R59- Industrial Automation and Process Control
Author: John Stenerson.
Publisher: Prentice Hall.
ISBN 0-13-033030-2.

R60- Industrial Automated Systems
Author: Terry Bartelt.
Publisher: Delmar Cengage Learning.
ISBN: 10-1-4354-888-1.
R61- Introduction to Fluid Power
Author: James L. Johnson.
Publisher: Delmar Cengage Learning.
ISBN: 10-0-7668-2365-2.

R62- Summary for Engineers
Author: Dr. Abdel Nasser Zayed.
Publisher: Dr. Abdel Nasser Zayed .
ISBN: 977-03-0647-9.

R63- Mechanics of Materials
Author: Ferdinand P.Beer, E. Russell Johnston Jr., John T DeWolf.
Publisher: McGraw Hill Publishing .
ISBN: 0-07-365935-5.

R64- Oil Hydraulic System, Principles and Maintenance
Author: S. R. Majumdar.
Publisher: McGraw Hill.
ISBN 10: -0-07-140669-7.

R65- Contamination Control in Hydraulic and Lubricating Systems
Publisher: Pall

R66- Diagnosing Hydraulic Pump Failure
Publisher: Caterpillar.

R67- Oil Service Products Catalog
Publisher: Schroder Industries.

R68- Industrial Fluid Power Volume 1
Author: Charles S. Hedges.
Publisher: Womack Educational Publication.
ISBN: 0-9605644-5-4.

R69- Industrial Fluid Power Volume 2
Author: Charles S. Hedges.
Publisher: Womack Educational Publication.
ISBN: 0-943719-01-1.

R70- Industrial Fluid Power Volume 3
Author: Charles S. Hedges.
Publisher: Womack Educational Publication.
ISBN: 0-943719-00-3.

R71- Electrical Control of Fluid Power
Author: Charles S. Hedges.
Publisher: Womack Educational Publication.
ISBN 0-9605644-9-7.

R72- Hydraulic Cartridge Valve Technology
Author: John J. Pippenger, P.E.
Publisher: Amalgam Publishing Company.
Post Office Box 617, Jenks, OK 74037 USA.
ISBN: 0-929276-01-9.

R73- Noise Control of Hydraulic Machinery
Author: Stan Skaistis.
Publisher: Marcel Dekker, 270 Madison Avenue, New York, NY 10016.
ISBN: 0-8247-7934-7.

R74-Solenoid Valves
Author: Hydraforce

R75-HF Proportional Valve Manual, Author: Hydraforce

Index

W

Y